Environmental Science

Environmental Science

An Empirical Approach

First Edition

Steve Fitzpatrick

Perimeter College at Georgia State University

SAN DIEGO

Bassim Hamadeh, CEO and Publisher
Skyler Bryant, Project Editor
Susana Christie, Senior Developmental Editor
Samantha Hansen, Production Editor
Monica O'Keefe, Editorial Associate
Jess Estrella, Senior Graphic Designer
Karisa Kampbell, Licensing Coordinator
Natalie Piccotti, Director of Marketing
Kassie Graves, Senior Vice President, Editorial
Alia Bales, Director, Project Editorial and Production

Printed in the United States of America.

Contents

Chapter 1

What Is Environmental Science?

A Simple Term for a Complex Planet

Objectives

- Become able to define various key concepts and terms included within the "environmental science" umbrella term.
- Gain a broader perspective of the breadth of scientific disciplines utilized in order to the study of Earth's environment.
- Understand the overall structure of your course.
- Understand the relevance of science in our daily existence.
- Reach an understanding of how vital Earth's environmental systems are in order to achieve carbon neutrality, utilize resources in a sustainable and renewable manner, and develop renewable energy resources.
- Understand that an implicit understanding of science is necessary in order to solve some of the greatest problems that our planet and its living inhabitants face.

KEY WORDS AND TERMS

Hypernym (n.): A word or phrase that references a broad set or subset of related ideas or objects. This is also known as a "blanket term" or "umbrella term." Examples include words like "rock," which encompasses many types of related things, including granite, shale, and marble, as well as phrases like "environmental science," which refers to many varied inter-related scientific disciplines like "geochemistry," "hydrology," and "wildlife biology."

Environmental science (n.): A hypernym that describes the range of scientific disciplines that analyze, codify, educate, investigate, and quantify descriptive phenomena concerning the interrelated abiotic and biotic systems of the planet Earth. Ecology, the study of the interactions between the abiotic (non-living) and biotic (living) environment is implicit in the study of environmental science. Emphasis is on the formal and natural sciences since objectivity is of importance.

Environmental studies (n.): A hypernym that describes the range of academic disciplines that analyze, codify, educate, investigate, and implement humanities relationship to the environment, economically, ethically, politically, and socially. There is an emphasis on the social sciences, informed by ecology, as well as the utilization of science to solve problems through engineering. Human society also deals with the environment in the context of economy; in fact, economics is a major driving force for human civilization. Emphasis is on the social sciences, although data and modeling are necessary; thus, one cannot ignore the utilization of the formal and natural sciences.

Environmental engineering (n.): A hypernym that describes and investigates the range of scientific disciplines utilized to analyze, codify, educate, and quantify mathematical and scientific methods, as implemented through the discipline of engineering, in order to explicitly manage Earth's environment for humanities purposes, which includes solving environmental problems that impact human society. Since there is political/societal context, economics is implicitly involved as there are costs to creating environmental systems in addition to the costs of mismanaging our environment.

Formal sciences (n.): *Academic disciplines that deal with analysis and deduction on a purely mathematical, or logical basis.* These disciplines are often abstract in nature and even removed from what we may call human experience at times. Nevertheless, they are vital as they allow us to gather and resolve data, model natural phenomenon, and solve problems on a quantitative level. These include mathematics, theoretical philosophy, computer science, formal linguistics, and others.

Natural sciences (n.): *Academic disciplines that deal with the analysis, codification, education, and quantification of factual systems and phenomenon.* The natural sciences, or "science," as it is generally known, are a group of disciplines dedicated to examining, exploring, and understanding everything that is objectively realized within the entirety of the universe itself (in some cases beyond this universe). These sciences are explicitly dependent on utilizing the formal sciences, especially mathematics and computer science, to measurably explain the phenomenon, as well as to create predictive models and solve problems through engineering solutions. These include the first-order disciplines of physics and chemistry, as well as the combinatory (combine understanding physics and chemistry) disciplines of biology and geology. Physics explores energy and energetic systems on a universal level, and chemistry explores matter, that which has mass and volume, on a universal level. Other combinatory disciplines include atmospheric science, climatology, forestry, hydrology, physical geography, soil science, and many others. Physics and chemistry are often thought of as the basis of all science. Engineering is a discipline that combines science in general with the formal sciences in order to solve systematic problems. It can be argued that the main scientific discipline of this textbook is ecology, the science of the relationship between living things and nonliving things.

Social sciences (n.): *Disciplines dedicated to dealing with analyzing and understanding human social interactions, human institutions, and institutions.* This also includes education of said disciplines. These disciplines can be either focused on the individual or on people as part of the greater human society. They are somewhat dependent on the formal and natural sciences, although there can be some degree of subjectivity regarding their application and understanding. Despite not technically grouped in the natural sciences, an understanding of the social sciences is vital to studying environmental science since environmental problems are inherently human problems. These disciplines include anthropology, economics, human geography, law, political science, psychology, sociology, and others.

Per capita (adj.): Per person when discussing some sort of capital resource, often used in economics and population studies.

Consider This . . .

What do you already know about **environmental science**? How does your environment affect you and how, in turn, do you affect your environment? While our actions within the environment help shape our viewpoint concerning it, the scientific phenomena around us, what we study in environmental science, has a big influence too. Ultimately, the environment that we all live in is part of the environment of the entire Earth, our planet, the only planet that we know of that has life.

Figure 1.1 From space the atmosphere, oceans, and life itself can be seen on Earth. No other planet, that we know of, contains an abundance of such dynamic systems

Introduction

What is environmental science? This is a question that you may have asked now that you are taking a course in it. The problem with answering this question is that the answer is manifold, meaning that there are many answers. It is not one scientific discipline or endeavor, but many different scientific disciplines. The term "environmental science" is a **hypernym**, a word or phrase that describes a broad range of related concepts or disciplines. Environmental science (ENVS) covers a variety of scientific and related disciplines.

As we explore ENVS, we'll discuss the total Earth environment within which we live, including the deep Earth, the solid Earth, the ocean abyss, Earth's ecological systems, and Earth's atmosphere (Fig. 1.1). One's environment is the space within which they live and operate on a day-to-day basis. This is true for all human beings, as well as all of the other organisms that live on Earth. In fact, those scenarios are, so far, unique to this planet because we have yet to find concrete proof of life on other planets.

Studying ENVS involves investigating all the sciences (**formal, natural,** and **social**) involved in understanding our environment in order to become the best stewards of this plan possible. The main emphasis of this textbook is on the natural and, to some extent, formal sciences, hence "environmental science" not "**environmental studies**." This means that we'll explore quantifiable, data-oriented phenomenon, and we will use information that has been time-tested utilizing the scientific method. The scientific method is explained in the next section. Science itself is not a "thing," nor is it a belief; it is a series of systems utilized to understand, well, everything!

FYI: ACRONYMS

Acronyms and abbreviations are widely used in science and technical writing. Many key words and terms are used repeatedly within science magazine articles, journal articles, and textbooks. The convention is to use acronyms with terms used multiple times, especially when the terms employ more than one word. This applies most often for the names of scientific disciplines, government agencies, chemical elements, and chemical compounds. Chemical elements and compounds are simplified to their atomic symbol or chemical formula: ENVS for "environmental science," USGS for the "U.S. Geological Survey," CO_2 for carbon dioxide, and S for sulfur. (Note: FYI stands for "for your information.")

Ecology

Much of what will be discussed in this textbook has to do with the science of ecology. Ecology is the interdisciplinary science that deals with the relation of living organisms to each other and their environment. Through ecology all organisms and relationships between organisms are explored, but the context of this textbook, and generally with ENVS as a whole, is focused on human interactions with other organisms and the environment. This, of course, includes human impacts on the environment. This is also known as "anthropogenic impacts." Although ecology is more or less the "backdrop" for our studies in ENVS, we will also study geology, soil science, forestry, economics, geography, physics, chemistry, hydrology, toxicology, and more! Always remember that science as a whole seems complex because it is. Real-world answers are rarely black and white, and observations as well as solutions are often multi-layered and tricky to work through.

Scientific Methods

As explained before, science is not an object or even a belief system; it is a series of methods that are used to objectively investigate, quantify, and understand phenomena on a universal scale. In practice, one must be methodical in order to investigate scientific phenomenon. To do this, organization is called for, a plan, and this plan is often called the scientific method. In truth, there is no one "scientific method"; rather, there is a generally agreed-on system that is utilized by scientists in order to investigate scientific phenomena and solve problems. These are plans based on a general outline and can be arranged or rearranged according to the work that must be done in order to discover new phenomena and investigate and analyze scientific problems. Using methodologies in this manner to solve or explain scientific, or any other, problems is not limited to people who work in the field of science. For that matter, science is not a prohibitive practice; anybody who is curious and wants to understand the world, even the universe, through reason and logic can engage in science!

Question

Using evidence of all sorts, from observation scientists initially start formulating questions about the nature and workings of the scientific phenomena in question. Scientists often start out on a path of scientific discovery of making observations or exploring phenomena within their chosen fields, at times when attempting to solve problems. This is when questions are asked, for instance, “Why do objects fall to the ground when let go?” or “How do rocks in streams become rounded pebbles?”

Research

Once the initial question is asked, one must conduct background research when attempting to investigate scientific concepts. There are many pathways to research, including personal observation, but the most rigorous takes the form of watching, word-of-mouth, reading, and studying. There are many pathways to gaining knowledge, including articles, books, film, videos, and maps. Some of the most rigorous research comes from studying the works of other investigators and scientists in peer-reviewed publications. These works are subject to scrutiny from experts in the field in order to provide for the most accuracy and honesty of content. Peer-reviewed publications often provide the most detailed and rigorous background information for all forms of scientific matters.

Hypothesis

A hypothesis is an explanation, or assumption, based on observations and reasonable assumptions. The explanation is somewhat tentative in that it must be proven, but it is usually framed as a statement. It is not a series of wild speculations, nor is the intent for it to be truth from the start. Ultimately, through the course of research, a hypothesis may be proven to be just plain wrong to a certain degree. That is fine. The scientific process of discovery is one of finding the truth behind the workings of the universe. Sometimes the truth is not what we desire. To fit the data to something that is less than completely accurate, precise, and, quite frankly, honest, is not the goal of science at all; it is biased. It is even deceptive. When a hypothesis is proven incorrect through rigorous investigation it can actually lead to some amazing discoveries! If a hypothesis is incorrect, it means that there is more work to do, either reformulating the hypothesis or even throwing it out and coming up with a new one. The universe has a way of surprising us.

Experimentation and Investigation

Once the hypothesis is formulated, scientists must come up with a way to test the hypothesis to see whether it is true. In order to do this, they must explore the phenomenon or problem through experimentation, investigation, or a combination of the two. The experimentation and investigation phases are meant to test the initial hypothesis, or hypotheses, to see whether they hold true. The approach to this depends on what is being investigated. Ultimately, data is gathered from observation and testing, which is used to assess the hypothesis and arrive at the truth. There are many ways to go about investigating scientific problems, but two of the main ones are through controlled or natural experiments.

Controlled Experiment

A controlled experiment is one in which predetermined variables are controlled by the experimenter, often just one. This is the classic laboratory experiment featuring set, controlled variables. Generally, only a few variables, or even only one, are different in the experimental design, thus introducing causality in the variable that is different in the experiment. The variable that is different in each set, the one that is being tested, is called the independent variable. It is the one that is being manipulated to check for an outcome. For instance, in an experiment involving the speed at which ball bearings travel down a series of ramps, all of the ball bearings could be the same size and travel down identical ramps with identical slopes, the only difference being the weight of each ball bearing. Since the weight is different for ball bearings, the weight is the independent variable. The result of the experiment due to the manipulation of the independent variable is called the dependent variable. The dependent variable is functional to the independent variable, so, mathematically, the independent variable would be plotted on the x-axis and the dependent variable on the y-axis, which is f(x), a function of the x-axis.

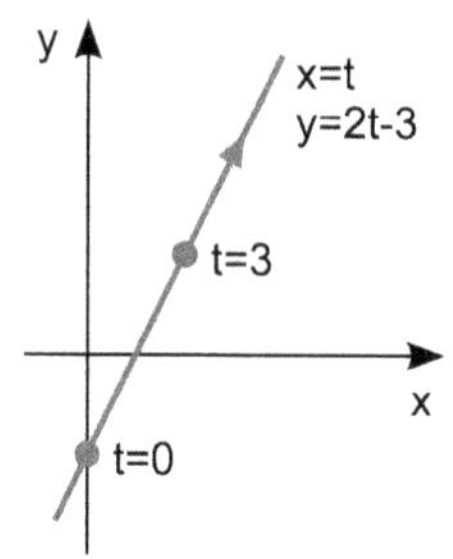

Figure 1.2 Relationship between an independent variable, X, and a dependent function, Y, or f(X)

Natural Experiment

A natural experiment, or a field experiment, as opposed to a controlled experiment, is not one based on controllable variables. It is mostly based on observation and testing natural parameters. There is a vast amount of scientific research that investigates scientific phenomenon and that is beyond our control, the workings of nature itself. For example, a scientist studying the amount of lava expelled by a volcano has no control over what the volcano does. They can only take careful measurements and record data. There are some controllable variables in field experimentation, such as taken-down data at precise, predetermined time intervals, but there is also a vast amount of data from uncontrollable natural responses. At the same time, nature follows predictable patterns that scientists can ultimately understand, such as the slow, steady movement of tectonic plates or the precise radioactive decay of radiometric isotopes. A lot of the scientific disciples under the ENVS umbrella depend on data taken from careful observation and testing, including atmospheric science, biology, geology, hydrology, and more.

Analysis

When exhaustive experimentation, investigation, and observations have been done, including all of the data gathered from that work, the scientist goes elsewhere to analyze the data that they have gathered. This can take the form of the results of chemical experimentation, creating models to understand field data, running samples through mass spectrometers, and looking at samples through a microscope, among others. This analysis usually results in numerical data, which is used, in turn, to formulate mathematical or statistical analysis. So far, all scientific endeavors have yielded mathematical results that have resulted in a deeper understanding of our universe. Even mathematical analyses that are poorly resolved may someday be solved. This type of rigorous analysis is used to discover whether a hypothesis is true or false.

Conclusion/Discussion

Once all of the data-driven analyses have been obtained, scientists can draw conclusions from the results and discuss them. If a hypothesis is true, it will continue to be tested, often by other investigators. If a hypothesis is false, one must go back and reformulate a new hypothesis for testing from the original research. Regardless of whether the hypothesis is true or false, it must be made available to not only the scientific community, but to all individuals. Even a false result bears reporting because it is still based on truth.

Reporting

Once a particular scientific study is concluded, it is made available. This often takes the form of submission to peer-reviewed publications with the goal of publication in mind. This need not always be the case though. Sometimes a scientific investigation is part of the work that scientists do with a governmental agency, corporation, or nonprofit organization. Sometimes the results never get published but are made available in a university or research center. Regardless, a faithful and honest account of the entire process behind an investigation must be made available in order for future work to take place, leading to further understanding of the phenomena.

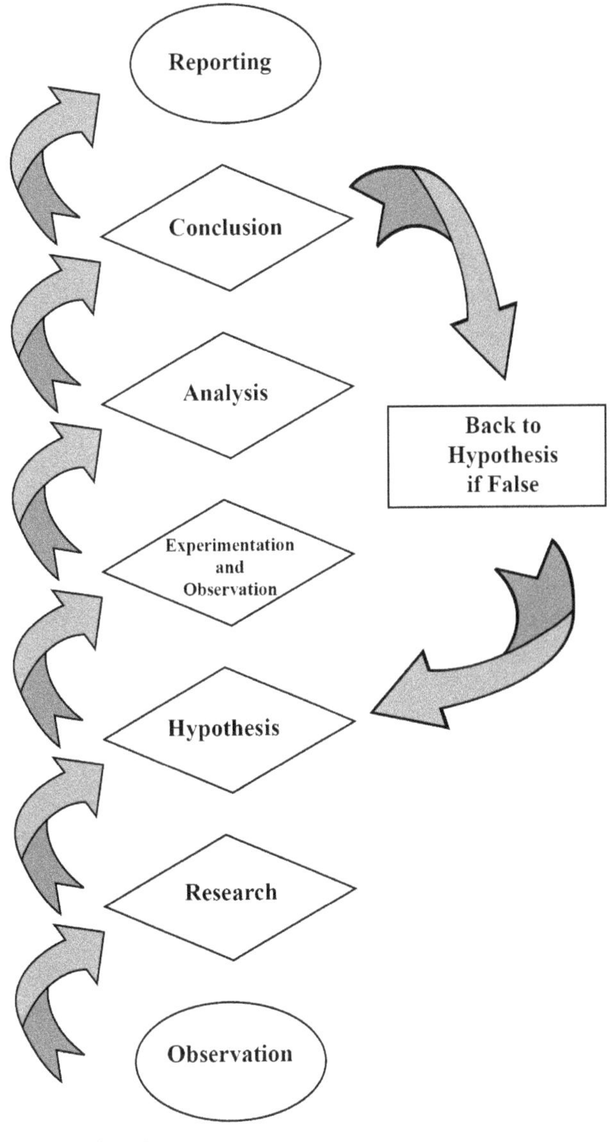

Scientific method conceptual flowchart

GROUND TRUTH

Example of an Environmental Challenge: Proctor Creek, Atlanta, Georgia

Proctor Creek is a 10,094-acre watershed that originates in downtown Atlanta, Georgia, and runs over 9 miles northwestward to the Chattahoochee River. Before 1927, Proctor Creek carried raw sewage from most of downtown Atlanta to the Chattahoochee River. Among the lasting effects of this practice was the devaluation of working middle class into low-income neighborhoods. In 1927, a combined sewer system was constructed in the 1652-acre Proctor Creek/North Avenue (PNA) watershed, headwaters of the Proctor Creek Basin. Large, combined sewer trunk lines were built (partially above ground), followed by low-cost housing, perpetuating what some have likened to "third world" conditions. Until recently, some of these conditions continued to exist in the PNA watershed. In fact, the work is ongoing.

This area has experienced historic, frequent, and repeated flooding with sewage and all of the health problems associated from combined sewer/stormwater systems, including higher incidences of West Nile Virus. Downstream from the combined sewer overflow facility, Proctor Creek is polluted with sewage, chlorine disinfection byproducts, untreated stormwater runoff, illegally dumped construction debris, and tires. The Proctor Creek Watershed is primarily composed of working-class African American citizens with little access to amenities and services. Many neighborhoods in the watershed are currently experiencing high rates of unemployment and underemployment. According to 2010 Census data, over 19,000 residents live within the community with a median income of $14,791.00, and 33% live below the poverty level.

Figure 1.4 Gathering data at the Proctor Creek Watershed, Atlanta, Georgia

The Proctor Creek Watershed is a real-world example of a series of challenges within the fields of the environmental sciences and utilizing the work of many people within a host of scientific and nonscientific disciplines. Work was done to clean up the watershed and provide a higher quality of life for the people who live in the surrounding area. In 2013, the U.S. Environmental Protection Agency (EPA) Urban Waters Federal Partnership named Proctor Creek as a priority Urban Waters site. Partnerships were formed with local leaders and organizations, including the Georgia Institute of Technology (Georgia Tech), Georgia State University, Morehouse College, and the Proctor Creek Stewardship Council, in order to investigate Proctor Creek's problems as well as to implement solutions for both the watershed and the community in order to revitalize the ecology and economics of the area. It was and continues to be a successful effort. The following solutions have been and continue to be employed: the creation of green space and green infrastructure, research on the impact of increased stormwater management, improvements in the overall water quality of the watershed, engagement with the local community to support stewardship, and the reduction of food deserts in the area by providing healthy nutritional choices for the community in the form of local vegetable gardens.

Earth's Systems

As we explore ENVS, we'll discuss the total Earth environment within which we live, including the solid Earth, the ocean's abyss, the atmospheric interface with outer space, the deepest rainforests, and more. In fact, the amount and diversity of Earth's ecosystems is staggering. On a personal level, one's environment is the space within which one lives and operates on a day-to-day level. Thus, a natural environment is not just spatial but oriented in time. This is true for all human beings, as well as all of the other organisms that live on Earth and the nonliving components of Earth's varied environments.

FYI: ECOSYSTEMS

An ecosystem is a functioning biological community within some given area, on some level. It is a living community that exists in relation to a nonliving environment, as well as in relation to all other organisms in an ecosystem. Ecosystems are singular and unique. There are a vast number of ecosystems across the planet, ranging from giant deserts and swamps to the bottom of the ocean, to soil profiles and subterranean communities in the mud-ringing river systems. The physical and chemical characteristics of an ecosystem, and the organisms that they support, define that ecosystem as unique, although any given ecosystem likely shares characteristics with others.

Abiotic Components

The Earth concluded its initial formation about 2.8 million years (Myr) after the formation of the solar system about 4.571 billion years (Gyr), making Earth 4.543 Gyr old. At this point in time, Earth was a mostly molten spheroid of chemicals in melt. As it cooled, Earth's outer layers formed the mineral-bearing rocks that would be the setting for everything that has ever happened or existed on, or within, the Earth. Earth's abiotic components had initially formed, although calling them "abiotic" was irrelevant at the time.

Abiotic components, or factors, of an ecosystem are the non-living chemical and physical portions of the environment. They basically compose the setting for life to take place. With all of the complex chemicals, energy, gases, rocks, and water that the Earth contained at that time, there was still nothing alive during those early days of the Earth,

Earth's Archean Eon, as if anything could live in or on the dangerous, poisonous inferno that Earth was at the time. There were no ecosystems in any sense of the word since there were no living things on Earth. It would take about 870 Myr, that we know of, for life to appear on Earth.

Figure 1.5 Artist's impression of the Archean Eon

Biotic Systems

Biotic components, or factors, are the living parts of the environment or an ecosystem. In fact, an environment cannot truly be called an ecosystem without the presence of life. From what is known, life appeared on Earth in the Archean Eon about 3.7 Gyr ago, but there is some evidence for the building blocks of life from much earlier in Earth's history. Biotic systems do not just live in the backdrop of abiotic systems. Life actually affects the environment itself, even the very rocks in the ground! Excess oxygen that was given off by photosynthetic bacteria started accumulating about 2.3 Gyr ago. Prior to this the excess oxygen was absorbed into iron-rich rocks until they could accept no more. This is the origin of Earth's breathable atmosphere, rich in oxygen (O_2). This is why there is not only air for us to breath, but why we exist in the first place!

Natural Resources

Natural resources are critical for all living things, but they are especially important to human beings and human civilization. Humans evolved the ability to use tools, and this proved important to the species' continued survival prior to the beginning of history. Since human beings are not especially well equipped with natural weaponry or the vast speed, or strength, of many other animals, they depended on tools and a knowledge of the environment to thrive. Once civilization, and history, began, around 10,000 years ago, humans had learned to utilize all sorts of natural resources for energy, goods, shelter, survival, and well-being. Humanity continues to rely on natural resources in our current post-industrial, global economy, but many of our valuable resources are being threatened with overuse and overexploitation.

The Limits of Resource Availability

Natural resources are rarely ever available in unlimited amounts. To us, Earth is huge and seems unlimited, even with our knowledge of current conditions and our ability to travel around and communicate across the globe. Yet it has its limits. Many natural resources, especially resources that are mined yielding rocks, minerals, and energy resources, are currently running out. Others, like solar and wind, are available in unlimited amounts but require technology to harness. Other resources like forests, soil, and water are variably available; they become nonrenewable if not carefully managed. Note that resources are also often renewable or nonrenewable according to time, such as groundwater, which takes many years to replenish. An extreme example of this is crude oil.

Solar power is considered a completely renewable energy source since its availability is unlimited

Renewable Resources

Truly renewable resources are available in unlimited abundance relative to the time frame in which we use them. For instance, solar energy is essentially available in unlimited amounts to us right now, but the Sun will eventually lose its ability to undergo the current style of nuclear fusion in about 5 Gyr. The truly renewable energy resources include solar power, wind energy, tidal energy, and wave energy, among others. In other words, of the various forms of energy resources, in fact, only certain forms of energy resources are completely renewable. One would think that crop-based agricultural resources would be available in unlimited abundance because of solar energy, but crops can become degraded due to incorrect agronomic (agronomy is the science of raising plant crops) practices. Thus, available plant life, as well as soil and water availability, is only variably renewable. They occupy the continuum between renewability and non-renewability. Most natural biochemical resources on, or near, Earth's surface are variably renewable.

Nonrenewable Resources

Nonrenewable resources include anything that is mined from the ground, as well as all fossil fuel resources. These are natural chemical resources that are produced within the Earth at rates in accordance with geologic time. Geologic time, as you can probably guess, is counted in thousands to millions to billions of years. These resources are replenishable within our human life spans. Once any given mineral, or fossil fuel energy resource, is extracted from the Earth approaching its available limit, there will simply be no more that we can easily use. Many mineral and fossil fuel resources are approaching this limit due to the rapid consumptive demands of a global civilization.

Earth's History

The planet Earth's age is 4.54 Gyr. It formed some 2.8 Myr after the entire solar system itself was established. The very early, primordial Earth was essentially a molten ball of magma and lava. It was at this point that most of the heavier elements such as iron and nickel sank to Earth's center of gravity, leaving most of the lighter (lower numbered atomic number) elements at or near the Earth's surface. A large planetoid collided with Earth during this primeval time, the Hadean Eon, which created an orbiting ring of debris that eventually accumulated to become Earth's moon. Outgassing from volcanoes and strikes from comets in that early, violent solar system contributed volatile elements and compounds to the Earth system, foremost among them water. This also contributed to Earth's early atmosphere, which contained no free oxygen but plenty of carbon dioxide, hydrogen sulfide, and methane. As age-long rains occurred and started to fill the lower parts of the topography, oceans started to form. The planet continued to cool.

Eventually, Earth had cooled down enough for solid land masses to form, and poked out over the primitive ocean. Complex chemicals were produced in these waters as well, the building for life. This was the Archean Eon, and it began some 4 Gyr ago. Ultimately, life itself emerged at around 3.7 Gyr, from what we know. This took the form of bacteria and bacteria-like organisms that formed enormous photosynthetic mats in Earth's shallow seas. Through photosynthesis oxygen began accumulating in Earth's atmosphere, after iron-bearing rocks absorbed it. This was during the Proterozoic Eon, which began around 2.5 Gyr ago. Life began to flourish during that time as free oxygen began accumulating in the atmosphere during the Great Oxidation Event (GOE) that lasted from 2.4 to 2.0 Gyr ago. This abundance of free oxygen, among other factors, led to an ever-increasing abundance of life until what can only be called an explosion of life occurred about 541 Myr ago, the beginning of the Phanerozoic Eon.

Life's Prehistory

The beginning Phanerozoic Eon at 541 Myr ago was also the beginning of one of its subdivisions, the Cambrian period (Figure 1.7). The beginning of the Cambrian is also called "the Cambrian explosion" because almost instantaneously, for geologic time, a massive overabundance of life appeared in the Earth's oceans. This was not only a major increase in the usual single-celled organisms of the time but also of multicellular, also known as metazoan, life. The Cambrian period was also the beginning of the Paleozoic Era. The Paleozoic Era, which lasted from 541 to 251.9 Myr ago, was an eventful time for Earth, especially when life itself is considered. Metazoan life thrived as invertebrates evolved alongside primitive vertebrates. Eventually animals began colonizing the land around 450 Myr, the end of the Ordovician period.

Land-based organisms, including animals, fungi, and plants, began to dominate Earth's continents and land masses during the Silurian period, which began around 443.8 Myr ago. Giant, primitive tree-like plants dominated the Earth during very warm time spans within the Carboniferous period, 358.9 to 298.9 Myr ago. These giant plants died in droves and formed thick beds within sediments. Once the Earth began to cool again and these plants became buried, they formed the basis for a large amount of the coal that we currently use as an energy source, and unfortunately as a source of greenhouse gases and pollution. Remember, fossil fuels are called fossil fuels because they are derived from the remains of ancient organisms. The remains of organisms throughout Earth's history were fossilized, and we utilize these fossils today to help us understand Earth's history of life.

Figure 1.7 Artist's conception of life during the Cambrian period

The end of the Paleozoic, at the end of the Permian period some 251.9 Myr ago, is marked by the largest extinction event the Earth has ever seen. During this Permo-Triassic extinction about 81% of marine and 70% of terrestrial species became extinct. It also coincided with the break-up of Earth's most recent supercontinent (when all of Earth's continental landmasses are joined) called Pangaea. This also marked a new beginning, the beginning of the Triassic period within the Mesozoic Era. The Mesozoic Era, from 251.9 to 66 Myr ago, is called "the age of dinosaurs" because that's what it was! The magnificent dinosaurs that continue to fascinate us to this day dominated the Earth during this era. It was during this time that angiosperms, flowering plants, evolved around 125 Myr ago. The Mesozoic came to an end when an asteroid crashed into Earth 66 Myr ago off the Yucatan Peninsula, marking the end of the dinosaurs and of the Mesozoic Era.

The Cenozoic Era (66 Myr ago to the present) has been called "the age of mammals." True mammals evolved much earlier, around 225 Myr ago, during the late Triassic period, but these were small shrew-like animals that did their best to stay out of the way of domineering dinosaurs. With the dinosaurs gone, and a cooler climate in place, the mammals, as well as the birds who evolved directly from dinosaurs, were set to take over the planet. At the beginning of the Cenozoic, land masses were formed into essentially the same set of continents we currently have. Toward the later part of the Cenozoic, a series of ice ages began, which aided in the evolution of a peculiar type of mammal, an ape, known as human beings.

Human History

Human beings (*Homo sapiens sapiens*) evolved in Africa during the Pleistocene Epoch around 300,000 years ago from earlier members of the genus Homo some 2.8 Myr ago during the late Pliocene Epoch of the Quaternary period in the Cenozoic Era. These first human beings were essentially like ourselves genetically, although the world that they lived in was vastly different from the one we live in. There weren't a lot of human beings in the early days, between

100,000 and 300,000 individuals. This number may have plummeted around 75,000 years ago when a supervolcano, Mount Toba, erupted throwing the planet into a volcanic winter. It is estimated that the human population may have been reduced to around 3,000 individuals because of this disaster, something known as a "bottleneck" event, although this theory has been successfully challenged recently. Regardless, survival was difficult for early human beings as they spread forth and diversified across Earth.

Human beings shared the world with other types of humans for a large part of its history, most notably the Neanderthals (*Homo neanderthalensis*, or *Homo sapiens neanderthalensis*). Neanderthals were a hardy human species who lived on the ice sheets of Europe during the Pleistocene Epoch until they went extinct around 40,000 years ago. There is strong evidence that they interbred with human beings who had migrated out of Africa into Europe and elsewhere.

For around 97% of their time on Earth, human beings lived as hunter/gatherers traveling the globe on foot in small bands of families or groups of families. Although not particularly well armed with the weaponry, strength, or speed that other animals had, human beings were successful because of their large, sophisticated brains, ability to utilize tools, and endurance (humans can easily walk 20 miles in one day in favorable terrain). During most of its history, humanity's activities had very little impact on the Earth's environment because human beings used very little in the way of energy resources, other than the occasional fire, and did very little harvesting of plants. That being said, there is evidence that human beings were in some part responsible for the extinction of many species of large land mammals, known as the Pleistocene Megafauna, due to overhunting. Humanity's impact on the Earth did not change much until the beginning of the Holocene Epoch some 11,650 years ago.

The Agricultural Revolution

The Holocene Epoch is beginning of what we call "recorded history." It is the epoch we are currently in (there is a proposed "Anthropocene Epoch," starting in the mid-20th century, but it has not been officially approved by geoscientists), and it encompasses two major changes in human civilization: the agricultural revolution and the industrial revolution.

The agricultural revolution began around 10,000 years BCE (before the Common Era) at the beginning of the Holocene Epoch. The human economy was a sustenance economy before that time, an economic system of simply living off the land and utilizing barter, the direct trade of goods. Conditions changed at the beginning of the agricultural revolution in part because conditions really did change. The planet had entered an interstadial period, or a period of warming during an ice age, which we continue to enjoy to this day. With a favorable climate, humans could relax their frequent seasonal migrations. Also, since they were staying in one geographic location at a time, human beings began cultivating plants and raising livestock to a greater degree than before, the beginnings of agriculture. (Note: Human beings had already learned about agricultural practices prior to the beginning of the agricultural revolution but did not practice it frequently or on a large scale.)

The beginnings of wide-scale cultivation did not start at the same time everywhere across the globe. The earliest known start of widescale agriculture was in the "Fertile Crescent," which is in part of the Middle East these days (Figure 1.8). This also coincided with the beginnings of some of the world's earliest civilizations shortly afterward such as Sumer, Mesopotamia, Assyria, and Egypt. These civilizations oversaw a break from sustenance economies with other means used for trade, market economies, and centralized economies in which military or religious leaders essentially owned all of the land.

Throughout the agricultural revolution, human beings continually improved on technological advances. Boats made it possible for people to slowly travel around the globe and take part in maritime trade. The smelting of metals, starting with bronze and then steel, made it possible to construct all sorts of tools, and primitive machines, as well as tools of war, an unfortunate byproduct of human "conflict resolution." Human history marched on, kingdoms rose and fell, the human population slowly grew to close to one billion people, and then another revolution took place around 250 years ago.

Figure 1.8 Map of the agricultural revolution in the Fertile Crescent

The Industrial Revolution

Human beings had always harnessed energy resources throughout history, including animal power and wind to turn windmills. Some Ancient Greek philosophers had even tinkered with using steam to power machines, although it did not catch on, but steam power came back around in the 1700s. By the mid-18th century, steam power through the burning of coal started to become used on a widescale basis to power the new invention known as the locomotive as well as factory and agricultural machinery. Human beings were discovering how to synthesize new types of chemicals for commercial and industrial use. Steel production became common and with it innovations in factory-based tool-making processes. The industrial revolution had begun, and it changed the world.

With all these advances came an exponential increase in the human population, and with that came the first signs of dissent at the new industrialized systems and the changes they brought about. Thomas Malthus was an economist who published "An Essay on the Principle of Population" in 1798. He was alarmed at the sharp increase in population as well as the dirty air that came from industrialization in England at the time. He predicted large-scale famine if people did not change their ways due to population overtaking adequate food supplies. This never happened because as mechanization increased, so did agricultural production rates.

As mentioned before, the industrial revolution brought about sharp increases in pollution, mostly atmospheric, but waterborne as well. Crowded cities in Europe and in the East Coast of the United States were infested with disease, crime, and poverty as well as the aforementioned pollution. As England continued its maritime expansion into colonies such as India, along with other European nations, the United States began its westward expansion into the heartland of America. These expansionary pushes did not come without a cost for the environment as well as human capital. In the United States millions of acres of old-growth (original) forests were clear-cut for timber and farmland. The United States also saw the shameful exploitation and genocide of its Native American population in the name of "progress." Similar forms of exploitation were occurring in European colonies in Africa, Asia, and South America. Meanwhile, wars raged across the landscape and technology continued to progress through advancements in science, some of which were beneficial.

Medical advances made it possible for humans to live longer, as did the production of greater amounts of food. Scientific advances also allowed us to gain a deeper understanding of our planet. Throughout the 19th century many intellectuals became alarmed at the problems that an industrial civilization was bringing to the table. In England some of the earliest environmental organizations began during this time, and in the United States John Wesley Powell, the second director of the United States Geological Survey (USGS), warned of the coming problems associated with untrammeled expansion into the wilderness and exploitation of land resources. The naturalist John Muir advocated for preservation of native lands, keeping them away from advancing industry. A personal friend of President Theodore Roosevelt, Muir's activism made good on his promises. Incidentally, the world's first national park, Yellowstone National Park, was established by the United States Congress and President Ulysses S. Grant in 1792. Thus began an awareness and reverence for the environment that continues to this very day.

Context: Environmental Studies

As mentioned, environmental studies is the umbrella term that includes the range of academic disciplines used to understand, utilize, and explain the environment through economics, ethics, politics, and sociology. This textbook covers these concepts contextually with reference to the natural environmental sciences, but a complete survey of environmental studies is beyond the scope of this work. Nevertheless, science has a responsibility for providing truth to humanity in general. Science is not divorced from human endeavor and must strive to protect and support humanity as a whole. We will return to discussing environmental studies in the last chapter of this book once we have covered the breadth of scientific disciplines that will provide us with tools for understanding our environment, for becoming stewards of this planet Earth. But there are a few factors to consider as we move forward.

Economics, Ecology, and the Environment

It is no coincidence that the root word for "economy" *oikos* is the root word for "ecology." Oikos is a Greek word that simultaneously means family, or the family home. A family's home must have an economy to thrive, but it also must exist in internal and external harmony. Although the current basis of the world's economy is wholly numeric and not based on any sort of precious metal standard, it saw its beginnings in the use of natural resources. We derive our original economy from agricultural, mineral, and energy resource products. Today the world is still dependent on coal and oil for much of its energy use. At this point in time, we must also use more sustainable, clean sources of energy to help stave off the environmental and human costs of fossil fuel use in industry and transportation.

Ecological Footprint

Figure 1.9 Fishing boat amidst pollution in Dungeness, Kent, England

Although it can be argued that we currently live in a "post-industrial" world, global-scale industry and transportation still accounts for enormous volumes of damaging greenhouse gases into the environment. Not only are these gases (carbon dioxide, methane, nitrous oxide, sulfur dioxide, etc.) the main cause of accelerated heat flux (warming) due to the greenhouse effect, they are also the main cause of air pollution and in many ways water pollution across the globe. Humans have also mined many of our mineral resources to the point that they will no longer be available with the next 100–200 years. These, along with crude oil, are the chemical components that were originally used for modern industry. Although we can now artificially synthesize many of these chemicals and materials, the bulwark for global industry is in mined materials.

Through global agriculture human beings have stressed soil- and water-based resources to their limits. Desertification is rapidly spreading to parts of the globe and soil in areas of intense agriculture is constantly having artificial fertilizer added to make it usable at all for growing. Widescale agricultural runoff is also part of the water pollution that threatens water supplies every day. Each year vast volumes of the Earth's oceans become hypoxic, depleted of dissolved oxygen, resulting in the destruction of marine life.

Animal and plant resources experience overexploitation through agriculture and other human activities. Fish in the world's oceans have experienced, and continue to experience, near extinction levels of fishing. Meanwhile cattle, chickens, and pigs are herded into feedlots experiencing horrific conditions. These livestock operations account for actual rural air pollution, not to mention diseases borne of animal waste entering the planet's waterways. Human tinkering with life itself may have led to the worst environmental crisis of our lifetime, the current COVID-19 pandemic.

All of this has contributed to what is called humanity's ecological footprint. The term "ecological footprint" was coined by the Global Footprint Network, an international think tank founded in 2003 by Mathis Wackernagel, PhD, and Susan Burns in Oakland, California. An ecological footprint is a metric that is used to describe human demand on natural resources. This includes the per-capita use of natural resources, geographic space, contribution to the waste stream, and more. The term ecological footprint can be applied to the specific use of vital chemical elements and compounds made available by our environment such as carbon and water. We will explore these concepts later in this book.

Conclusion

Moving forward we will focus on the science behind the environment, and in some cases engineering, based on said science. Ecology is a large part of what we will explore, but the Earth's environment is much more than that. It encompasses the very rocks we live on through geology, studying the air we breathe through the geographic science of the atmosphere, relating our energy use to fundamental analysis of energy dynamics and chemical production, knowing about the versatility and strength of life itself through biology, and so much more. All of our actions have consequences. No one living thing, including we humans, live in a bubble on this planet. On our planet Earth, we are all interconnected on very fundamental levels. Let us begin our exploration of the sciences behind the environment in which we all live.

Discussion Questions

Directions: Review the chapter in order to completely and correctly respond to the questions and prompts.
Keep this in the back (or maybe even front) of your mind: Science is a system of methods that allows us to understand not only our planet, and ourselves, but of the entirety of the universe. Answering the following questions correctly will lead to a more robust understanding of how to apply science to ENVS problems.

1. In what ways did you come to an understanding of the fact that understanding science is necessary to understanding Earth's environment?
2. What are some of the scientific disciplines that must be taken into account in order to understand environmental processes?
3. What are some of the challenges that we face from the overexploitation of Earth's resources?
4. What are some of the costs to human well-being that have historically been brought about by overexploiting the Earth environment?
5. What are ecosystems, and why is it important to understand how they work?
6. Are some resources variably renewable and nonrenewable and, if so, how is this possible?
7. What sorts of environments did human beings live in prior to 10,000 BCE, and how are they different from environments today?
8. Why did humanity's population see an exponential increase in numbers starting in the 18th century?
9. What is the difference between environmental science and environmental studies?
10. Why is an ecological footprint a good measure of a person or group's impact on Earth's environment?
11. What sort of impact did you make on your local environment today? It's OK to make a positive impact!

Bibliography

Barker, G., 2006. *The Agricultural Revolution in Prehistory: Why Did Foragers Become Farmers?* Oxford: Oxford University Press, 616 p.
Global Footprint Network, 2003. *Tools & Resources*: https://www.footprintnetwork.org/resources/ (accessed November 2021).
Levin, H., and King, Jr., D., 2017. *The Earth Through Time*, 11th ed., Hoboken, NJ: Wiley, 590 p.

Lowe, B., 2002. *The Formal Sciences: Their Scope, Their Foundations, and Their Unity: Synthese*, v. 133, n. 1/2, p. 5–11.

Nityasya, M.N., Mahendra, R., and Adriani, R., 2018. *Hypernym-Hyponym Relation Extraction from Indonesian Wikipedia Text: 2018 International Conference on Asian Language Processing (IALP)*, p. 285-28.

Sharma, J.P., 2016. *Environmental Studies: For Undergraduate Classes in Science, Commerce, Humanities, Engineering, Medicine, Pharmacy, Management and Law: Strictly as Per Model Curriculum Prescribed by UGC*, New Delhi: Laxmi Publications, 316 p.

Weiner, R.F., and Matthews, R., 2003. *Environmental Engineering*, 4th ed., Oxford: Butterworth-Heinemann, 484 p.

Withgott, J., and Laposata, M., 2018. *Environment: The Science Behind the Stories*, 6th ed., San Francisco: Pearson Benjamin Cummings, 784 p.

Credits

Chapter 2

Science Basics, Part 1

Data, Analysis, and Matter

Objectives

- This chapter introduces you to the overall discipline of science, including the multitude of subdisciplines, starting with chemistry and physics.
- Once the basics of science are mastered, you'll understand that all of the disciplines are connected. Science is a supportive endeavor dedicated to understanding not only our planet, Earth, but the universe itself through logic, reason, and cooperation.
- It will be apparent that science is not a series of beliefs, or a static object, but various systems of methodologies that are employed to rationally understand, again, not only Earth but the universe itself.
- This chapter gives you the basic tools for understanding, through science, what is required for exploring the vast range of mechanisms the govern the operations of the natural world: the environment of the planet Earth.
- This chapter contains certain facts and first principles that form the bulwark of the scientific understanding of this planet, as well as all life on it.
- This chapter will impress upon the reader the fact that many of the current environmental challenges and problems that we face have been, and will always be, solved using reason, logic, and science.

KEY WORDS AND TERMS

Science (n.): Branches of study and disciplines that employ bodies of demonstrated truths and observed facts that are systematically and understood through a series of general laws and rigorous methods of analysis. This includes the use of the scientific method incorporating falsifiable hypotheses in order to discover various truths within the specific domain. Essentially, it is a series of methods used to uncover the truth about the physiochemical and quantifiable nature of the entire universe.

Chemistry (n.): The branch of science dedicated to the study of matter as well as transformations of matter in all its forms.

Ecology (n.): The branch of biology dedicated to the study of how organisms relate to each other and their environment.

Engineering (n.): The scientific and technological discipline dedicated to the design, construction, and use of engines, machines, structures, and systems.

Anthropogenic (adj.): That which is produced through human economic and social activities or development. It often refers to the negative effects of human activity on the environment or health.

Greenhouse gases (n.): Gases that absorb and reradiate some of the longwave, or infrared, solar energy that is reflected off of the Earth's surface. Some of this infrared radiation is directed back to Earth and acts to increase the heat of Earth's surface.

GROUND TRUTH

Example of Scientific Inquiry: Svante Arrhenius

Global warming has become a cultural term within popular media, implying that humanity's rampant use of certain gases is rapidly causing the Earth to become warmer. In broader terms, "global climate change" became a suite of cultural, economic, political, and social problems related to overall **anthropogenic** climate change. All of these factors have only added to its complexity as an issue, due to divisive sociopolitical issues concerned with possible impending global climate disasters. Regardless of the impact on human civilization, global warming is not new. The temperature of Earth's atmosphere since the origin of Earth has fluctuated greatly throughout time. This is also the case for our current Phanerozoic Eon, the ~541 Myr period that saw the explosion of multicellular life on Earth, average temperatures varied greatly. We now know that one of the main reasons for climate change throughout Earth's geologic history, regardless of human activity, is atmospheric **greenhouse gases** content especially with regards to **carbon dioxide** (CO_2), but how do we know this?

Greenhouse gases, such as carbon dioxide, **methane** (CH_4), **sulfur dioxide** (SO_2), and even water vapor, were primordial constituents of Earth's atmosphere, having been present on and in the Earth since around the time when it was a new planet, over 4 Gya (billion years ago). Earth's climate has always been dynamic, and Earth's average temperature has fluctuated over geologic time, ever since the planet formed, albeit fairly gradually and uniformly (Figure 2.1). There are several reasons for these fluctuations in temperature: tectonic activity, volcanism, and the chemical composition of atmosphere. Of this chemical composition, greenhouse gas content is the strongest line of evidence for warming on a global scale. These gases, especially naturally occurring CO_2, and CH_4, have warmed the planet out of ice ages in the past. CO_2 is responsible for the intensely hot temperatures on the planet Venus. Yet how do we know that these gases warm things up?

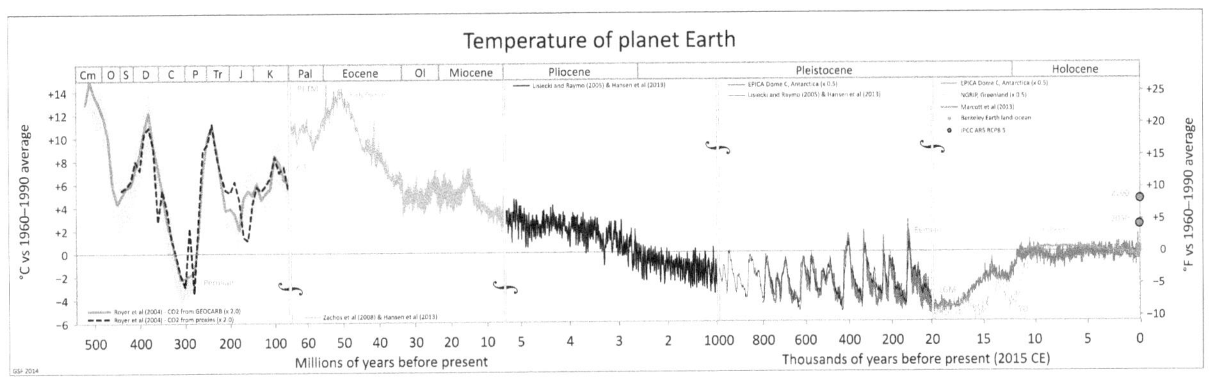

Figure 2.1 Temperature of planet Earth during the Phanerozoic Eon, based on geological evidence

We have been told, by the vast majority of scientists for over 30 years, that the accumulation of these gases, notably CO_2, was a cause for concern. Many people, not just scientists, know that these gases absorb **electromagnetic radiation (EM)** from the Sun and re-emit it as **infrared radiation (IR).** It occupies the same electromagnetic spectrum of solar emissions as visible light, ultraviolet light, microwaves, and others. IR is a type of electromagnetic radiation that has a longer wavelength than microwave radiation, which in turn has longer wavelengths (hence lower frequencies) than visible light. One of the properties of IR is that it causes an excitation of atoms and molecules in matter; in others words, it increases temperatures. So, when were greenhouse gases first investigated? As it turns out, several scientists including **Svante Arrhenius** wrote about this phenomenon during the late 19th century, long before the modern, cultural, and sociopolitical concept of global warming started become used (Figure 2.2).

Arrhenius, a Swedish scientist, published "On the Influence of Carbonic Acid in the Air upon the Temperature of the Ground" in Series 5, Volume 41, of *Philosophical Magazine* and the *Journal of Science* in April 1896. He based much of his analysis on the works of the American astronomer **Samuel Langley**, who had observed that the temperature of the Moon responded very differently to **insolation**, through exposure to solar rays, as opposed to the surface of the Earth. This was due to the presence of what is now known as greenhouse gases, particularly "carbonic acid," a term for CO_2 at the time. Similar to Langley, Arrhenius himself was more of a physicist, yet his work, including this one, resulted in him becoming one of the founders of the science of physical chemistry. After the publication of this paper, he went on to win the Nobel Prize in Chemistry in 1903 for the electrolytic theory of dissociation, which basically dealt with the way molecules break up into charged particles in mixtures.

Figure 2.2 Svante Arrhenius

Svante Arrhenius came to his conclusions through the examination of how air conducts heat due to the properties of the constituents of air in Earth's atmosphere: its various gases. One of the ways atmospheric retention of heat occurs is through diffusion as heat is transmitted through the air. The other way is through molecular absorption. Some gases absorb more heat than other gases. Nitrogen (N_2) and oxygen (O_2) are both present in the atmosphere as tightly bound diatomic molecules. These molecules do not absorb much heat since they do not vibrate enough to do so. On the other hand, molecules like methane (CH_4), water vapor (H_2O), and carbon dioxide (CO_2) do vibrate upon heating due to IR because of the loose arrangement of the different bound elements in each molecule. Arrhenius used mathematics to calculate the heat absorption coefficients of the two most prevalent greenhouse gases, carbon dioxide, and water vapor. He then created a table that detailed the atmosphere's ability to absorb heat with regard to the concentration of these gases. It almost goes without saying that the higher the concentrations of carbon dioxide and water vapor, the more heat gets absorbed.

Since he came from Sweden, a country that is at a very high latitude, Arrhenius's paper on carbon dioxide as a forcing mechanism for warming the Earth's surface may have actually been written out of an interest in the promotion of warming the Earth's surface. In fact, heating from greenhouse gases was seen as somewhat beneficial at this time because these gases help make our planet habitable and not unbearably cold. It is now well known that the cyclical pulses of Earth's Ice Ages throughout geological time are tied to atmospheric greenhouse gas content, especially CO_2. Arrhenius was also aware that the burning of fossil fuels, mostly coal during that time, contributes to the overall atmospheric CO_2 budget. Arrhenius probably did not foresee the wide-scale use of fossil fuels that was to come in the 20th and into the 21st centuries.

Consider This . . .

Examples of the end products of scientific inquiry are abundant in even our day-to-day lives. For instance, when you wake up on certain mornings and look out the window you may see fog. This is easily explained as the dew point temperature exceeding that of the condensation point near Earth's surface, but why is it explainable at all? It is because some individual or group of individuals in the past investigated that phenomenon based on a hypothesis that they developed. They then utilized a series of tests, employing the scientific method and experiments, to determine the validity of the hypothesis. Such is the case for water vapor condensing onto surfaces. This phenomenon, something television meteorologists discuss almost every day, was originally explained by the individuals who developed **air conditioning**, technology used to modify the temperature and/or humidity of an enclosed space. In the early 20th century, **W.N. Shaw**, **Willis Carrier** (the inventor of air conditioning systems), and others had created the world's first air conditioner, which was actually initially used in the agriculture industry (Figure 2.3). Condensation was a byproduct of its operation, and thus it led to an understanding of condensation. This was a testament to the close relationship between science and **engineering** since indoor air conditioning is as much a feat of engineering as it is science.

Science is a cumulative process in which collaboration is encouraged. In fact, collaboration is optimized through a process known as peer review in which scientists test the findings and theories of other scientists. This keeps the findings of the scientific community focused on uncovering the truth and weeding out findings that stray from rigorous analysis. Since so much of the infrastructure of our current civilization is directly dependent on the findings of science, the constant critical analysis of the scientific process is vital.

Collaboration is also prevalent in a lot of work done through the various fields of science. Many scientific problems and investigations require the work of multiple scientists across multiple disciplines. For instance, there is a branch of geology called geophysics, a discipline that is concerned with the physical properties of the Earth as well as its physical, energetic processes. It also covers the use of quantitative analysis of Earth's properties utilizing signaling and sensing instrumentation. Many biochemists work alongside geologists and other Earth scientists to solve environmental problems. For instance, biochemistry has been employed in the **remediation** (basically, cleaning up) of oil spills in the ocean. Bacteria that consume the organic matter comprising crude oil are let loose, in a controlled manner, into the volume of ocean water containing the spills. Through digestion, the bacteria break the crude oil down into components that are safe for the environment, including the wildlife that is often devastated when such disasters occur.

Figure 2.3 Willis Carrier with colleagues

Introduction

In order to gain an understanding of the natural environmental systems that we all live in, interact with, and depend on it is important to understand, in an objective manner, the mechanisms behind these systems. It should also be understood that conclusions are arrived at through rigorous testing and analysis. The practice of **science**, and all of the other disciplines related to it, provides the means to understanding our environment here on Earth. In fact, science provides us means to understanding the entire universe, and a greater part of our existence within it.

As we have established, science is a complex and multiscale practice. The phenomena that can be described and understood through science are generally just as complex. Science is utilized through practice to develop our understanding of everything that can be described as chemical and physical. In turn, we use that information to create practices and systems that help us to engineer means to utilize resources, develop structures that we require, and, hopefully, live in harmony with the environment. The latter is vital because, as a civilization, it has become apparent our human population and Earth's ability to sustain it has definite limits.

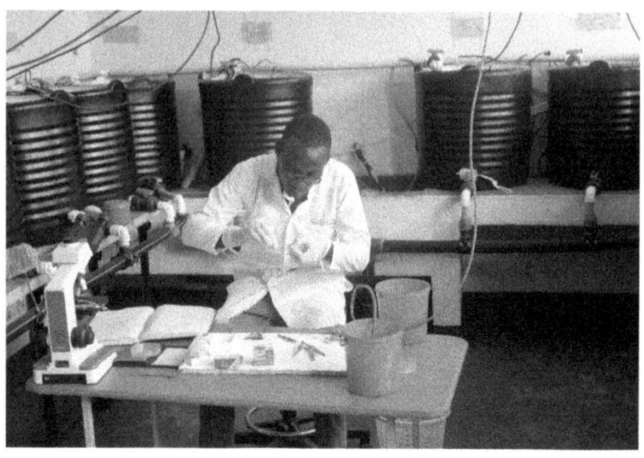

Figure 2.4 In order to test hypotheses, scientific procedures are often conducted in a laboratory setting

Scientists use the scientific method to gain knowledge of natural systems. This includes hypotheses that are developed through observation and logical inference. These hypotheses are tested through experimentation either in a controlled laboratory setting or in the field (Figure 2.4). Often, though, it is a combination of the two, especially when doing work in the fields that comprise ENVS. Data and samples are gathered from the field and brought back to the laboratory to be analyzed using any number of powerful instruments. Instrumentation is often used out in the field as well, especially when a given sample may breakdown or become altered through its transport. This work is done with rigor in order to achieve a robust result.

FYI: INTERNATIONAL SYSTEM OF UNITS

The measurement of any given set of parameters is vital to all scientific disciplines. Whether it is determining the number of moles of a substance that is produced in a chemical reaction, an accurate assessment of a population of wolves in a national park, the precise measurement of tectonic plate movement, or anything else being done in the vastness of scientific endeavor, correct measurements are important. Since science seeks to uncover truths about our world, exact measurements must be made as often as possible. Also, in order to determine the causality in scientific phenomenon through mathematical calculations and statistical analysis, accurate assessments of data must be made.

Two aspects of measurement are often discussed: **accuracy** and **precision**. They are not the same. Accuracy is the extent to which a given measurement, or calculation, is closest to the correct standardized amount. Precision refers to the consistency of the measured or calculated amounts. Both are important for scientific measurements, and both are important when reaching conclusions that are as close to the truth as possible, especially when proving the validity of a hypothesis.

The **International System of Units** is essentially the **metric system**. The metric system is a system of measurement based on log-10, or decimal components. This means that each "jump" in units is based on multiples of 10. The current International System of Units, or SI system, is the refinement of the base metric system that was introduced in Europe in the late 18th century. As of this time, the metric system is the predominant system of measurement on Earth. Currently there are only three countries that have not adopted the metric system as the official standard of measures: Liberia, Myanmar, and the United States of America. It is not entirely clear why this is the case other than the costs that would be incurred if maps and publications were changed, mainly in the United States.

There are seven base units of measurement in the SI system from which nearly all the other SI units can be derived (Figure 2.5):

- **Ampere** (A): The basic SI unit of electrical current, or flow of electrons, defined as charge per unit time
- **Candela** (cd): The base SI unit for luminous intensity, or wavelength of visible light, per section of an illuminated sphere
- **Kelvin** (K): The basic SI unit for temperature, the movement of atoms and molecules as influenced by energy or heat
- **Kilogram** (kg): The base SI unit for mass, the amount of matter in a body or object
- **Meter** (m): The base unit for length in SI
- **Mole** (mol): The base SI unit for the amount of substance, or elementary particles, in matter
- **Second** (s): The SI unit for time and the only SI unit that is not log-10

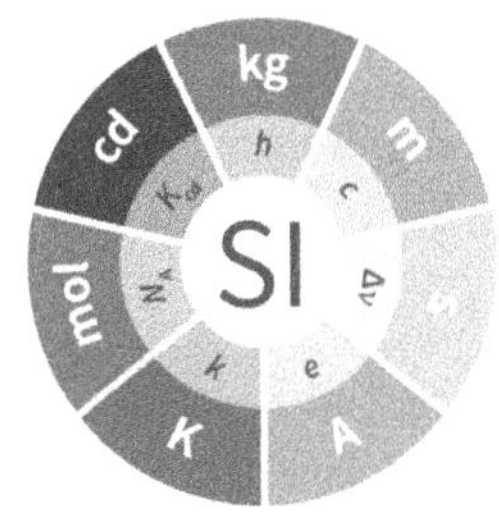

Figure 2.5 Logo of the International System of Units

Mathematics: Data and Analysis

In order to explore phenomena through the discipline of science, one has to have a means in order to interpret precise, analytical data. One would require methods that utilize numbers and the transformation of those numbers, representing actual values. This is what we call **mathematics**. The importance of mathematics, or math, to science as a whole cannot be overstated. At its very base, the representation of all chemical and physical singularities use math in its purest form. It may seem silly to say this, but math is the "language" of science. In fact, math itself is a science, one of the formal sciences, which includes other fields like analytical linguistics and computer science.

Through its application, math is a way to investigate and represent the real world. Science is the objective investigation of real-world phenomena, so math acts a language of sorts for the purpose of explaining the real world. Engineering is a discipline that utilizes science and math to solve real-world problems. In order to study engineering and, in the case of this environmental science course, science, a basic understanding of mathematics is required.

Data

The melting point of water ice is 0° Celsius (32° Fahrenheit), the Earth is approximately 4.54 Byr, Cesium (Cs) atoms vibrate at 9,192,631,770 cycles per second (used for precision in atomic clocks), and any organism consuming another in a lower trophic level will receive only about 10% of the energy content of its food. All of that is information, or **data**, based on real-world phenomena. Data is gathered in science through precise measurement and observation. Then through analysis, often using very complex mathematics, this data yields equations that allow us a deeper understanding of our world than before. For instance, Albert Einstein's famous equation, $E = MC_2$, gives us information about the fundamental nature of matter and energy. It is a derived formula that was calculated using information Einstein gathered through experimentation, observation, and calculation. Einstein's formula tells us that when mass (M) is accelerated to the speed of light (C) squared (2), it yields an equal amount of energy inherent within the system.

USING EXPONENTS AND LOGARITHMS

An **exponent** is a value raise to a "power" such that each consecutive number is a multiple of the previous result by the original number. For example, 22 = 4, 210 = 1024, 310 = 59,049, and so on. With variables the form is generally x2, x3, x57, and so on.

The **logarithm** refers to the exponent itself and the base of the logarithmic function is the variable. For example, regarding the exponential value 103 = 1000, the logarithm of 1000 to base 10 is 3. The logarithm is 3 and the base is 10. Log base-10 is commonly used for values in multiples of ten for anything from population statistics to measuring the acidity of solutions (Figure 2.6). Using a logarithmic scale on a graph can be helpful when the data covers a large range of values that increase exponentially in value over given ranges.

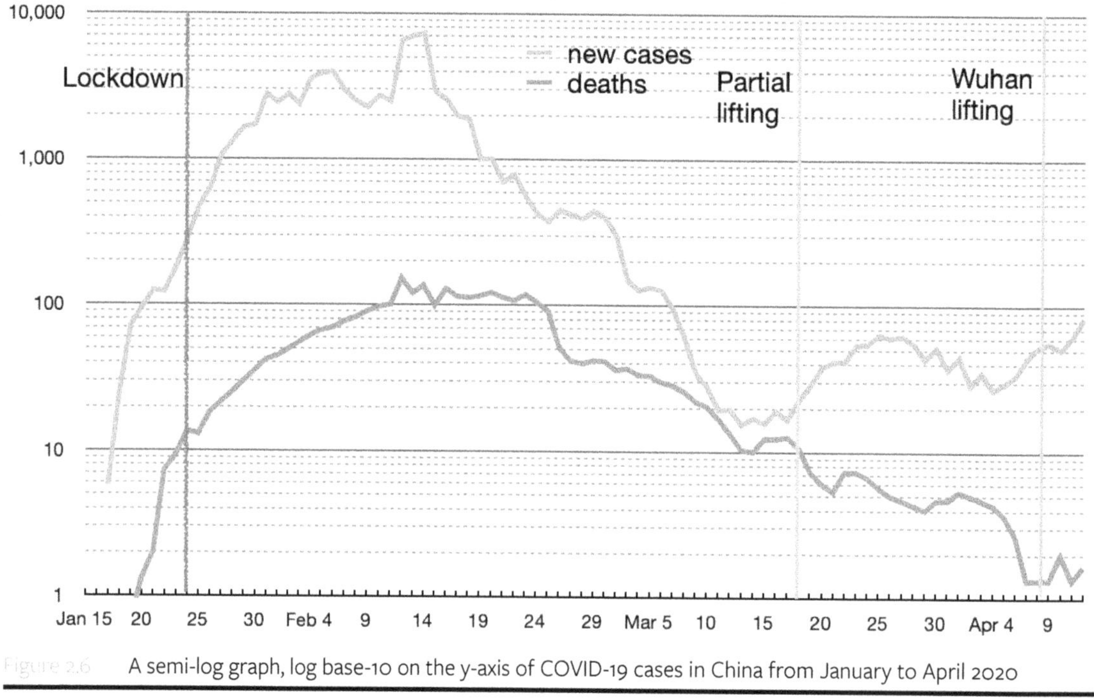

Figure 2.6 A semi-log graph, log base-10 on the y-axis of COVID-19 cases in China from January to April 2020

Scientific Models

Scientific **models** are used to represent events, ideas, processes, or systems. They can also be used to make predictions of ongoing processes in time. These models can either be mathematical or physical, although they often are mathematical in order to foster reproducibility. Scientists use models to identify and understand patterns in order to predict phenomenon. For instance, engineers and scientists use hydrological models to help predict when and where floods may occur. Scientific models have to be consistent and based on proven scientific understanding. Models also utilize rigorous and often complicated mathematical functions because nature is complicated. Models are tools for predicting and understanding science; they are not factual statements. Scientific models can be either physical or mathematical. Mathematical models are usually run as computer programs.

Mass and Volume

Chemistry is the science that deals with the behavior and nature of matter, everything in the universe that has mass and occupies volume. **Chemists** are the scientists who work within this field. It is a fundamental science that is utilized in many other disciplines, such as biology, geology, and chemical engineering. Through chemistry, the working of atoms and chemical reactions are understood, as well as the essential nature of matter.

Matter

Matter is all material in the universe that has mass and occupies space and volume. There are four natural phases of matter: solid, liquid, gas, and plasma. Of these four, three—solid, liquid, and gas—are of primary interest to us in our survey of ENVS.

Solids are composed of particles, like atoms and molecules, that are tightly packed together. This doesn't give particles much chance to move, and all particles move. Even in the somewhat rigid solids, atoms and molecules vibrate. Volumes of solids have their own shape. Solids "rest" in their shapes; in other words, solids have both a fixed shape and volume.

The particles that make up **liquids** are more loosely packed than that of solids. Liquids do not rest in their shapes; they take the shape of their containers or surfaces that they rest on. They are not necessarily less massive than solids. You know this if you've ever carried two full pails of water up a hill! Liquids have no fixed shape but a do occupy a fixed volume.

Gases are very loosely packed assemblages of atoms and/or molecules. Gases are very energetic states of matter and are only bound if they are placed into some kind of sealed container. Gases are extremely low density, and the space between particles of gas is less than liquids and solids. Gases have no fixed shape or volume.

Plasma, like gas, has no fixed shape or volume. It is similar in lack of density to gas, yet it is a different state of matter. Plasma consists of charged particles, negatively charged electrons and positively charged ions, within a sort of an energized "miasma." It is the most energetic of the states of matter. It generally exists at extremely high temperatures. The Sun and other stars are largely composed of plasma. The conditions that form plasma are rarely found on Earth; lightning generates it and is part of the phenomenon of atmospheric auroras.

Most substances, whether singular or as components of something else, are capable of changing from one state of matter to another. This is called a **phase change**. Phase changes are brought about due to changes in energy, usually heating or cooling. This induces change in temperatures, which leads to the critical points necessary for a mass of substance to change phase. For instance, on Earth, water exists in the three main environmental states of matter: solid, liquid, and gas. Heating ice higher than 0° Celsius (32° Fahrenheit) causes it to melt into a liquid. Heating water further to 100° Celsius (212° Fahrenheit) causes it to boil and become vapor (gas). Under certain conditions, ice can undergo a phase change to water vapor; this is called sublimation. The opposite is also true, and water vapor can change from gas to solid without becoming liquid; this is called deposition. All substances have unique phase change properties.

Basic Components and Properties of Matter

Matter can be transformed from one type of substance into another, through phase changes or other means, but it cannot be destroyed or created within the universe. Matter simply becomes something else. This is known as the law of conservation of matter. This first principle of chemistry was not discovered by one person, but generally accepted as a guiding precept by researchers as far back as the 18th century. The law of conservation of matter helps us understand that the amount of matter stays constant in all systems. It is vital for understanding how matter is cycled through the complex systems that make up Earth, from sinking slabs of oceanic crust becoming consumed by the Earth's mantle through subduction to the nutrients that are exchanged annually between and within ecosystems.

Atoms and Molecules

Atoms are the smallest components of matter that maintain an element's chemical properties (Figure 2.7). Atoms comprise **elements**. Elements are a fundamental type of matter, with given sets of properties based on their component atoms. Even down to one atom, an element will always retain its particular set of properties. In turn, the fundamental nature of an atom, the basic characteristics of an element, are mainly governed by the number of protons in the nucleus. The **Periodic Table of Elements** is used to categorize elemental properties (Figure 2.8).

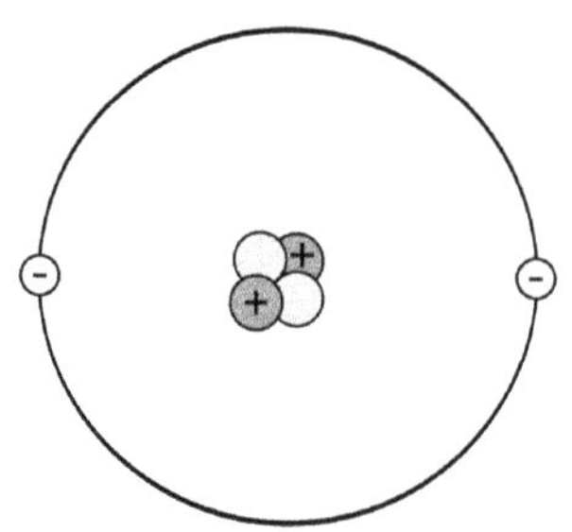

Figure 2.7 Diagram of an atom (not to scale). This is an atom of element 2, helium (He)

The nucleus is the central part of an atom containing protons and neutrons, other than hydrogen, in which only a proton, under normal conditions, occupies the nucleus. The number of protons in an atom impart its characteristics. This is an atom's **atomic number**. For instance, the gaseous nitrogen, element number seven has one extra proton in its nucleus compared to its neighbor, solid carbon, yet nitrogen is vastly different from carbon. Both have very different properties, yet the difference is down to one proton.

The atomic nucleus contains both positively charged protons and, with the exception of hydrogen, neutrons that carry no charge, hence they are "neutral." Surrounding the nucleus are electrons that orbit the atom in discrete levels called shells. Electrons are negatively charged particles of a magnitude that are exactly the opposite of the positive charge that protons contain. Electrons are also very small; a proton has 1,836 times the mass of an electron. Electrons moving between atoms results in what we know of as electricity. Generally speaking, each atom contains the same number of electrons as protons, but there are exceptions.

Again, with the exception of the basic hydrogen atom, all atoms of every element contain neutrons. Neutrons are fundamental particles that are similar in size to protons, albeit slightly more massive, yet again carry no charge. The number of neutrons that are normally present in atoms is the same as the number of protons and electrons. The number of neutrons and protons measured together is called the **mass number** of an atom, or its **atomic mass**. When there are more neutrons than protons in the nucleus of an atom, it can result in a different sort of atom with different sorts of properties.

Period																																
1	1 H																															2 He
2	3 Li	4 Be																									5 B	6 C	7 N	8 O	9 F	10 Ne
3	11 Na	12 Mg																									13 Al	14 Si	15 P	16 S	17 Cl	18 Ar
4	19 K	20 Ca															21 Sc	22 Ti	23 V	24 Cr	25 Mn	26 Fe	27 Co	28 Ni	29 Cu	30 Zn	31 Ga	32 Ge	33 As	34 Se	35 Br	36 Kr
5	37 Rb	38 Sr															39 Y	40 Zr	41 Nb	42 Mo	43 Tc	44 Ru	45 Rh	46 Pd	47 Ag	48 Cd	49 In	50 Sn	51 Sb	52 Te	53 I	54 Xe
6	55 Cs	56 Ba	57 La	58 Ce	59 Pr	60 Nd	61 Pm	62 Sm	63 Eu	64 Gd	65 Tb	66 Dy	67 Ho	68 Er	69 Tm	70 Yb	71 Lu	72 Hf	73 Ta	74 W	75 Re	76 Os	77 Ir	78 Pt	79 Au	80 Hg	81 Tl	82 Pb	83 Bi	84 Po	85 At	86 Rn
7	87 Fr	88 Ra	89 Ac	90 Th	91 Pa	92 U	93 Np	94 Pu	95 Am	96 Cm	97 Bk	98 Cf	99 Es	100 Fm	101 Md	102 No	103 Lr	104 Rf	105 Db	106 Sg	107 Bh	108 Hs	109 Mt	110 Ds	111 Rg	112 Cn	113 Uut	114 Uuq	115 Uup	116 Uuh	117 Uus	118 Uuo

Alkali metals | Alkaline earth metals | Lanthanides | Actinides | Transition metals

Poor metals | Metalloids | Nonmetals | Halogens | Noble gases

State at standard tempurature and pressure

Atomic number in red: gas

Atomic number in blue: liquid

Atomic number in black: solid

solid border: at least one isotope is older than the Earth (Primordial elements)

dashed border: at least one isotope naturally arise from decay of other chemical elements and no isotopes are older than the earth

dotted border: only artificially made isotopes (synthetic elements)

no border: undiscovered

Figure 2.8 Current, widely used Periodic Table of the Elements

Isotopes

When atoms have differing numbers of neutrons they are called **isotopes**. This results in an atom having a higher than normal mass number. Isotopes of elements behave differently than normal atoms. When isotopes become very heavy they become unstable and begin to decay. This is called **radioactivity**. These unstable isotopes decay into more fundamental particles until they are transformed into stable products of that isotope, often called **daughter products**. Radioactive decay is not necessarily a linear process. For instance, uranium-238 (U-238), an isotope of the element uranium, undergoes a decay chain. The final stable daughter product of U-238 is lead-206 (Pb-206). Through the process, U-238 decays into many unstable daughter products such as radioactive isotopes of Bismuth (Bi) and Thorium (Th).

An isotope's half-life is the amount of time it takes for one-half of the atoms of a radioactive element to give off radiation and decay into some other daughter product. **Radioactive decay**, in this manner, occurs spontaneously but also at a remarkably regular rate. Due to this fact, **radioactivity**, the property of exhibiting radioactive decay, a term coined by the pioneering scientist **Marie Curie**, can be used for radiometric dating. The age of ancient materials can be very accurately assessed. **Radiometric dating** is used by scientists such as astronomers and geologists to calculate the age of rocks, both terrestrial and extraterrestrial. Radiometric dating has allowed scientists to discover the age of the Moon from lunar rocks collected by asteroids. The Moon is 4.51 Gyr. Uranium-235 (U-235), used in commercial nuclear power, has a half-life of around 700 million years, which is slightly older than Earth's current geologic Phanerozoic Eon. Rocks containing U-235 and other radioactive isotopes, or radioisotopes, allow scientists to determine the age of Earth materials and even the age of the Earth itself, which has been determined to be 5.543 Gyr.

Tritium, also known as hydrogen-3, is another radioisotope. It is an isotope of hydrogen. Tritium is unstable, hence radioactive, with two neutrons in the nucleus along with one proton. There is another isotope of hydrogen called deuterium, also known as hydrogen-2, that has one neutron in the nucleus. Deuterium, unlike tritium, is stable. Tritium is used in commercial, industrial, and technical applications such as luminous dials and as an environmental tracer for determining water flow. When unstable isotopes like tritium decay, they emit varying levels of energy or radiation. Much like tritium, radioactive elements have properties that are widely used productively for a variety of reasons, such as energy production, like with U-235. Despite their benefits, these radioactive materials often constitute hazards to the environment and to human health and well-being. We all know that radioactivity is the basis for atomic weaponry.

Chemical Bonding

Atoms are bonded together through outer valence shell electrons and through charge interactions. Molecules are combinations of two or more atoms bound covalently. A **covalent bond** occurs when two atoms share available valence electrons. Valence electrons are electrons in the outer shell. In covalent bonding these available electrons are shared in the outer shell of all atoms in the molecule. Atoms of the same element, such as atmospheric free oxygen (O_2) and nitrogen (N_2), as well as atoms from different elements, such as water (H_2O), form molecules in this manner. When atoms come from different elements, those molecules form compounds. Other examples of compounds are carbon dioxide (CO_2), methane (CH_4), and glucose, a type of sugar ($C_6H_{12}O_6$).

Covalent bonds tend to be quite strong. For instance, it takes a fairly large amount of energy to split atoms apart in most molecules, such as in the compound we know as water. The water molecule also exhibits a large negative charge due to the size difference between the oxygen and the two hydrogen atoms. The oxygen atom is much larger than both the hydrogen atoms combined, so the overall charge of the covalent water molecule is negative. This is known as a **polar covalent bond**.

One of the properties of the water molecule is that in its liquid phase water is composed of many independent, individual molecules that are bound together through electrical attraction. This, again, is due to the overall negative charge site of the water molecule. The larger amount of exposed negative charge readily yet tenuously attracts the positively charged hydrogen atoms in water molecules, but these sites also attract anything else with a positive charge. This is called a **hydrogen bond** and gives water the electrical property to bind to other atoms or molecules that have an overall positive charge.

Depending on the element and the conditions, atoms ionically bond to form compounds rather than covalently bond. An **ionic bond** occurs when atoms are bound through the exchange of positive and negative charges. Atoms with charge differences due to extra or missing outer valence electrons are called **ions**. Ions that lose electrons are positively charged; these are cations. Ions that gain electrons and are negatively charged are anions. Since nonmetallic elements have a negative charge, these tend to be anionic. Metallic elements are generally positively charged and tend to be cationic. If one looks at the periodic table of elements, one notices that metallic elements occupy the left side of the table, and nonmetallic elements occupy the right side. The bond that results between cations and ions results in full valence shells for both atoms.

Unlike covalent bonding in which electrons are shared, electrons are completely transferred in ionic bonds. This is what occurs when sodium (Na) and chlorine (Cl) bond, forming the mineral halite. Halite is also called salt, although it's only one kind of salt, our common table salt. Ionic bonding forms salts rather than molecules, and these, in turn, comprise ionic compounds. Examples of salts forming ionic compounds are common table salt, which is the mineral halite (NaCl), potassium chloride (KCl), and calcite ($CaCO_3$). Ionic compounds may be formed from elements that are ionic or from charged covalently bonded atoms.

Some covalently bonded atoms carry a positive or negative charge because the number of electrons do not match the number of protons in the bond, so they form an ion rather than a molecule. These are known as **polyatomic ions**. Ammonium (NH_4) and sulfate (SO_4) are examples of polyatomic ions. This can be seen in the chemical formula of these ions with the appearance of a positive (+) or negative (-) symbol, as well as the number of charges preceding the symbol in the upper right of the formula. The aforementioned calcite (also known as calcium carbonate) is composed of an elemental cation (Ca^{2+}) and a polyatomic anion, the carbonate ion (CO_3^{2-}). These polyatomic ions often form compounds that are important to Earth's ecosystems, such as calcite, which is used by many marine animals such as coral to make their homes, and naturally occurring ammonium, which is vital for plant growth.

Metallic bonding is the third major type of chemical bond. In metallically bonded compounds the valence electrons travel freely around the cationic nuclei within the compound. This gives metallic substances significant properties such as ductility (the ability to deform something with it breaking), strength, and electrical and thermal conductivity. All of the metallic elements that we know of, such as iron (Fe), copper (Cu), gold (Au), and mercury (Hg), are bonded metallically. Steel is a common alloy. It is iron that is often alloyed with tungsten (W) and vanadium (V), which are also metallically bonded elements. Interestingly enough, Fe is often bonded with C, a nonmetallic element, which increases the overall strength of steel.

If one, again, looks at the Periodic Table of Elements in Figure 2.8, one notices that the nonmetals on the right side are generally covalently bonded into molecules, the metals on the left side (most of the table really) are metallically bonded, and a bond between a metallic and nonmetallic element is ionic. This generally holds true in nature and human engineering, although there are many exceptions and permutations of this general rule. For instance, many complex substances exhibit combinations of two or more types of bonds. As previously mentioned, calcite is both covalently and ionically bonded because it is a formed from an element (Ca) and a polyatomic ion (CO_3).

Solutions are composed of atoms, molecules, salts, metals, or some combination of two or more substances exhibiting these bonds. A solution is a homogenous mixture of two or more substances. These mixtures involve one or more components to be dissolved; these are called solutes. The component or components that solutes are dissolved into are called solvents. Solutions can be quite complex and often form very important substances in the environment. Examples of solutions are blood, crude oil, and milk.

Conservation of Matter

The 18th-century French chemist **Antoine Lavoisier** discovered that the total mass of the products of a chemical reaction have the same total mass of the reactants. Conservation of matter is, simply put, the fact that matter is neither created or destroyed chemical reactions, only changed. This is one of the fundamental laws of chemistry. It can be extended to all chemical reactions on Earth, including those that occur within the environment. Even the combustion of materials like wood does not result in less matter, only the transformation of matter. The char and vapors given off by chemical reactions once the wood has been burned are exactly equal to the mass of the original wood. The implications of this law extend to all chemical reactions in Earth's environment, including anthropogenic activity, such as energy production, pollution, crop fertilization, water treatment, and so on.

FYI: STOICHIOMETRY

What happens during the chemical reactions that form compounds? Chemical reactions occur because compounds or elements of some sort come into contact with each other, but this is not enough. Energy (heat) must also be present to drive the reaction forward. If energy cannot drive a reaction forward, it will probably not occur. Chemical reactions are often complex, but in general energy and proximity are the two main determining factors. The chemical components, or **chemical species**, that react with each other are called the **reactants**. The results of the chemical interactions are called the **products**. Chemical reactions are conservative, meaning that the mass of the reactants is going to be the same as the mass of the products. This is in accordance with the law of conservation of mass. Since mass is conserved, chemical reactions can be written as balanced equations.

Stoichiometry is the relationship between products and reactants, in which balanced equations are used to determine what occurs within a chemical reaction. Through these equations, chemical reactions can be understood and even predicted, especially when experiments are conducted in a laboratory setting. Things get more complex in the natural world, of course, but stoichiometry can still be applied in order to understand chemical reactions in the natural world. Our environment is, in many ways, driven by chemical reactions such as photosynthesis, which essentially provides us with food and oxygen, and combustion, which provides us with energy and results in wildfires. Many of our environmental problems are the result of chemical reactions due to human activity, such as pollution, and acid rain deposition. Chemical reactions can also help solve environmental problems such as the catalytic convertors in most automobiles that reduce the amounts of carbon monoxide, nitrogen oxides, and non-combusted hydrocarbons that enter the atmosphere.

Organic Compounds

Chemical compounds that contain C-C or C-H bonds are called **organic compounds**. These are, in turn, bonded to a limited number of other elements, commonly N or C. Sometimes P, S, or a few other chemical elements are bound within organic molecules as well. Organic compounds tend to be more complex than inorganic compounds, and organic molecules can be massive. There are millions of organic chemicals. These compounds, as one can deduce, are vital to life and biological processes. Examples of organic compounds include the sugar sucrose, vitamin B-12, and deoxyribonucleic acid (DNA).

Hydrocarbons are organic compounds that contain only C and H. The simplest hydrocarbon is methane (CH_4), and hydrocarbon molecules get larger as well as more complex from there. There are hundreds of thousands of these types of compounds. Hydrocarbons comprise all of the compounds that we utilize as energy, including methane, which formed over time within the Earth from prehistoric organisms.

All-Encompassing Water

There is one simple chemical compound above all others that is central to our existence on Earth. This compound is **water**. Water is not only present on Earth, it is present on other planets in the solar system, albeit in the form of ice. In fact, water is ubiquitous throughout the universe.

Saying that water is essential for life on Earth is an almost absurd understatement. In fact, life as we know it could not exist without water. Life itself had its beginnings in Earth's waters. In addition to that, the presence of water in all three phases—solid, liquid, and gas—on Earth is characteristic of our planet, the only planet that we know of that has life. Water is essential to Earth's unique weather and climate patterns. It flows across and under the Earth, transporting chemical species and sediments. It is the main solvent for both biochemical and geochemical reactions. Water is part of the very molecular structure of many of Earth's rocks and minerals. In other words: water is vital.

Water has several key properties giving it unique characteristics:

- The angle between the hydrogen atoms attached to the central oxygen atom is 104.5°, an oblique angle (greater than 90°) imparting solid ice with the following properties; it is less dense than liquid water, and water actually expands in volume when frozen.
- Water molecules are held together through hydrogen bonding, giving liquid water its unique flow properties.
- Water's strong cohesion through its hydrogen bond feature a large negative site. This is due to the large size of the oxygen atom. It allows for the transport of chemical species, nutrients, sediments, and waste products.
- Water absorbs heat with only small changes in its temperature; it has a high specific heat capacity, which acts to stabilize temperatures on Earth such as near coastlines. Note that specific heat capacity is the amount of energy (or heat) per unit mass required for raising or lowering the temperature of a substance by 1°C.
- Water dissolves other molecules, in solution, in its capacity as a solvent.

Water has many other properties that make it the main reason for Earth's characteristically exceptional environment for harboring life. Water covers nearly 71% of Earth's surface. It is part of the very structure of the Earth. It is found not only in the empty spaces between rock and soil particles, but it is also in the very molecular structure of many minerals that make up rock units. Water is present on Earth's surface, mostly in the ocean, and it is part of the atmosphere. The presence of water in all of its phases has a significant effect on Earth's weather and climate. Water has shaped Earth's geological history and continues to be essential to life's ongoing evolution.

Acids and Bases

Acidity and basicity are important chemical properties grounded in the properties of water. **Acids** are chemical compounds containing a hydrogen cation (H+), called either **hydronium** or **protium**. They are essentially just a proton, a hydrogen atom stripped of its electron. **Bases** are chemical compounds containing a hydrogen-oxygen anion (OH-) known as **hydroxide**. Basicity, the amount of base concentration, is also sometimes called alkaline, but that is a bit misleading. Alkalinity is a subset of basicity. Alkaline compounds are basic salts containing an alkali metal or alkaline earth metal, usually calcium or magnesium.

Hydronium and hydroxide ions are formed from water molecules. In any given volume of water, some H_2O molecules will spontaneously dissociate into H+ and OH- depending on various conditions. The amount of hydronium and hydroxide ion concentration is expressed in terms of the negative logarithm of hydrogen (H+) content, or **pH**. pH is defined as the negative logarithm of proton concentration because of product over reactants calculations.

Measuring pH yields a measure of proportional hydronium cations to hydroxide anions in a given solution of water. This type of solution, acid and base with water as the solvent, is called an aqueous solution. There are many different aqueous solutions in the environment, from sulfuric acid formation at volcanic vents in the deepest parts of the ocean to extremely basic alkaline lakes such as Lake Natron in Africa. Aqueous solutions also result from anthropogenic activity, like acid mine drainage due to the improper management of coal mines.

The pH scale generally runs from 0 to 14, although values less than 0 and greater than 14 are possible. Pure water contains equal parts hydronium cations and hydroxide anions. Pure water has a pH of 7. Acidic solutions (high H+) have a pH less than 7. Basic solutions (high OH-) have a pH higher than 7. pH uses a logarithm based on multiples of 10: log base-10. Thus, a pH of 5 is equivalent to 10 times more hydronium ions than a pH of 6. For that matter, a pH of 4 is equivalent to 100 times the number of hydronium ions than a pH of 6. The same is true for bases, just on the opposite end of the scale from 7. A basic solution with a pH of 10 has 100 times more hydroxide ions than a solution with a pH of 8. Natural waters contain both hydronium and hydroxide ions in varying proportions based on the environment in which they are found. It is rare to find natural waters that are completely neutral, with a pH of 7. A pH of around 8.2 is essential in ocean water in order to sustain healthy and functional aquatic ecosystems. This pH has dropped slightly lower recently in many parts of the ocean to around 8.1. This small difference is of great concern and is one of the major environmental challenges of our time.

ENVIRONMENTAL SCIENCE ENCOMPASSES MANY FIELDS OF SCIENCE

As mentioned in Chapter 1, ENVS is not a science per se, but a blanket term for the scientific fields that enable understanding of our planet, its mechanisms, its living organisms, its ecosystems, and the unique challenges that are presented to us in order to maintain its systems. In addition to the aforementioned primary scientific fields of chemistry, biology, physics, and geology, the following are some of the main fields employed in environmental scientific endeavor. Engineering is included here since there is a strong analytic and quantitative basis for engineering in its employment of the natural and formal sciences to solve environmental problems. Fields like sociology, economics, and political science are not included because those, and other similar fields, fall under the heading of social sciences. At the same time, as we move further into our study of ENVS, we will explore some of the social, political, cultural, and economic aspects and implications of our relationship with Earth's environment. Following is a list of disciplines beyond basic biology, chemistry, geology, and physics that are within the ENVS umbrella. Most of the following are multidisciplinary in nature, meaning they combine several academic disciplines into one due to the nature of the work. Note, the following list is by no means complete:

- Agronomy: The subdivision of agricultural science concerned with applications and principles of soil, water, and crop production
- Biochemistry: The study of the chemistry of living organisms
- Botany: The subdivision of biology concerned with the study of plants
- Cartography: The subdivision of geography concerned with creating maps
- Civil engineering: The branch of engineering dedicated to the creation and maintenance of public works
- Climatology: The study of weather patterns over extended, usually 30 years or more, time
- Computer science: A formal science dealing with machine logic, computation, and information
- Demography: The study of populations, particularly that of human populations
- Ecology: The science of the relationships between organisms as well as the relationship between organisms and the nonliving world
- Environmental engineering: The applied science of improving environmental quality
- Epidemiology: The study of the occurrence and distribution of health patterns, such as communicable disease, within populations
- Food science: The science dealing with all aspects of food and nutrition
- Forestry: The interdisciplinary science of the study and management of woodlands and wild spaces in general for the better of human life and environmental quality
- Geography: The study of Earth and planetary surfaces, including Earth's atmosphere and surface water
- Geotechnical engineering: The branch of engineering concerned with utilizing Earth materials for developmental purposes
- Hydraulic engineering: The branch of engineering concerned with the development of water and sewage structures
- Hydrology: The multidisciplinary study of water, as well as water movement, occurrence, and management
- Marine biology: The study of life within the ocean
- Meteorology: The science of weather forecasting and weather over short timescales
- Microbiology: The subdiscipline of biology dealing with microscopic life
- Oceanography: The science of the oceans and all that occurs within them
- Soil science: The study of soil formation, properties, classification, and occurrence with an emphasis on soil conservation and management
- Statistics: The mathematical study of the analysis, organization, and use of data
- Zoology: The subdiscipline of biology dedicated to the study of animals

Conclusion

Scientific processes require the analysis of data both in the laboratory and in the field. In order to do this mathematics is required in addition to primary accurate measurements observations. The scientific method is actually a series of methods that make use of data, mathematics, and logic to reach verifiable conclusions about our Universe and our planet, Earth. One of the fundamental natural sciences, chemistry, the science of that which occupies volume and has mass, requires the implementation of data analysis through both accuracy and precision.

Discussion Questions

Directions: Review the chapter in order to completely and correctly respond to the questions and prompts.
Keep this in the back (or maybe even front) of your mind: Science is system of methods that allows us to understand not only our planet, and ourselves, but of the entirety of the universe itself. Answering the following questions correctly will lead to a more robust understanding of how to apply science to ENVS problems.

1. Why was Svante Arrhenius's study of the effects of carbon dioxide, or carbonic acid, on surface warming as relevant back in 1896 as it is today?
2. What is a biological solution for the remediation of oil spills?
3. A kilogram (kg) is 1,000 grams (g). How many kilograms is 100 kg?
4. In keeping with question 3, 1 kg is approximately 2.2 pounds (lb.). Is the answer heavier than 1 lb.?
5. Look at the graph (Figure 2.9). Is the difference in average global temperatures between 1920 and 2000 exponential or linear (more steady)?
6. What other sorts of phase changes occur in nature, other than the melting or evaporation of water?
7. Would you expect elements with large atomic numbers, greater than 10, to have a greater tendency toward radioactivity than elements with atomic numbers smaller than 10? Why?
8. What happens when halite, otherwise known as common table salt, goes into solution when mixed into boiling water on a stovetop?
9. Why are organic compounds so common on Earth? Think about what these compounds contain.
10. Why does water expand in volume when it freezes?
11. Why do marble (high pH) monuments and headstones become worn away when exposed to weather outside, particularly to rainwater?

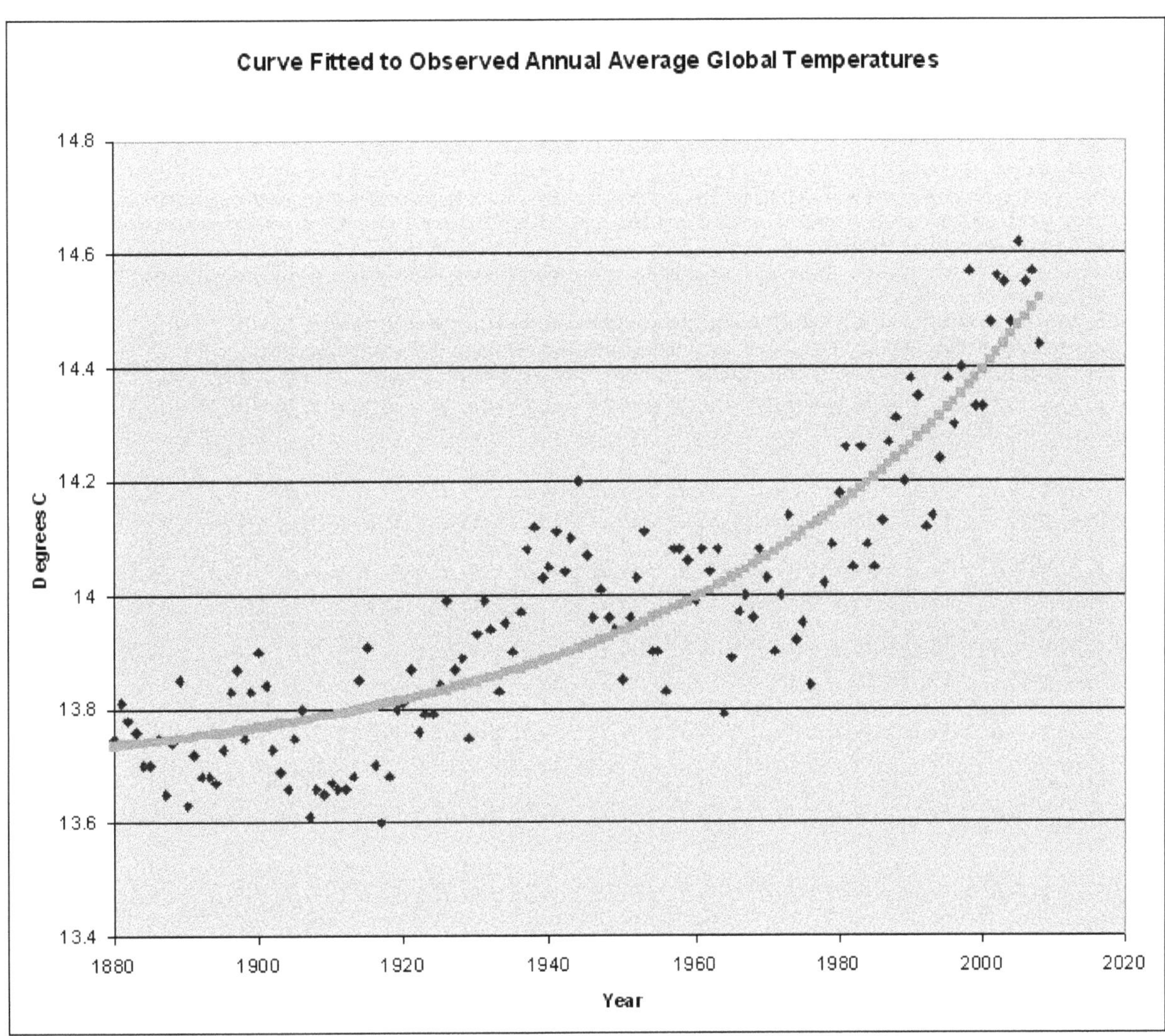

Figure 2.9 Graph of global average temperatures

Bibliography

Arrhenius, S., 1896. *On the Influence of Carbonic Acid in the Air upon the Temperature of the Ground: Synthese*, v. 133, n. 1/2, p. 5–11.

Doll, J.E., and Baranski, M., 2011, *Greenhouse Gas Basics: Michigan State University, Climate Change and Agriculture Fact Sheet Series*, E3148, 2 p.

Gilbert, J., and Boulter, C.J., 2000, *Developing Models in Science Education*: Boston, Kluwer Academic Publishers, 387 p.

Houghton, J.T., Jenkins, G.J., and Ephraums, J.J., eds., 1990, *Climate Change: The IPCC Scientific Report*: Cambridge, Cambridge University Press, 364 p.

Kuhtz, R., 2015. *Chemistry: Understanding Substance and Matter*: New York, *Britannica Educational Publishing*, v. 1, 138 p.

Obaya, A.V., Barocio, Y.R., and Rodríguez, Y.M.V., 2021, *Online Simulators for the Teaching of the Law of Conservation of Matter and Chemical Reactions in High School: Science Education International*, v. 32, n. 3, p. 209–19.

Ouellette, R.J., and Rawn, J.D., 2015. *Principles of Organic Chemistry*: Amsterdam: Elsevier, 498 p.

Shaw, W.N., 1888. *Report on Hygrometric Methods; First Part, Including the Saturation Method and the Chemical Method, and Dew-Point Instruments: Philosophical Transactions of the Royal Society of London*, v. 79, p. 73–149.

Credits

Fig. 2.1: Copyright © by Glen Fergus (CC BY-SA 3.0) at https://commons.wikimedia.org/wiki/File:All_palaeotemps.svg.
Fig. 2.2: Photogravure Meisenbach Riffarth & Co. Leipzig, "Svante Arrhenius," https://commons.wikimedia.org/wiki/File:Svante_Arrhenius_01.jpg, 1909.
Fig. 2.3: Carrier Corporation, "Willis Carrier," https://commons.wikimedia.org/wiki/File:Magna_Mine_A-C_System_made_large-scale_air_conditioning_feasible.jpg, 1902.
Fig. 2.4: Copyright © by Bruno ashimwe winks (CC BY-SA 4.0) at https://commons.wikimedia.org/wiki/File:Fish_scientist_in_a_laboratory.jpg.
Fig. 2.5: Copyright © by International Bureau of Weights and Measures (BIPM) (CC BY-SA 3.0 IGO) at https://commons.wikimedia.org/wiki/File:SI_Logo_with_defining_constants.png.
Fig. 2.6: Copyright © by Chris55 (CC BY-SA 4.0) at https://commons.wikimedia.org/wiki/File:Covid_19_new_cases_and_deaths_China_showing_Hubei_lockdown.png.
Fig. 2.7: Copyright © by Svdmolen/Jeanot (CC BY-SA 3.0) at https://commons.wikimedia.org/wiki/File:Atom.svg.
Fig. 2.8: Copyright © by AnthonyDuo122 (CC BY-SA 4.0) at https://commons.wikimedia.org/wiki/File:PtableWide.png.
Fig. 2.9: Conhegarty, "Graph of Global Average Temperatures," https://commons.wikimedia.org/wiki/File:CNH_Figure_3.jpg, 2009.

Chapter 3

Science Basics, Part 2

Life, Energy, and the Earth

Objectives

- This chapter introduces you to the concept of life on Earth.
- When this chapter is completed, you will understand the structure of the cells that are the basic components of life.
- You will understand the interrelated nature of organisms through evolution once this chapter is completed.
- The foundational chemical components of all organisms are covered in this chapter.
- The foundational concept of energy, that it moves and transforms everything, will become apparent after this chapter is read.
- Life itself is dependent on energy; that is discussed in this chapter.
- Earth and all of its living things are dependent on thermodynamic system, and those properties are covered here.
- An insight into the enormity of Earth's geological history will become apparent toward the end of this chapter.
- You will know that Earth is a dynamic planet with regard to plate tectonics after this chapter is read.
- Rocks and minerals as the basic components of the solid Earth are discussed here.

KEY WORDS AND TERMS

Biology (n.): The branch of science dedicated to the study of matter as well as transformations of matter in all its forms.

Physics (n.): The branch of biology dedicated to the study of how organisms relate to each other and their environment.

Geology (n.): The science of the Earth that deals with its physical and chemical characteristics, history, and processes; also includes the study of other planetary bodies in the universe.

Physiochemical (adj.): Relating to both chemical and physical (energetic) properties.

Consider This . . .

As stated before, life on Earth is unique. It is not just "life on Earth," it is the only life that we know about! Currently, over 5,000 exoplanets (planets that exist outside of our Solar System) have been confirmed by astronomers and other planetary scientists. Although water and other chemicals that form the building blocks of life have been detected on these exoplanets, through spectroscopic analysis and other means, no incontrovertible proof of life outside of our planet has been confirmed. Even Mars in our Solar System, one of the most likely candidates for harboring life, has not yielded any concrete proof of life, although NASA's Perseverance rover has sent back data that indicate life could have been possible in the past. Earth currently remains the one planet in the Universe, that we know of, that bears the necessary **physiochemical** conditions for life to exist on its surface.

Introduction

Life has existed on Earth, in some form or another, for around 4 Gyr. A form of graphite, a type of carbon, that is associated with living processes has been found in 4.1 billion-year old zircon crystals found in the Jack Hills region of Australia. This predates the Earth's second eon, the Archean, placing it squarely within what was originally thought to be the completely inhospitable first eon of Earth, the Hadean. This means that conditions were favorable for life very early on in Earth's distant geological past. This is truly remarkable but life itself is remarkable, as is Planet Earth's propensity to favor it.

Astronomers and other planetary scientists refer to Earth as being in the "Goldilocks Zone." Goldilocks Zones are also known as "habitable zones," orbital regions around a star in which conditions are favorable for liquid water to exist on a planetary surface. Earth certainly features an abundance of surface water, but there are other factors favoring life on its surface as well.

Earth's atmosphere moderates its surface temperatures, as does the presence of its vast ocean. Plate tectonics has also led to the formation of continental landmasses that jut above the ocean. These landmasses, slowly move into, past, and apart from one another as Earth's outer surface glides across its, unimaginably hot and energetic lower depths. This creates the surface topography in which lifeforms ultimately came to dwell. In the distant past, these landmasses began shedding sediments that provided some of the chemicals that life itself depends on, because life began in the sea.

Within the darkest depths of the oceans, volcanic vents have been providing even more chemicals for living processes for eons. These are powered by the immense heat that comes from within the Earth itself. At its distance from Earth, the Sun provides the right amount of energy on Earth's surface for, well, basically everything else, from weather to the water cycle to life itself! For if life is to exist at all, it must have energy.

The Complexity of Life

Life, as far as we know it, is unique to only one planet, Earth. Probing the solar system and beyond, explorers and researchers have found some of the basic components of life, mostly amino acids in meteors, but no true signs of living things. Whether there is life beyond Earth has yet to be determined, but **biologists**, scientists who study living things, do have basic criteria for what is living and what is not living. A living thing is called an **organism**. In order for something to be classified as an organism it must, at some point in its development, exhibit the **characteristics of life**. For something to be called an organism it must respond to its environment in some manner, be capable of change and growth, have the potential to reproduce, have offspring that have the potential to pass their traits to their offspring, and be composed of at least one **cell**. A cell is the smallest unit of life that can either live on its own or be part of the composition of a larger, multi-celled organism. All living things are composed of at least one cell.

Characteristics of Cells

A cell is the basic unit of life's organization. Cells are capable of fully functioning within the parameters of the characteristics of life, whether the organism is unicellular (a one-celled organism) or multicellular (a complex organism containing tissues and possibly organs made of many cells). All organisms on Earth are comprised of either one of two types of cells, the concept known as **cell theory**.

Prokaryotes are ancient; they evolved over 3.5 Gyr ago. This makes them, as far as we know, the earliest forms of life. Prokaryotes are all unicellular organisms, and prokaryotic cells are the simplest sorts of cells. These cells lack two basic features that are part of more complex cells: an abundance of **organelles** and a **nucleus**. Organelles are small structures in cells that perform specific tasks. Prokaryotes have organelles but on a limited basis. They contain basic organelles such as ribosomes, which create proteins but not much else. Another important characteristic of a prokaryotic cell is that it does not contain a nucleus. The biological meaning of nucleus is different from the chemical definition. In **biology** a nucleus is a membrane structure in a cell that contains genetic material. Bacteria and archaea are prokaryotes.

Eukaryotes are organisms that are composed of the other type of cell, eukaryotic cells. Eukaryotic cells are more complex than prokaryotic cells. Eukaryotic cells contain many more specialized organelles than do prokaryotic cells, such as **mitochondria,** which extracts energy from sugars and fats, and **chloroplasts,** which allow plants to manufacture their own food (Figure 3.1). Eukaryotic cells also tend to be much larger than prokaryotic cells, but the major difference between the two types is that eukaryotic cells always contain a membrane-bound nucleus containing genetic material. Also, though there are many types of single-celled eukaryotes, there is a multitude of multicellular eukaryotic organisms containing any number of specialized eukaryotic cells.

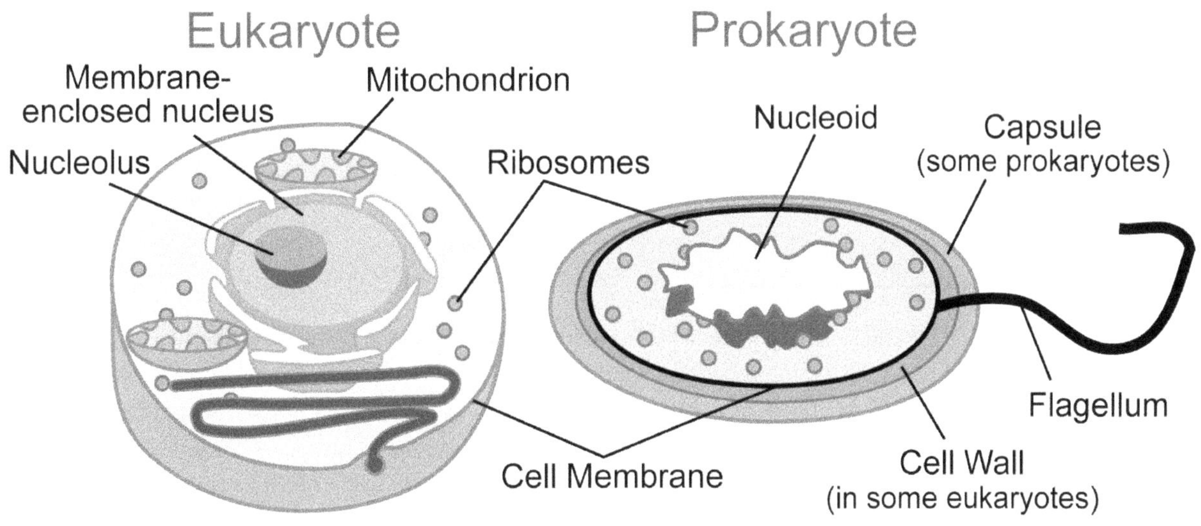

Figure 3.1 Basic differences and similarities between prokaryotic and eukaryotic cells

Classifying Life

As with most processes in science, at some point it became apparent that a system of classification was necessary for organisms. Carl Linnaeus, an 18th-century Swedish scientist, proposed a system of **taxonomy** that in essence continues to be used today, although in some ways that system is outdated. Taxonomy in biology is the practice of naming and classifying organisms based on common characteristics. The original system was somewhat hierarchical; since then other systems have come into use.

You may be familiar with the classic hierarchical classification of life. This system is based on appearances and characteristics. The following ranking system, from broadest to most specific, in this original taxonomic system may be familiar to you: kingdom, phylum, class, order, family, genus, species, and even subspecies. Recently, though, these categories were assembled within the largest groups of all, above kingdoms: **domains**. There are three domains of life: **bacteria, eukarya,** and **archaea** (Figure 3.2). Bacteria include all the single-celled prokaryotic organisms that we know, as well as, bacteria. Eukarya includes all eukaryotic organisms: single-celled **protists**, multicellular **fungi** (except for single-celled yeast), multicellular **plants**, and multicellular **animals**. Archaea, organisms of the third domain, are prokaryotic, single-celled organisms that seem similar to bacteria although that is not the case. While they are similar in size and structure to bacteria, their biochemistry is very different. It is generally accepted that archaea and eukarya are actually related. Archaea are also known as "extremophiles." Various types of archaea can live in very extreme environments: subzero temperatures, boiling water, extreme salinity, and extremely acidic waters.

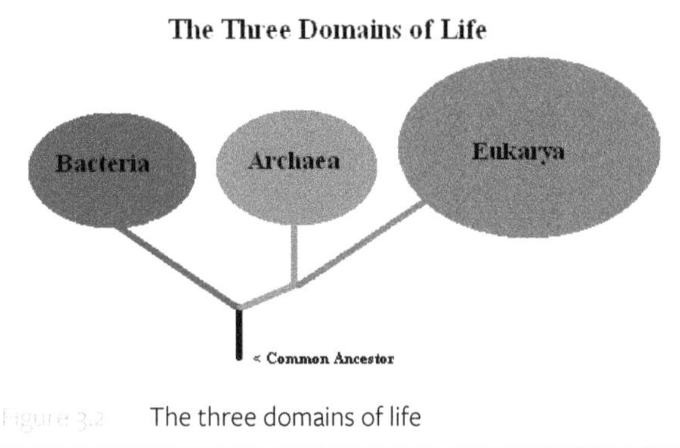

Figure 3.2 The three domains of life

Another system of classification is based on relatedness, **cladistics**. Cladistics has no defined levels such as kingdom or class like the older system; it is based on shared characteristics of ancestral relations. A species is classified according to shared common ancestors as are the species that came before it. What leads to the development of new species are new traits that help to define the emergence of new species that diverge from their common ancestors. Note that the older system is still used to name organisms, such as genus and species. This system is based on the fact that organisms all evolve from previous ones. Organisms are grouped into broad clades, organisms with a common ancestor, reflecting evolutionary precedent.

Evolution

Evolution is a process proposed by **Charles Darwin** and **Alfred Russel Wallace** in the early 19th century. It is based on the observations and principles that competition, as well as sheer survival in Earth's environment are based on successful biological characteristics. These characteristics or traits are passed down from generation to generation through reproduction of some sort. The means of transferring these characteristics was not known until the results of Gregor Mendel's experiments became published in the early 20th century. Mendel, a European monk, experimented with plant growth through succeeding generations. This work led to him founding the modern science of genetics.

The mechanism by which evolution works is called **genetics**. In all organisms, each individual passes on traits to their offspring. Through sexual reproduction with two parents, multiple traits are passed down to offspring. These traits can be anything, from the color of a plant to horns on a gazelle, to the types of hands a human has. It doesn't matter; it all gets passed down in some form. These genetic traits are often important for survival. Genetic changes occur due to mutations, random alterations in genetic material that are often due to changes in an organism's environment. Major successive changes, or adaptations, lead to the formation of new species. The actual biochemical means for genetics is contained in an organism's genetic materials, complex molecules that are essential to life.

Atoms and Molecules Integral to Life

Even the components of the simplest cells, such as bacteria, are complex. There are a few compounds composed of key elements that are essential for life on Earth. Water, H_2O, is one of them. Water is vital to all life on Earth. Even though there are organisms that can exist with much less water in their bodies and surroundings than others, water is still needed.

Six elements are key to the formation of the biochemicals (the chemicals of biology) that comprise Earth's living things: carbon, hydrogen, nitrogen, oxygen, phosphorus, and sulfur. In turn, these elements, in varying proportions, are integral to four distinct types compounds that are essential to life itself on Earth: carbohydrates, lipids, nucleic acids, and proteins. These four types of compounds form **macromolecules**, very large molecules that are important to biological function.

Carbohydrates

Carbohydrates are a group of compounds formed from hydrogen (H) and oxygen (O), proportionally to H and O in water (H_2O), bonded to carbon (C), hence "carbohydrates." These compounds are formed when plants undergo **photosynthesis**, the process by which plants produce nutrients using sunlight as a catalyst. Carbohydrates form **polymers,** which are long chains of molecules that have a repeating structure. There are thousands of different types of carbohydrate molecules. **Autotrophs**, like plants and certain types of bacteria, form carbohydrates when they undergo photosynthesis. Autotrophs are "self-feeders," organisms that can utilize environmental C to create what is essentially food for themselves.

Starches, cellulose, and sugars are all carbohydrates. Sugars include fructose, glucose, and sucrose. Carbohydrates store energy and provide essential structure in living organisms. In addition to cellulose in plants, the carbohydrate **chitin** reinforces the exoskeletons of **arthropods** (insects, spiders, and the like) as well as the cell walls of fungi. Carbohydrates also function to store energy in organisms, which can later be broken down for use. There are commercial and industrial applications for carbohydrates, such as fabrics, photographic film, plastics, and paper products.

Nucleic Acids

Another group of molecules vital to biological function are **nucleic acids**. Nucleic acids contain structures formed by the bonds between five key elements: carbon, hydrogen, oxygen, nitrogen, and phosphorus. Nucleic acids function to transmit hereditary features to the next generation; they also control the manufacture of specific proteins. There are two basic types of nucleic acids that occur naturally: DNA and the generally less complex ribonucleic acid (RNA). Like carbohydrates, nucleic acids are polymers consisting of often very large molecules in the form of a helix double-strand with various, regularly spaced attachment sites. The attachment sites are called **nucleotides** and are composed of a sugar, a nitrogen base, and a phosphate: C, N, and P, the three macronutrient elements.

Proteins

Proteins are complex macromolecules that are essential for life and govern a wide range of cellular activities. Like carbohydrates and nucleic acids, proteins are polymers. They are largely composed of carbon and nitrogen, as well as hydrogen and oxygen. Proteins are responsible for most of the chemical reactions that occur in cells. These reactions are driven by **enzymes**, proteins that cause and regulate chemical reactions to occur in cells without the enzyme being altered. Proteins also provide structure in cells and provide binding for tissue formation. They also help protect organisms from disease, control gene activity, and often take the form of hormones. Protein is an essential nutrient for animals. Human beings require protein-rich foods such as beans, fish, and meat.

Lipids

The fourth biologically essential macromolecule, **lipids**, are molecules that are insoluble in water. Unlike carbohydrates, nucleic acids, and proteins, lipids do not form polymers. Lipids include fats, waxes, and steroids. Lipids have various functions such as energy storage, structural support, and **cell signaling**. Cell signaling is the ability of cells to process, receive, and transmit signals. This can be done through physical means such as pressure and temperature, or biochemically. Fats are a part of human dietary consumption, including meat and fish, as well as vegetables and other sources.

FYI: SYNTHETIC POLYMERS

Polymers are a highly diverse group of chemicals that have been used by people for many reasons, for building materials, energy production, fabrics, food, and a lot more. Prior to the industrial age, people used polymers from naturals sources, mostly animal and plant products. That changed in 1907 when the Belgian American chemist **Leo Baekeland** invented "Bakelite," the world's first fully synthetic polymer. Baekeland experimented with formaldehyde and phenols (a group of organic compounds) in order to create an electrical insulator that was more effective than shellac. The result was Bakelite, a versatile polymer that proved to be an excellent insulator for electrical components. Today, human civilization relies on a vast number of different synthetic polymers providing us with a multitude of uses. These are often known by their brand names such as Kevlar, nylon, and Teflon. Plastic, which includes a wide range of different compounds, is a synthetic polymer.

Figure 3.3 Seal entangled in plastic netting

These polymers are often very durable and resistant to chemical breakdown. This is also problematic because when disposed of these compounds take a long time to naturally break down. Polymers such as plastic can break down into potentially harmful chemicals. The result is long-lasting pollution, and when this pollution breaks down it often results in environmentally harmful chemical species that leads to health problems with both wildlife and human beings (Figure 3.3). These polymer pollutants are especially harmful to marine organisms. Many of these polymers, again mostly plastics, are derived from petroleum hydrocarbons, which adds to their impact to the environment, as well as to Earth's climate. It should be apparent that less harmful substitutes for plastic need to be used. For instance, the non-plastic straws that are currently being used in many restaurants are a good start, although more powerful and timely solutions are still needed.

Characteristics of Life

There are certain criteria for something to be called living, an organism: (1) they need to use energy, (2) they must respond to their environment in some way, (3) they must be able to reproduce in some way, and (4) they must be composed of at least one cell. Anything that meets at least those four criteria is an organism and is **biotic**. Biotic refers to anything that is an organism. More specifically, **biotic living** refers to currently living organisms such as a frog or a tree, and **biotic nonliving** refers to any organism that is deceased, such as a deer that has been hunted and killed. It can also refer to parts of an organism such as beefsteak, coal, or the wooden leg of a chair. Anything that has never lived and never met all four of the basic criteria for the characteristics of life is called **abiotic**. For instance, a chunk of granite, salt crusts in a desert, or the water molecules in a pond are all abiotic.

Autotrophs and Heterotrophs

One of the characteristics of life is the need for organisms to use energy. In this regard there are two types of organisms: **autotrophs** and **heterotrophs**. Autotrophs are self-feeders that utilize external means to produce their own food, a form of stored energy, within their bodies. Examples of autotrophs are plants and algae. Heterotrophs gain energy by feeding on other organisms. All animals and fungi are heterotrophs. This makes human beings heterotrophs. There is an interdependency between autotrophs and heterotrophs with regard to cycling energy and nutrients in ecosystems.

Photosynthesis and Chemosynthesis

The type of autotrophic system that we are all familiar with is **photosynthesis**. Photosynthesis is the process utilized by plants, and many kinds of single-celled organisms, to make food within their bodies. This food contains stored energy in the form of a carbohydrate, such as a starch or sugar. In photosynthesis, carbon dioxide and water undergo a chemical reaction in which the product is a carbohydrate of some sort. This process is initiated by light from the Sun. In plants and other autotrophs, photosynthesis takes place within cellular organelles called **chloroplasts**. A green-tinted protein in chloroplasts called **chlorophyll** absorbs light at the beginning of the photosynthetic process. Photosynthesis is responsible for the basis of all the foods that heterotrophs consume. In addition to that, the waste product from photosynthesis is something very important to life on Earth: oxygen. Here's the basic stoichiometry for photosynthesis: $6CO_2 + 6H_2O + \text{light} = C_6H_{12}O_6 + 6O_2$.

Chemosynthesis is similar to photosynthesis in that the product is a carbohydrate, but the way the process is initiated is very different. Chemosynthesis only takes place deep in the ocean beyond the reach of the Sun's rays. Through volcanic activity at the bottom of the ocean, hydrogen sulfide (H_2S) is released into seawater. Certain types of autotrophic microscopic organisms use the energy of reaction with H_2S to produce food in the form of carbohydrates. This is similar in manner to surface autotrophs using carbon dioxide and water to produce carbohydrates. The difference is the driving mechanism: solar energy for surface autotrophs, a chemical reaction with hydrogen sulfide for deep sea autotrophs. The waste product from hydrogen sulfide–based chemosynthesis is some form of sulfur or sulfur compound, usually sulfuric acid (H_2SO_4), which thankfully dissociates completely within the ocean's vast amount of water. Here's the basic stoichiometry for chemosynthesis: $6CO_2 + 6H_2O + 3H_2S\,(\text{energy}) = C_6H_{12}O_6 + 3H_2SO_4$.

Cellular Respiration

When organisms use carbohydrates that they obtain for energy, whether through autotrophic or heterotrophic behavior, the mechanism is **cellular respiration**. Cellular respiration is, more or less, the reverse of chemo- or photosynthesis in that carbohydrates are broken down using oxygen, which produces energy. The breakdown products of this process are carbon dioxide and water. All organisms use cellular respiration for energy, even autotrophs. Once plants, and other autotrophs, produce their own food, through chemo- or photosynthesis, the food is either broken down for energy or stored for later. Here's the basic stoichiometry for cellular respiration: $C_6H_{12}O_6 + 6O_2 = 6CO_2 + 6H_2O + \text{energy}$.

Energy Moves Everything

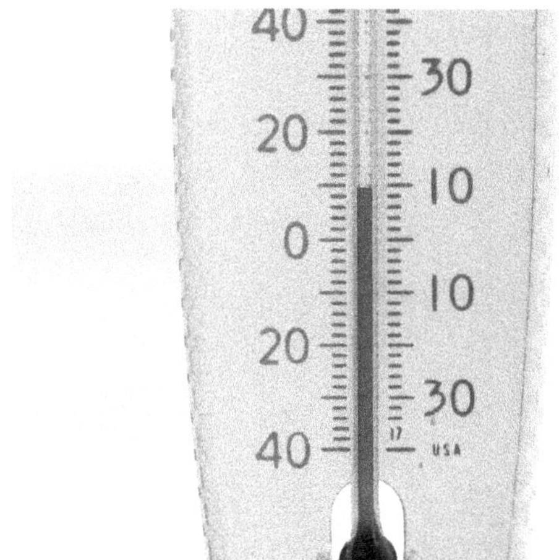

Figure 3.4 A change in temperature is a measure of the change in movement of atoms and molecules due to the application of heat, or energy

Energy is the ability to do work and affect change. **Physics** is the study of energy, as well as the interaction between energy and matter. Scientists who study and apply these principles are **physicists**. As a discipline, physics has a highly interdisciplinary relationship with mathematics, with math acting as the basis behind most of the principles of physics. Physics is a foundational science that is behind nearly every scientific discovery and principle.

Two essential manifestations of energy are **kinetic energy**, the energy of motion, and **potential energy,** which is the energy of position or storage. Energy stored in atomic and molecular bonds within matter and living organisms is called chemical potential energy. When potential energy is harnessed, it is transformed into kinetic energy. Heat is also a form of energy that changes the temperature of matter. Temperature and heat are not the same. Heat, as energy, causes the atoms and molecules in matter to move more rapidly in a system. This movement of atoms and molecules causes an increase in temperature. When heat leaves a system, the opposite happens and the temperature of that system drops (Figure 3.4).

The basic formula for kinetic energy is $KE = \frac{1}{2}mv^2$, where *KE* is kinetic energy, *m* is mass, and *v* is velocity. This is energy of motion.

The basic formula for gravitational potential energy is $PE = mgh$, where PE is potential energy, m is mass, g is gravity, and h is height. Other forms of energy storage have their own equations, but the principle remains the same: energy transfromation within a system. For instance, a meal that a person prepares contains stored energy within chemical bonds. Once that person eats that meal, the digestive process breaks down the chemical bonds in the food, releasing stored energy that is transformed into kinetic energy used for any number of actions, such as talking or walking.

There are many other ways potential energy is transformed into kinetic energy in the world at large; in fact, this is how nearly everything happens. Examples include wind dislodging a rock, causing it to roll downslope; someone lighting a campfire, causing the wood to burn and release heat; gravity causing a meteorite to slam into the Earth at high speeds; and food that someone eats being transformed into energy through cellular respiration. Please note that energy, both kinetic and potential, in the metric system is measured in **joules**.

Energy in Earth Systems

There is a difference between energy that is stored, causes motion, or affects change deep within the Earth and near, or on, Earth's surface. Earth's internal energy is generated through (1) the fact that the Earth's **core**, the super-heated center of the Earth, is constantly cooling down and releasing heat; (2) the radioactive decay of minerals containing elements such as potassium (K), thorium (Th), and uranium (U); and (3) heat induced from pressure and friction. This internal heat can be harnessed for electricity and heating at certain places on Earth in the form of **geothermal energy**, energy that originates from within the Earth used by people.

Earth processes at the surface of the Earth are powered by the Sun. The Sun provides energy for everything on and near Earth's surface. For instance, solar energy drives heating that changes Earth's temperature on the surface and drives wind patterns. Also, evaporation through solar heat is a main driver of Earth's water cycle, the **hydrologic cycle**. The Sun provides warmth for living things and warms soil. Even materials just beneath the surface of the Earth were originally provided energy from the Sun. **Coal** is the altered remains of woody plant material buried beneath the ground in low to no oxygen conditions. The potential energy in coal originated from plant life that depended on solar energy to grow.

The Four Laws of Thermodynamics

Nature itself is, in great part, governed by a set of physical properties known as **thermodynamics**. Thermodynamics is the study of heat, temperature, and work, and their effects on matter. There are four laws of thermodynamics that have been developed by scientists over the years. They have broad, universal applicability to all of the natural sciences:

1. *The zeroth law of thermodynamics*. The basis of this law is that of thermal equilibrium. Simply put, when one system comes into contact with another, both systems will eventually equilibrate at the same temperature regardless of the initial temperature of either. If a third system comes into contact with the two joined systems, it will equilibrate to the same temperature as the other two eventually. If all three systems have the same temperature, nothing changes.
2. *The first law of thermodynamics*. This is also known as the law of conservation of energy. Essentially, it is similar to the law of conservation of matter. In a closed system, energy is neither created nor destroyed; energy can only be transformed. Energy cannot spontaneously be emitted from nothing. In this regard, the entire universe can essentially be thought of as a closed system.
3. *The second law of thermodynamics*. The second law deals with **entropy**, the tendency for all systems to eventually reach maximum disorder and the lowest energy state over time. Energy eventually dissipates evenly within a system over time, whether it be the heat generated from the internal Earth or the universe itself, as some scientists have postulated. None of it matters. Entropy will eventually reach a maximum. For

instance, before a log is put on a fire it is an orderly piece wood replete with cellulose and well-structured plant cells. After the wood has been completely burned, though, all that remains is a burnt remnant of ash, basic random atoms and molecules, incapable of further providing heat, of providing energy. That log has reached a higher state of entropy.

4. *The third law of thermodynamics.* The third law of thermodynamics is, from a subjective viewpoint, a rather exotic one. This law basically states that entropy approaches a constant state as the temperature of a system lowers. Eventually the temperature of system, say a cold distant planet, approaches, but does not achieve, **absolute zero,** which is zero degrees on the Kelvin scale. This is the point at which, theoretically, all motion ceases. Absolute zero has never been observed in either nature or in a laboratory setting.

The Earth Itself

The Earth is our home, and it is unlike any other planet that we know of. Through current research, many new planets outside of the solar system are being discovered. None so far have what Earth has and that is life. Life itself has altered the very make-up and properties of the Earth, as well as the atmosphere surrounding it. **Geology** is the study of the Earth itself. It is the science of all the principles governing the functioning and properties of Earth. It is also the science of the history of the Earth, including the history of its life. Scientists who study geology are called **geologists**. The structure of the Earth, and all of the processes that occur upon and within the Earth, have framed our existence on this planet. Should humans inhabit other planets, then much the same can be said for those new planets because geology is also involved in the study of the functioning and properties of other planetary bodies as well.

Earth's Origin and Structure

The Earth formed from primordial dust and gas 4.546 Gyr within a spinning disc of materials generated by the concurrently forming Sun. Earth was originally a molten spherical body. Within that sphere there were different sorts of elements and Earth materials with differing densities. As the planet formed, the heavier materials sank to the gravitational center of the Earth while the lighter ones remained atop to "float" on top the denser ones. As Earth cooled and became more solid in its outer layers, eventually three somewhat distinct, chemically differentiated layers formed: the **crust**, the **mantle**, and the **core**. The crust, where we reside atop, is composed of mostly lighter elements such as aluminum (Al), calcium (Ca), silicon (Si), and sodium (Na). Below the crust is the mantle composed of some of the elements in the crust along with a greater abundance of heavier elements, primarily iron (Fe) and magnesium (Mg). Dense yet super-heated rocks under high pressure give portions of the mantle a semi-liquid property. This is where **magma**, molten rock, is generated. Magma that erupts onto the surface of the Earth is **lava**. The Earth's center, the core, is extremely heavy and dense. It is composed of mostly iron (Fe) and nickel (Ni).

Tectonic Plates

As the Earth's layers were set in place, a lot of energy continued to be generated. Heat from within radiated, and continues to radiate, outward to a discrete layer of the mantle called the **asthenosphere**, the second layer of the mantle. This is the portion of the mantle that is in a semi-liquid state, or a state of **plasticity**. Due to this property, radiative heat from deep within the mantle causes the asthenosphere to **convection**: heating of the asthenosphere is driven upward then in turn, cooling downward, forming a continuous cycle. This "conveyor belt" of heat has enormous, almost unimaginable, magnitude; because of that it moves the layer above it around in a continuous manner.

The layer above the asthenosphere consists of the entire crust and the first layer of the mantle. This layer is stronger than the asthenosphere; it is in a brittle, not plastic state. It is known as the **lithosphere**. The lithosphere is not a single continuous layer, but it is broken up into seven very large plates and many smaller ones. These **tectonic plates** have been reconfiguring themselves since **plate tectonics**, the phenomenon of movement and occurrence of Earth's lithosphere, initiated around 3.2 Gyr ago (Figure 3.5).

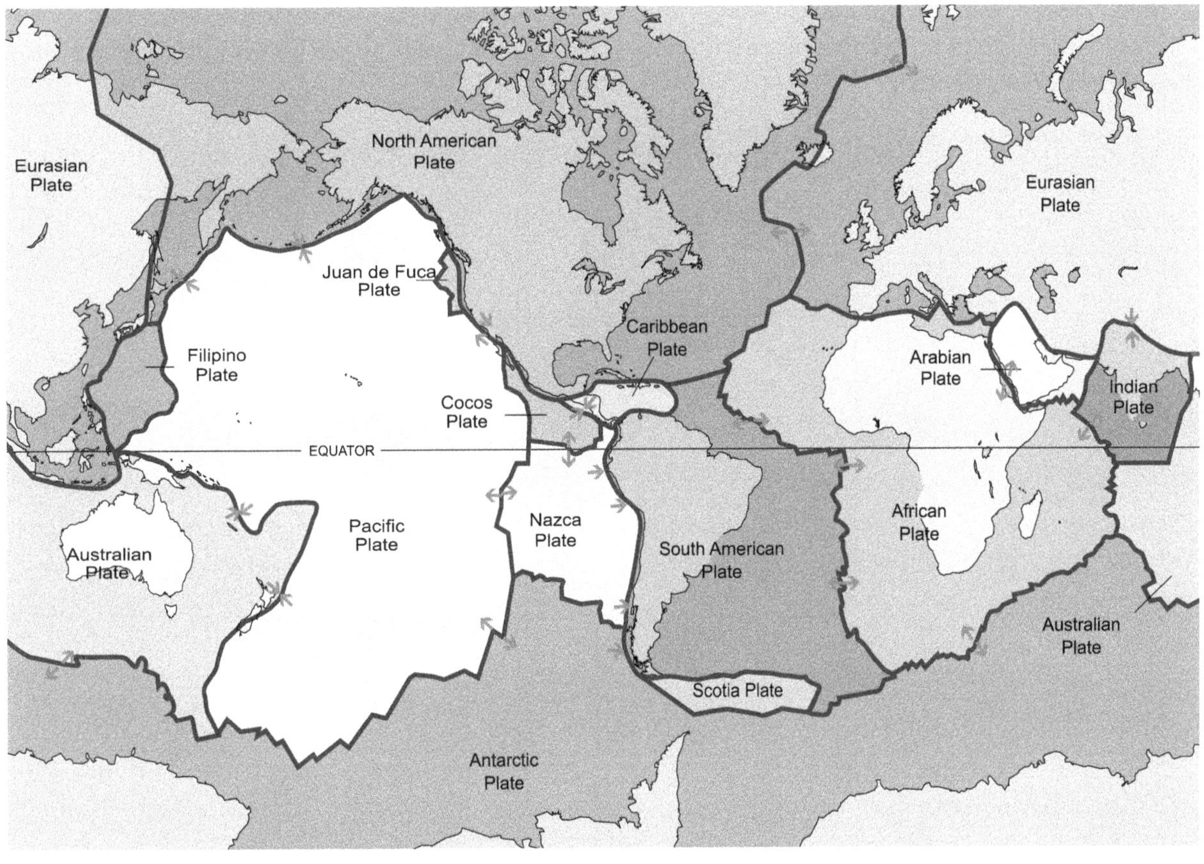

Figure 3.5 Earth's principal tectonic plates

Plate tectonics is the setting for Earth's natural history. It is responsible for the formation of mountains and most volcanoes, as well as the basins that contain Earth's oceans. Plate tectonic activity alters climate and weather patterns; it is the phenomenon behind most earthquakes. It also frames the ecosystems that contain life.

Evidence of life has been found as early as 3.7 billion years ago (Gya), some 500 Myr prior to what is thought to be the activation of full-scale plate tectonic activity. The earliest lifeforms were all prokaryotic. It wasn't until around 1.85 Gya that more complex eukaryotic life evolved, and it took until 541 Mya for the rapid expansion of multicellular life, during the **Cambrian period**, called the **Cambrian explosion** to occur. All of these events may seem disconnected because of the enormous spans of time separating them, yet they are all connected because they are part of Earth's total history. Processes that have shaped the history of Earth occur slowly and steadily over time spans that are far beyond that of human life spans, even that of human history itself. This is the concept known as **deep time**. It helps us understand the rate at which geological processes such as plate tectonics and rock formation occur, as well as the implications behind those processes.

The Rock Cycle

The solid Earth is composed of Earth materials known as **rocks**. Rocks are hard, dense materials that are usually composed of **minerals** and sometimes other components such as organic matter. Naturally occurring minerals are defined as (1) being essentially solid in their natural state, (2) having been formed naturally through geologic processes, (3) having an orderly atomic or molecular structure composed of crystals, (4) having a distinct chemical formula, and (5) being inorganic in nature. That last one means that coal, which is a rock, is not composed of minerals, since it is not compositionally crystalline. Those five defining principles also mean that water ice is actually, by definition, a mineral since it meets all of the criteria!

Figure 3.6 These features in Arches National Park, Utah, are composed of sandstone, a sedimentary rock

There are three different types of rocks in the Earth, formed from assemblages of minerals under three basic types of conditions. Rocks and the minerals that compose them are important because they have been used by human civilization for a multitude of important purposes: building materials, energy resources, medicine, and more. **Igneous rocks** form from Earth materials that were in the liquid phase at some point in time, molten. There are two types of igneous rocks, **intrusive**, which formed from slowly cooling magma deep in the crust, or **extrusive**,which formed from rapidly cooling lava on Earth's surface. Rocks that form at or near Earth's surface are known as **sedimentary rocks** (Figure 3.6). These rocks form from fragments or preexisting rocks that cement and compact into new forms.

They can also form from chemical residues that originate in biological or chemical activity near the surface. Fossils, the evidence of remains of ancient life, are found in sedimentary rocks. Deep rocks that never melted yet formed through intense heat and pressure are called **metamorphic rocks**. These rocks form from **parent rocks**, which can be any of the other three types of rocks: igneous, sedimentary, or even metamorphic. Many of the characteristics of metamorphic rocks are dependent on the characteristics of their initial parent rocks. Metamorphic rocks are often found near the Earth's surface as the remains of large mountain chains after those mountains have eroded away.

Conclusion

Physics is a primary natural science that addresses the phenomenon of energy, that which does work. The secondary scientific disciplines of biology and geology, as with all other scientific disciplines, are dependent on an understanding of physics, as well as chemistry, in order to implement solutions.

Discussion Questions

Directions: Review the chapter in order to completely and correctly respond to the questions and prompts.

Keep this in the back (or maybe even front) of your mind: Science is system of methods that allows us to understand not only our planet, and ourselves, but of the entirety of the universe. Answering the following questions correctly will lead to a more robust understanding of how to apply science to ENVS problems.

1. What are some reasons why muscle cells in your body are eukaryotic rather prokaryotic?
2. From an evolutionary standpoint, why are birds the direct descendants of dinosaurs?
3. What are the three macromolecular polymers necessary for life?
4. During cellular respiration, what is broken down into carbon dioxide and water?
5. A calorie is a unit of stored energy, or heat. What are the calories in your last meal allowing you to do?
6. What is the kinetic energy (KE) in joules (j) of a 500 kg boulder falling off a mountain at a velocity of 10 m/sec?
7. Why does firewood, once completely burned, have higher entropy than before it was burned into char?
8. What elements would you expect to find in the minerals that make up granite?
9. Walk around your yard or some other natural space. Do you see any rocks? Why are they there?
10. Why do tectonic plates move?
11. What impact does rock formation have on Earth's environment? Think about things that we do every day.

Bibliography

Atkins, P., 2010, *The Laws of Thermodynamics: A Very Short Introduction*: Oxford, Oxford University Press, 106 p.
Benfield, R., 2009, *Biochemistry: Delhi*, Global Media, 96 p.
Doll, J.E., and Baranski, M., 2011, *Greenhouse Gas Basics: Michigan State University, Climate Change and Agriculture Fact Sheet Series*, E3148, 2 p.
Levin, H., and King, Jr., D., 2017, *The Earth Through Time: Hoboken: Wiley*, 11th ed., 608 p.
Ouellette, R.J., and Rawn, J.D., 2015. *Principles of Organic Chemistry*: Amsterdam: Elsevier, 498 p.

Credits

Fig. 3.1: National Center for Biotechnology Information, “Differences and Similarities between Prokaryotic and Eukaryotic Cells,” https://commons.wikimedia.org/wiki/File:Celltypes.svg, 2005.
Fig. 3.2: C3lin, “The Three Domains of Life,” https://commons.wikimedia.org/wiki/File:3_domains_of_life.png, 2008.
Fig. 3.3: Copyright © 2019 by Mmnsr (CC BY-SA 4.0) at https://commons.wikimedia.org/wiki/File:Rescue_with_Ocean_Conservation_Namibia.jpg.
Fig. 3.4: Copyright © by Don W. Stucke (CC BY 2.0) at https://commons.wikimedia.org/wiki/File:Thermometer_-_by_Don.jpg.
Fig. 3.5: United States Geological Survey, “Earth's Principal Tectonic Plates,” https://commons.wikimedia.org/wiki/File:Plates_tect2_en.svg, 1996.

Chapter 4

Earth

History, Systems, and Resources

Objectives

- Understanding Earth processes as well as the history of Earth is integral to studying the environmental sciences.
- It must be understood that everything that occurred on and in the Earth prior to our current time led to conditions on our planet as we know it.
- One must grasp the fact that geologic time is vast and, historically, Earth processes have constantly changed the composition, configuration, and even the nature of the Earth, albeit extremely slowly, and deliberately according to the principle of uniformitarianism. These processes are ongoing.
- Know that Earth's climate, the hydrologic cycle, and even life itself are directly and profoundly influenced by Earth's geologic surface, which is, in turned, controlled by plate tectonics.
- Geologic hazards exist due to the nature of processes on and within our planet. These are generally unavoidable, but we must learn to live with them through best practices.
- Understand that the Earth provides vast natural resources, such as minerals, building materials, and energy sources, for all living things, including human beings, but there are costs to mining and overusing Earth's resources.

KEY WORDS AND TERMS

Earth: The third planet from the Sun in the solar system and the planet that we call home, the only planet known to harbor life.

Geology: The science of the Earth that deals with its physical and chemical characteristics, history, and processes; also includes the study of other planetary bodies in the universe.

Geography: The study and science of Earth's physical features, outer surfaces, and atmosphere, as well as human activity that affects and is affected by them, including population and resource distribution, industries, and land use.

Planet: A celestial body that orbits a star (in Earth's case the Sun) that is big enough to have gravity force it into a spherical shape (the lower limit is around 400 km) and also big enough to have gravity clear its orbit around its star of any other sizable objects.

Systems: A set of processes or things that work together as parts of an interconnected mechanism or network.

Consider This . . .

Our very existence throughout human history has been defined by our relationship with our environment, **Earth**'s environment, and its **systems**. This includes the food we eat, the materials that we use to build things, our artistic achievements, the water that we use, the other organisms that we interact with, and even the biochemical composition of our bodies, all of which are the result of the composition, even the essence, of this **planet** we live on, Earth. Every event on Earth's past, every landform, every environment, every organism, has led us to the world we currently live in. If there is any guiding principle behind **geology** it is this one: The past is the key to the present, and the present is the key to the past.

Introduction

Earth, the only planet that is known to harbor life, operates, as a whole, under a vast set of unique chemical and physical characteristics (Figure 4.1). The entire Earth system is a complexity of subsystems and **emergent properties**. Emergent properties of a system are unique characteristics that function somewhat independently of a larger system yet are still components of that larger system. For instance, ocean water is a system that is independent of volcanic island arcs, yet ocean water is also a component of volcanic island arc formation, something that will be discussed later in this chapter.

In keeping with this discussion of systems, the solid Earth, including all the rocks within and above the Earth, as well as the sediments and soil at the Earth's surface, are part of the lithosphere of solid materials. The lithosphere, along with the other conceptual spheres of the Earth, the atmosphere, biosphere, and hydrosphere, is not an isolated series of systems but interacts with all of the other spheres of the planet. As an example, vast tracts of bedrock are composed of the sedimentary rock **limestone**, part of the lithosphere, yet many of the rocks are biological in origin. Many microscopic organisms in the ocean form exoskeletons, or **tests**. Much of Earth's limestone is composed of rocks made of these tests, an interaction of the lithosphere and biosphere.

The composition of the Earth is that of an enormous array of naturally occurring chemicals, including minerals that comprise rocks, hot liquids such as magma, and groundwater containing soluble chemical species. The study of the Earth's chemical composition is called **geochemistry**. The Earth is also an energetic planet. Radiative heat from within the planet causes convective energy to move Earth's outer layers constantly and slowly. At times those layers lock up due to friction. When that friction is overcome, earthquakes occur, sending forth massive acoustic signals known as seismic waves. The study of Earth's energetic systems, as well as the practice of utilizing energy signals to study the inner Earth, is called **geophysics**.

Water moves and is present within and atop the Earth; it also resides in the atmosphere. When water condenses in the atmosphere it eventually falls to the Earth as part of the hydrological cycle that shapes the very rocks on Earth's surface. This **hydrologic cycle** is powered from outside the planet, by the Sun. Some of this water sinks into the depths, forming groundwater, some of which becomes superheated near bodies of magma forming chemical-rich hydrothermal deposits.

The first photograph of Earth taken from space, October 24, 1946, by the V-2 rocket at an altitude of 105 km (65 miles)

Origin of Earth

Earth's history began with the formation of its elemental components, beginning with the formation of neutrons and protons about two minutes after the cataclysmic beginning of the universe known as the **Big Bang**. The Big Bang occurred some 13.7 billion years ago, give or take maybe 200 million years. It was a singularity of unimaginable heat and pressure that expanded within a very small amount of time into a field of fundamental energies and particles, the beginning of our universe. The fundamental components for atoms, basically everything, was shaped within those first few seconds. Some time later, anywhere from 200 to 500 million years after the Big Bang, hydrogen, the most basic element, accumulated in the first stars, then the first galaxies. These were the first stars, sometimes called **popula-**

tion III stars. They are hypothetical because no known example has been observed of a population III star but there is strong evidence for their existence.

It is theorized that population III stars were composed of mostly hydrogen since there were no other types of atoms that existed in the Universe at that time. Over time, after some of these primordial stars had exploded into massive supernovae, the energy caused by these explosives formed heavier elements, including metals. These newer stars were still metal poor. They are called **population II stars**, and there is evidence of their existence. When some of these newer stars had gone supernova, the 94 or so naturally occurring elements were formed. One such explosion happened in the volume of space near our own a little over 4.54 billion years ago. Once the clouds of cosmic dust and gases started coalescing, under gravity and momentum, the foundation for a metal-rich **population I** star was set in place. This star was our own Sun.

The Protostellar Disc

Our Sun formed from an enormous spinning, gravitationally clumped mass of matter of material left over from a **supernova** that occurred in our "local" volume of space a little over 4.54 Bya. Supernovae, exploding massive stars, are energetic enough to cause the formation of elements much heavier than iron due to nuclear fusion and the shock-waves that are part of a supernova. Thus, all of the elements that make up the Earth were in that cloud of material. Our Sun began as a cloud of dust and gases that became "clumped" together due to energy coursing through this assembly. Once a certain amount of mass was reached, this body of primordial elements attained a center of gravity and began to collapse on itself.

As the mass of dust and gases began to spin due to angular momentum, **centripetal force** (spinning that causes mass to move inward toward a central axis) initiated rotation, increasing the compaction of that mass through both gravity and the aforementioned centripetal force. Meanwhile, portions of dust and gas were flung outward in a protostellar disc that surrounded the newly forming star. This disc was spinning along with the newly formed Sun. The reason for this was likely due to the enormous magnetic field generated from the electrical currents coursing through the entire system. Within this **protostellar disc** of dust and gas, other masses began forming disturbances or eddies, if you will, due to the mathematics of the energy, as well as the position of these disturbances, that was present in the early **solar system**. This was the origin of the planets within the newly established solar system. Of course, Earth was one of these newly forming planets.

Collision and Creation

As one can surmise, the very early solar system was a very different place than the solar system we currently live in. As the new Sun began collapsing in on itself, initiating nuclear fusion, the material in the surrounding protostellar disc began collapsing into discrete volumes that we know of as planets. A lot of energy was coursing through the disc, and overall it was very hot. Because of this energetic formation and resultant heat, the new planets, including Earth, were molten bodies of superheated chemicals and gases. In addition to this, many other objects, including planetary bodies that no longer exist, were present, some of them with erratic orbits. Remnants of these ancient objects still crisscross the solar system to this day in the form of objects like comets and meteors. Meteors are some of the oldest objects in the solar system. When a meteor strikes the Earth's surface it is known as a **meteorite**. The oldest meteorite found on Earth is 4.6 Gyr old, predating the age of the Earth by 60 million years!

Beginning in 1969, astronauts with the United States' Apollo space program began bringing back samples of the Moon's surface in the form of rocks. When analyzed, it was discovered that these rocks had chemical and isotopic signatures that were oddly similar to that of Earth rocks. Through more analysis and testing of multiple parameters, a startling conclusion was reached: Earth's Moon was once part of Earth itself! The theory that was formulated is that a Mars-sized body, posthumously named **Theia**, collided with Earth around 100 Myr after Earth's formation. Some of the debris from this colossal event that did not escape into space became consolidated into one mass, the Moon. This new body assumed a spherical shape because it had a diameter greater than 400 km, which is the general threshold for a body to become spherical.

Deep Time

The concept of **deep time** simply refers to the vastness of Earth's history, which dwarfs human life spans and even the history of human civilization itself. It encompasses all of the complex changes that have occurred in the history of the planet since its formation 4.54 Gya. Throughout this almost unimaginable range of time, the Earth has gone through many changes, and in fact has been a very different place.

Earth's Timeline

Geologists sub-divide Earth's history into discrete intervals called the **Geological Time Scale (GTS)**. This is done through **geochronology**, the science of determining the age of Earth materials; **paleontology**, the study of life that existed prior to human history; and **stratigraphy**, the science concerned with the study of Earth's layers in determining conditions of the past.

All divisions within the GTS are determined by boundary conditions set by the development of life. A group of scientists that belong to the International Commission on Stratigraphy (ICS), which is part of the International Union of Geological Sciences (IUGS), oversee the development of the GTS, which changes slightly as new geological discoveries are made.

The largest divisions of the GTS are **eons**, each of which spans anywhere from around 0.5 billion years to around 2 billion years. The Earth's first eon, the **Hadean**, lasted half a billion years, yet it is an "unofficial" span of time because as far as we know it there were no living things on the Earth during the Hadean. Regardless, that eon is as significant as the other three because the development of our planet has been one of continuity. We know most about the current eon that we all live in, the **Phanerozoic**, meaning "visible life" in Latin. The Earth's first three eons, the Hadean, **Archean**, and **Proterozoic**, are often lumped together under the term **Precambrian** because they all occurred before the **Cambrian period** of the Phanerozoic Eon. Also, we know much less about the Precambrian than we do our current eon because it was so long ago.

The next smaller span of time is the **era**. We have more information on the eras within the Phanerozoic than eras within Precambrian time. The Phanerozoic Eon's three eras are the **Paleozoic era** when multicellular life first began to flourish; the **Mesozoic era**, otherwise known as "the age of the dinosaurs"; and the **Cenozoic era**, when mammals began to dominate the planet's surface. The next smaller unit is the **period**, of which we are in the **Quaternary period**, and then even more brief is the **epoch**. Two epochs are of concern to us, the **Pleistocene epoch**, often called the **Ice Ages**, and the epoch that followed, the one in which we live in, the **Holocene epoch**, which began around 10,000 years ago. We all currently live in the Phanerozoic Eon, Cenozoic era, Quaternary period, and Holocene epoch!

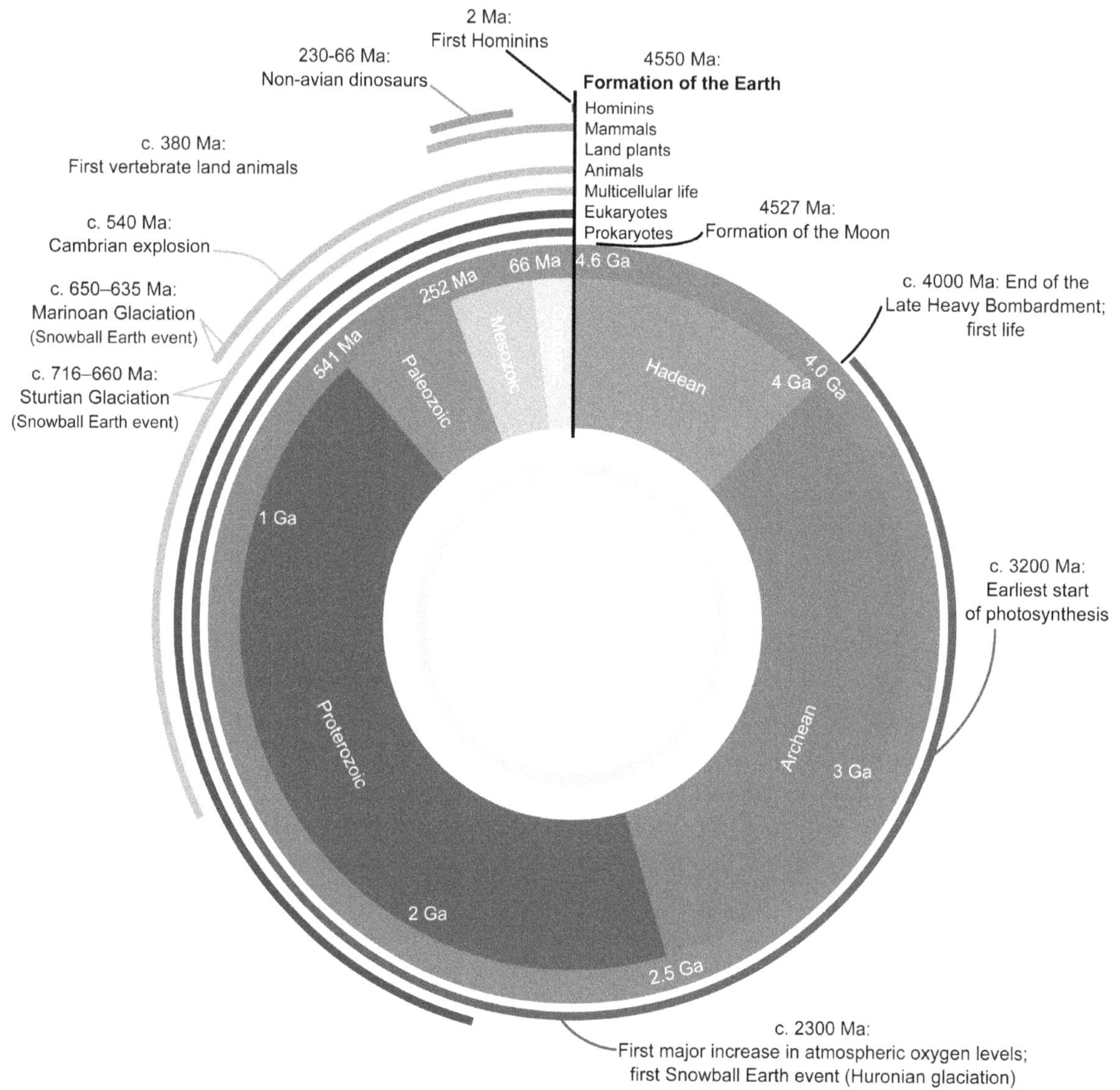

Figure 4.2 The geological clock, a projection of Earth's 4.54 Gyr history onto a clock. Note the tiny spans of time of recent history

GROUND TRUTH

Origins of Geology and Early Geologists

Throughout known history, many people, from scholars to those who were just curious, have explored and investigated the history and workings of Earth. Even Ancient Greek philosophers, like **Aristotle**, made observations and theories concerning our planet. **Eratosthenes** of Cyrene was a Greek scholar who also became the chief librarian of the Library of Alexandria. Over 2,000 years ago in 240 BCE, he successfully measured the circumference of the Earth using calculations based on the distance from Alexandria to Syene (which is now Aswan, Egypt) and the angle of sunlight in a well. His measurement of Earth's circumference was 96% accurate. This was possibly the earliest proof that the Earth is spherical.

During the Middle Ages, other researchers across the globe made significant discoveries in geology. Scholars in the Middle East made advances during this time. Between the 10th and 11th centuries CE, the physician **Ibn Sina,** known as Avicenna in the West, wrote with accuracy about erosion and mineralogy. Contemporaneous to Ibn Sina, the Persian poet **Omar Khayyam** made observations on the rising and falling of sea levels, which is integral to the modern science of stratigraphy. Around the same time, a naturalist named **Shen Kuo** made some significant observations concerning erosion, sediment deposition, and fossilization in China. He used the Taihang Mountains of northeastern China as a natural laboratory to formulate his theories. Shen Kuo also did some of the earliest research in climate change, surmising that climate changed over time using changing patterns of bamboo growth as evidence.

Niclaus Steno was a Danish physician living in Italy during the 17th century CE. He founded a series of geological principles, based on his observations, that ultimately formed the groundwork for modern sedimentology and stratigraphy. These are known as Steno's laws: (1) **the principle of superposition**, which states that sedimentary layers are laid down in determinable order with the oldest at the base unless overturned; (2) **the principle of original horizontality**, which refers to sediments being lain down in horizontal or nearly horizontal layers, and (3) **the principle of original lateral continuity**, which states that sediments are deposited over vast areas during long periods of time that can be traced laterally. In addition to sedimentology and stratigraphy, these laws form the basis for the modern science of geology.

The Scottish scientist **James Hutton**, often called "the father of modern geology," made significant further advances in geology during the 18th century. He advanced the notion that Earth processes occur slowly yet predictably according to physical and chemical laws. He stated that rocks, including the rocks that comprise mountains, weather and erode over vast periods of time. In fact, whole landscapes change over time because natural forces work on them, leading to our present landscapes. Thus, we arrive at the foundational statement "the present is the key to the past" because current features in the Earth can be used to interpret what came before them. Although Earth systems are dynamic, change is often slow and predictable in a uniform manner. In the 19th century, another geologist, the Englishman **Sir Charles Lyell** further wrote along these lines as an argument against catastrophism, which was the idea that changes in the Earth resulted from rapid catastrophic events. While certain isolated aspects of change do occur over small spans of time, often disastrously so, the guiding principle behind most of the development of the planet is due to uniformitarianism, gradual and predictable change over long periods of time.

A British mine and canal surveyor, **William Smith**, collected fossils and made drawings of rock layers, also known as strata, during the late 18th and early 19th centuries. He began publishing his observations and drawing detailed maps of what he found, culminating in the "Delineation of the Strata of England and Wales with Part of Scotland," a geologic map of England and one of the first real geologic maps ever. A working man, Smith, also known as "Strata Smith," was of humble background and education, and thus largely ignored by the academic community for much of his life. His efforts were not truly recognized until 1831 when the Geological Society of London honored him with its Wollaston Medal for his achievements. Four years later he was awarded an honorary doctorate of law degree at Trinity College in Dublin, Ireland. Smith is now known as "the father of English geology." His work is the basis for another important geological law, the **principle of faunal succession**, which states that sedimentary rock containing organisms will succeed each other vertically in predictable order over large horizontal areas. In other words, organisms evolve and become extinct in order, older below younger unless overturned. For instance, a Tyrannosaurus rex bone will never be found in the same rock strata as a wooly mammoth bone because those organisms did not exist in the same time period.

Figure 4.3 William Smith's A Delineation of the Strata of England and Wales with Part of Scotland, the earliest known actual geological map

By the mid-20th century, the theory that Earth's continents moved around slowly was being rediscovered from earlier works. Along with some partners, **Marie Tharp**, a geologist who worked for the Lamont-Doherty Geological Observatory at Columbia University in New York State, had drawn up maps of the Earth's oceans from data gathered by the United States Navy. Much of the more detailed work was done by Tharp herself, who also developed a crucial idea that would later prove to be true. She hypothesized that the Mid-Atlantic Ridge, which cuts through the bottom of the entire Atlantic Ocean from north to south, was actually a large crack in the Earth and on each side were two massive plates slowly pulling apart. This is true, and her conclusions are now integral to the advancement of geology as a whole, but at the time, her findings were essentially dismissed, unfairly so. She did not get credit for her efforts until later when it became known that hers was the work that truly broke ground!

Figure 4.4 Marie Tharp, geologist

Plate Tectonics

Throughout history we humans have thought the Earth to be a very stable place, unchanging even, but in the early 20th century a scientist challenged that assumption. **Alfred Wegener**, from Germany, was primarily a meteorologist but became curious about the Earth's continents from looking at their shapes. He noticed that some of Earth's landmasses fit together like a puzzle despite often being over a thousand miles apart. For instance, it doesn't take a huge intuitive leap to see that the western margin of Africa fits that of eastern South America almost perfectly. From this, as well as given other lines of evidence, such as the fact that the same sorts of fossils are found on continents thousands of miles apart, Wegener came up with his theory of **continental drift**.

Continental drift theory, simply put, states that Earth's continents are not stationary but actually slowly moving around across the surface of the planet. Wegener did not have much of an idea of any sort of mechanism behind such enormous movements. Continental drift was met with a lot of skepticism from the scientific community in general, and sadly, Wegener perished during an expedition in Greenland in 1930. Twenty years later, new evidence from new research and technology led to the confirmation that many of his ideas were in fact true, particularly that of continental drift theory.

Movement Under Foot

During the 1950s, researchers and scientists began using sonar to map the ocean floor. In fact, the world's first map of the ocean floor was produced during this time. Geologists noticed that features at the bottom of the ocean coincided with areas prone to earthquakes. An **earthquake** is a local, sudden, violent movement of the Earth due to movement within the Earth, volcanic activity, or some other disturbance. These earthquakes were especially numerous at trenches and at mid-ocean ridges. A trench is an especially deep part of the ocean parallel to island arcs or the edge of continents, and **mid-ocean ridges** are very long mountain chains in the deepest part of the sea, with trenches and other ridges parallel to the entire system.

In the 1960s, researchers like **Harry Hess**, a professor of geology at Princeton University, explored the ocean floor using the latest technology to further study mid-ocean ridge systems. Hess had used sonar while serving in the

U.S. Navy during World War II and became interested in marine geology. Hess wrote a paper in 1962 called "The History of Ocean Basins" in which he proposed his sea floor spreading theory as a mechanism for continental drift, utilizing his work in the mapping of the ocean floor. It was also found that the rocks on either side of one particular mid-ocean ridge, the **Mid-Atlantic Ridge**, had different successive magnetic signatures (Figure 4.5). The magnetic minerals in these rocks, primarily iron-bearing minerals, would alternate their alignment north or south in regular succession. It was also found that the youngest rocks could be found in the center of mid-ocean ridges, with those on either side becoming successively older as one gets farther from the ridge. In fact, the rocks at the very center of mid-ocean ridges are zero years old; they are being continuously produced. Sequences of alternating rock magnetism, known as **magnetic striping**, along with ocean floor mapping was some of the first direct evidence proving that Wegener's continental drift theory was correct: The outer surface of the Earth is indeed continuously moving!

We now know that Earth's magnetic field periodically reverses, once every 450,000 years or so, on average. One can think of the Earth as a gigantic bar magnet as far as the magnetic field is concerned wherein the polarity, north and south, flips from time to time. When this happens, rocks being produced at mid-ocean ridges, technically lava, cool down with the orientation of iron bearings aligning preferentially with the magnetic field. When reversals occur, these minerals orient in the opposite direction. Two researchers, Vine and Mathews, tied the discovery of magnetic stripes to Hess's sea floor spreading concept. New oceanic crust is continuously being produced at mid-ocean ridge systems, but what is happening to the Earth's crust elsewhere? Well, crust elsewhere is either moving into other portions of the crust atop the lithosphere, or crust riding the lithosphere is moving side by side in places like the **San Andreas Fault** in California. The Earth does not expand but maintains a fairly constant diameter while all of this complex movement takes place. One of the most powerful lines of evidence for plate tectonic movement is the fact that computerized satellite systems take precise measurements of plate movement on a constant basis.

The Inner Earth

Using technology, geologists have been able to study not only the Earth's surface, as well as the deepest parts of the oceans, but the interior of the Earth itself. Any disturbance of the Earth, such as an earthquake, explosion, or large shock causes a series of signals that reverberate throughout the planet. These signals can be used to understand what's in the Earth.

Exploration Through Invisible Signals

Using technology, geologists have been able to study not only the Earth's surface, as well as the deepest parts of the oceans, but the interior of the Earth itself. Any disturbance of the Earth, such as an earthquake, explosion, or large shock causes a series of acoustic waves, sound-like waves, to be produced. Depending on their intensity, these waves, known as **seismic waves**, propagate throughout the Earth, even to the other end of the planet. The type of material that seismic waves travel through, as well as changes in materials, determines the speed at which the waves travel as well as their general direction. Scientists can use a piece of equipment known as a **seismograph** to capture and analyze the signals produced by large disturbances in the Earth, such as earthquakes, in order to determine the properties of rocks deep within the Earth, even down to Earth's core! Geologists often use earthquakes to study the inner Earth because they occur constantly.

Seismic waves can be divided into two categories, body waves and surface waves. The fastest type of waves are body waves known as **p-waves**, also known as pressure waves or primary waves. These are the first waves that arrive at a seismograph from earthquakes, the first arrivals. P-waves propagate through matter in much the same way as sound does through compression and rarefaction. Rarefaction is essentially the relaxation phase of the sound waves cycle. The other type of body wave is the **s-wave**, otherwise known as secondary or shear waves. S-waves are transverse waves that generate perpendicular to the direction of propagation. This means that s-waves do not transmit through liquids, a good indicator of liquids in the Earth. S-waves die out once they get to the outer core of our planet, which is strong evidence that the Earth's outer core is liquid and not solid. S-waves are slower than p-waves and arrive at a given seismograph after p-waves.

Surface waves are found along Earth's surface mainly generated from shallow earthquakes. These types of waves have a lower velocity than s-waves, although they exhibit greater amplitude than either p- or s-waves. As the name implies, surface waves are generated at the boundary between two surfaces, in the case of seismic waves, the boundary between the Earth's surface and the atmosphere. Surface waves are fairly complex compared to body waves but worth studying because these are the seismic waves that cause the most damage to human life and property. There are two major types of surface waves: **love waves**, and **Rayleigh waves**. Love waves, or L-waves, are essentially waves that propagate in a side-to-side motion. Rayleigh waves are standing waves that propagate through materials in an elliptical fashion, similar to water waves.

Geologists and geophysicists have used other signals to explore the inner Earth. Other methods that have been used to explore the inner Earth include electrical conductivity and resistance, magnetism, and even gravity. Gravity itself is a useful method for understanding the density of materials of the Earth. Although the differences are slight, gravity is not the same across the Earth, and gravity maps have been created of our planet that help us understand this fact.

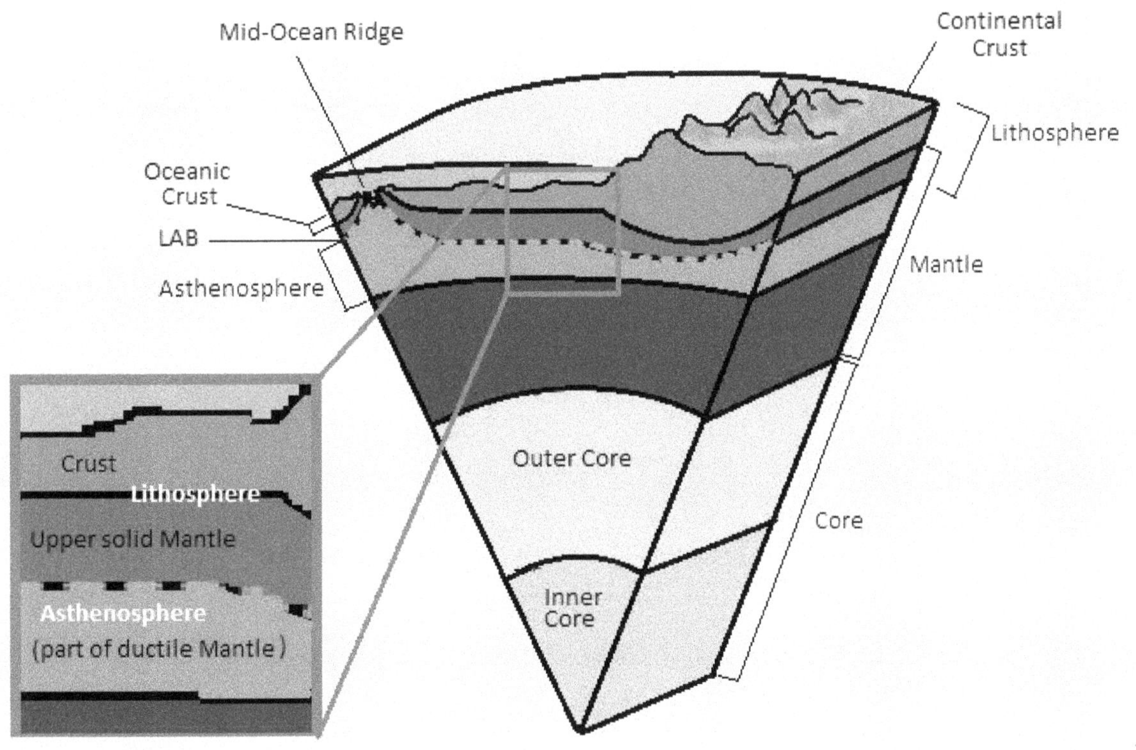

Figure 4.5 Diagram of the inner Earth (not to scale).

Compositional Layering

Through the use of geophysical means and other lines of evidence, geologists have developed a robust understanding of the differences between Earth's **compositional** or **chemical layers** (Figure 4.6). The Earth's outer layer is the **crust**, which varies in thickness from around 3 km (2 miles) in some ocean basins to 70 km (40 miles) in some mountain belts. There are two types of crust, **continental crust** and **oceanic crust**. Continental crust is thicker and less dense than oceanic crust with an average density of 2.7 g/cm3. Continental crust "floats" on top of thinner, denser oceanic crust. Oceanic crust, making up the floor of our planet's great oceans, has an average density of 3.3 g/cm3. Most continental crust is above current sea level but has been flooded in times past. Minerals in the crust are very similar in composition, whether continental or oceanic, and contain large amounts of oxygen and silicon. Continental crust is generally composed of **granitic** rocks, whereas oceanic crust is composed mostly of the rock **basalt** and its equivalent.

The Earth's **mantle** makes up around 83% of the planet's volume. It is directly below the crust and has an average thickness of around 2,900 km (1,700 miles). Minerals that make up rocks in the mantle have much less oxygen and silicon, and greater amounts of iron and magnesium, compared to rocks in the crust. Below the mantle, at the center of the Earth, is the **core**. The core is composed mostly of iron and nickel, which sank to the center of the molten Earth as it was forming 4.54 Byr ago. There are two layers in the core, the **outer core**, which is around 2270 km (1411 miles) thick, and the **inner core**, the central part of the planet, which has a diameter of around 2,432 km (1511 miles). The outer core is known to be composed of mostly liquid iron that convects, generating an enormous electrical current, which in turn generates Earth's magnetic field. The magnetic field helps protect us from highly energetic particles coming from outer space.

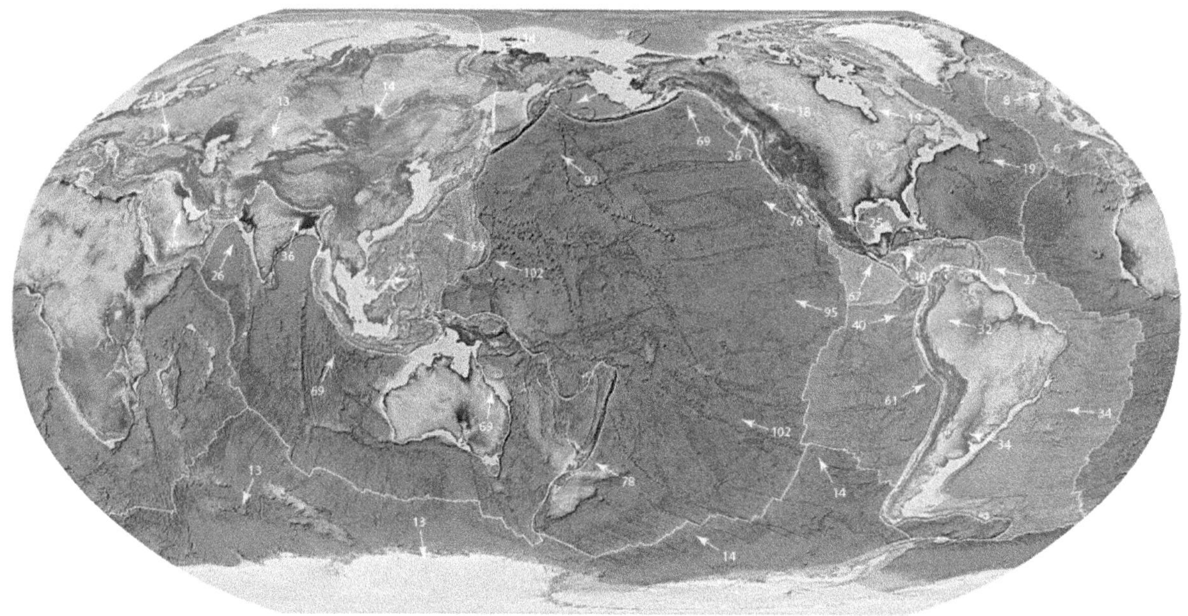

Figure 4.6 Map of major tectonic plate motion in mm/yr. relative to the African plate

Plate Tectonics

Another way to look at the inner Earth is through its physical layering, the way Earth materials behave mostly due to heat and pressure. There are two main physical, or rheological, layers: the **lithosphere** and the **asthenosphere** (Figure 4.5). The lithosphere is composed of the entire crust and the topmost layer of Earth's mantle. It is somewhat rigid in its physical behavior and is broken up into many sections known as tectonic plates. These are portions of the Earth that are constantly moving, slowly; this is what Alfred Wegener first wrote about when he published his theory of continental drift. Little did Wegener know, but plate tectonics, the affirmation of his continental drift theory, was to become the great unifying theory of geology!

The mechanism for plate tectonic movement occurs in the second layer of the mantle called the asthenosphere. Heat is generated in the inner Earth due in large part from radioactive decay of isotopes. This radiative heat from the layer of the mantle beneath the asthenosphere, the mesosphere, generates convective heat in the asthenosphere. The asthenosphere is composed of rock that is very close to melting, giving it an almost liquid-like, or plastic, consistency. Heating liquids cause them to **convect**, meaning the heat rises upward, then cools and drops down. The tremendous heat that is generated in this manner affects the lithosphere above, causing it to move slowly. The lithosphere is broken up into different tectonic plates, causing these plates to interact with each other and forming the basis of the Earth's entire physical topography, from the ocean's deepest depths to the tops of the highest mountains (Figure 4.6).

Boundary Issues

Earth's tectonic plates interact with each other depending on the relative motion of tectonic plates as well as, to some extent, the composition of the overlying crust. There are three main types of interactions between tectonic plates, in addition to cracks in the crust within the interior of lithospheric (tectonic) plates, often forming fractures (Figure 4.7). The boundaries between tectonic plates are called faults, although there are intraplate faults as well. Generally speaking, all interactions are related across the surface of the Earth in differential motion. If one thing occurs on one side of a tectonic plate, there will be some sort of compensation elsewhere. Earthquakes are associated with all tectonic plate boundaries.

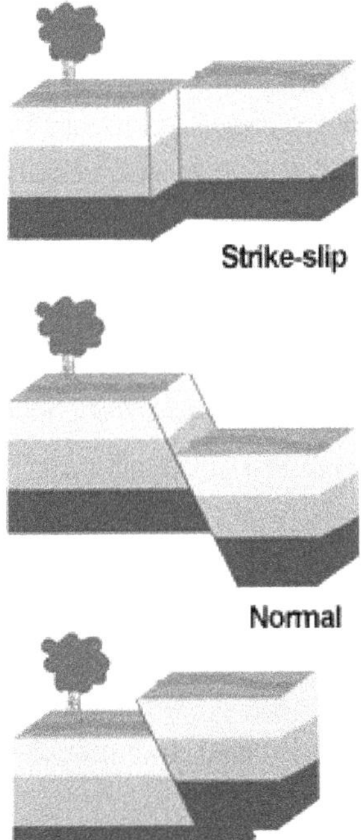

Figure 4.7 The three types of faults: strike-slip (transform), normal, and thrust (reverse)

The first type of boundary is composed of **normal faults**, which result in extension from tension. This is when two tectonic plates pull apart from each other. This mainly occurs at mid-ocean ridges where convection causes thinning of the crust at the boundary, resulting in lava welling up to form new crust at either side of the boundary. Given that fact, these are productive boundaries. Intraplate normal faulting occurs when crust is thinned from differential motion and results in fault blocks, which can form basins and other features.

Reverse, or **thrust faults**, result in compression. This results in crust becoming shortened, or even becoming destroyed. These are destructive margins. The deepest parts of the ocean, oceanic trenches, are the result of reverse faulting between two different plates. This leads to subduction: One plate disappears under another as if it was some sort of gigantic conveyor belt. This leads to the formation of island arcs and volcanoes. The Andes Mountains in South America and the Japanese Islands are examples. When two lithospheric plates carrying thick slabs of continental crust interact with each other, neither can truly subduct under the other. Thickening occurs, the boundary rises, and the result is the production of mountain chains, called an orogeny. This is currently occurring with the Himalayan Mountains in Asia.

Transform faults result in horizontal motion that is parallel to the fault, side-to-side motion. Nothing is created or destroyed, although these faults can cause great destruction. The San Andreas Fault in California is an example of a transform fault. It is a region that is prone to earthquakes because friction causes the two plates to lock up near the surface of the Earth. When that friction is overcome, the motion can cause great destruction. Transform faults that occur with plates are called **strike-slip faults**.

Basic Materials

The main body of the Earth is composed of an enormous variety of chemical compounds and elements called **minerals** that mainly originated from differentiation in the early Earth. These comprise Earth's **rocks**. In other words, rocks are made of minerals. The very foundation of the geographic landscape and ecosystems that all living creatures, including us humans, reside in has at the base solid rock. This geology is the framework of our existence. Rocks and minerals themselves are valuable materials that have been used by human beings to build and power entire civilizations.

Minerals

As a caveat, Earth's rocks are usually aggregates of minerals; there are exceptions. Some minerals can be found in the Earth on their own, although this is somewhat rare. For instance, sulfur, a mineral, can be found on its own under certain conditions, but usually sulfur is part of some other compound found in rocks. The basic definition of a mineral is as follows: (1) It occurs naturally under conditions found on and in the Earth, (2) it is an inorganic solid, (3) it has an ordered crystalline structure, and (4) it has a definite chemical composition. Given these sets of conditions, water ice is actually a mineral!

There are close to 6,000 naturally occurring minerals in the Earth. Around 92% of all minerals are **silicates**. Silicates are a group of minerals based on silica, the chemical bonding of oxygen and silicon. **Plagioclase feldspar** is a silica and is the most abundant mineral in the Earth's crust. Another silicate is quartz, which is simply SiO_2 and comprises most of the sands that occur in soil and on Earth's beaches. Other silicates include the micas and clay minerals. Clay minerals are also a major component of soil.

Around 8% of all minerals are composed of **non-silicates**. The most common, and arguably important, of these are the **carbonates** such as **calcite**, which comprise many different layers of rock such as limestone, chalk, travertine, and others. Other non-silicate minerals include oxides such as hematite, sulfates like gypsum, and phosphates such as apatite. Native elements are minerals that contain only one kind of atom: elements. These include valuable metals such as copper, gold, and silver, as well as non-metals like sulfur.

The Rock Cycle

Rocks are defined as a solid, naturally occurring geologic mass, or a solid aggregate of mineral grains that form under natural conditions within the Earth. With this in mind, a rock need not be composed of mineral grains, or crystals. For instance, obsidian is an igneous rock composed of glass; since glass is not crystalline, obsidian contains no minerals, yet it is still a rock. Coal is composed of fossilized plant fragments and does not contain crystals, yet it is still a rock. Rocks like obsidian and coal could contain mineral grains, though, because processes within the Earth are not constrained by our definitions. They just occur.

There are three main types of rock that are found in the Earth: **igneous**, **sedimentary**, and **metamorphic**. These rocks are both defined by the processes that are responsible for their formation as well as the physical and chemical composition resulting from said processes. The **rock cycle** is just that, a cycle, meaning that Earth materials, rocks and the like, are slowly but constantly being transformed though steady, uniform processes on and in Earth into some other form, or other type of rock (Figure 4.8).

Rocks that solidify from lava or magma become igneous rocks. When rocks become weathered into sediment, then eroded, ending up in a basin and "glued" together, these sediments form sedimentary rocks. Rocks that are buried deep in the Earth due to tectonic forces, subject to intense heat and pressure, change into new types of rocks: metamorphic rocks. These processes are continuous and will never end.

Igneous Origins

Igneous rocks form from molten Earth materials. These melts are either called **magma** when they occur under the Earth's surface or **lava** when they occur on the Earth's surface. Igneous rocks form when these materials melt, cool, and become solid. There are two basic types of igneous rocks depending on the origin of the melt:

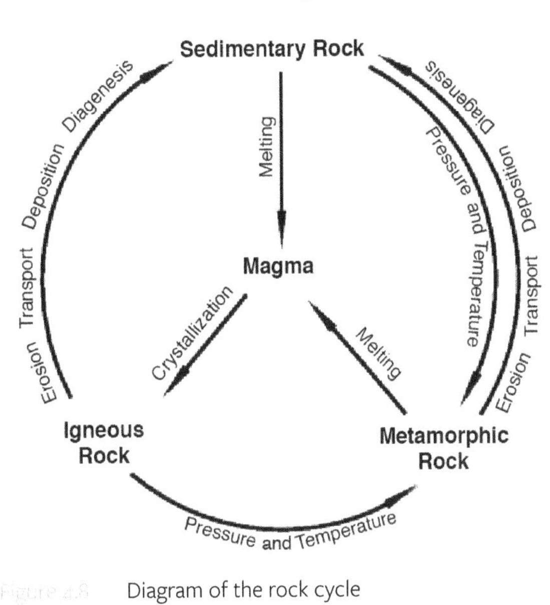

Figure 4.8 Diagram of the rock cycle

- **Plutonic or intrusive**: These rocks form under the Earth's surface from melt. Plutonic rocks may contain a wide range of mineral crystal sizes depending on the depth of formation and the rate at which they cool. Intrusive rocks that cool very slowly with cavities present in surrounding rock may form very large crystals. Since Earth was originally composed of only molten materials, one could conclude that all rocks ultimately originate from molten Earth materials. Granite is the most common type of intrusive igneous rock.
- **Volcanic or extrusive**: These rocks form at the Earth's surface or very close to the Earth's surface from melt. Extrusive rocks typically contain very small to microscopic-sized crystals because the crystals cool very quickly. In fact, some melts quench so quickly that no crystals from and the rock becomes composed of glass, such as obsidian. As the name implies, volcanic rocks erupt out of volcanoes and can pose quite a hazard depending on the type of volcano as well as the type of melt. Basalt is the most common type of extrusive igneous rock.

In addition to the two basic types of igneous rock determined from their origin, there are also four types of igneous rock determined from their mineral composition due to the chemical origin of the melt and the rate of cooling:

- **Felsic**: A combination of feldspar and sialic (silicon), light- and pink-colored minerals
- **Intermediate**: Literally a combination of felsic and mafic rocks, often dull or gray colored

- **Mafic**: A combination of magnesian and ferric (iron), dark- and black-colored minerals
- **Ultramafic**: Dense, deep Earth rocks, often with a greenish tint due to the presence of an iron and magnesium bearing mineral called olivine. Ultramafic rocks originate deep in the Earth and contain much less silica than rocks in the crust

Sediments and Sedimentary Layers

Rocks that form from loose mineral crystals, rock particles (**clasts**), and chemical or biochemical processes that occur at or near Earth's surface are called sedimentary rocks. Particles and crystals are deposited in low-lying basins on the Earth's surface. As these particles remain stationary and become buried by more layers of sediment, they become cemented or fused together by chemicals usually transported to the basin by water. Other sedimentary rocks are formed from chemicals present in natural waters, or from the residues left behind by living organisms. Sedimentary rocks are the rocks that contain evidence of past lives, **fossils**. These fossils may take the form of remains of organisms, such as bones and shells, or imprints of the activity of past life called **trace fossils**. Fossils tell us a lot about past life and in relation to with past climates and environments, which helps us understand current life, climate, and environments.

There are three types of sedimentary rocks depending on the processes that formed them:

- **Clastic**: These rocks form as aggregates of mechanically altered bits of rock, minerals, and other debris. Clastic sedimentary rocks form very close to the Earth's surface and are usually formed of particles cemented or fused together. As such, the original particles are unaltered. Sandstone is a type of clastic sedimentary rock and shale, composed of tiny particles of mud (clay), and is the most common sedimentary rock on Earth.
- **Chemical, or biochemical**: This type of sedimentary rock forms from chemicals that are present in Earth's natural environment, or from biochemical processes, usually in water. These chemicals can form in many different ways from solution, often with water as the solvent. Limestone is the most common type of chemical sedimentary rock. Some types of sedimentary rock, such as limestone, are formed from the minute exoskeletons of marine plankton. Other types of sedimentary rock are formed from the alteration of microorganisms, or secretions of marine organisms.
- **Organic**: The tissues of organisms, usually plants, formed directly by either biological processes or the alteration of biological material into rock material. Coal is the main organic sedimentary rock, and it is formed from the chemical alteration of woody plant material.

Sedimentology and Stratigraphy

Sedimentary rocks are found in sequences of layers in Earth. These layers, known as **strata**, hold direct evidence for past environments, also known as **paleoenvironments**. This evidence comes in the form of sedimentary structures such as mud cracks, scour marks from moving water in the past, and fossils, among others. The sedimentary record is important because it can teach us about processes that are currently occurring. The very sequences of mineral size in sediments in the ocean gives geologists information about global environments through **sequence stratigraphy**, which tracks oceanic **transgressions** and **regressions** (Figure 4.9). As ocean water levels lower, called an **offlap**, coarser materials like sands deposit at the top of a sequence of strata. Lowering sea levels indicate periods of glaciation, or even Ice Ages, wherein enormous amounts of Earth's liquid water become locked into ice. This is called a regression. When the sea rises over land surfaces, which is also called an **onlap**, it is called a transgression. This is due to increased warming of the ocean from excess water from ice melt and the expansion of water. Currently, all evidence points to a global transgression in our time.

Another important aspect of sedimentary rocks is that they often contain economically important metal, mineral, and energy deposits. These deposits include crude oil. Unfortunately, the extraction of these materials as well as crude oil causes stress to Earth's environment. In addition, the processing and use of many of these resources causes pollution, and, in the case of fossil fuels like coal and oil, causes the build-up of planet-wide warming greenhouse gases.

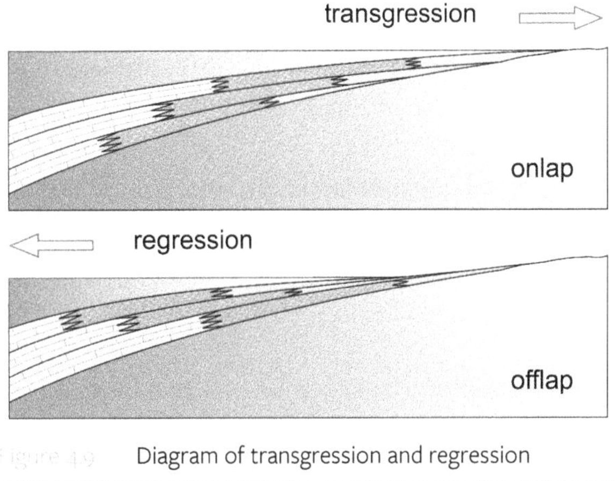

Figure 4.9 Diagram of transgression and regression

When sediments do not accumulate on the surface of the Earth, this becomes part of the sedimentary record known as an **unconformity**. Unconformities often occur due to the presence of erosive surfaces. They can also tell us a lot about what happened in the distant past, environmentally and geologically. What happened in the past can also tell us about the present because, as we now know, "the past is the key to the present, and the present is the key to the past."

There are three basic types of unconformities:

- **Angular unconformity**: This unconformity is formed when younger, horizontal strata is laid down on tilted rocks or any sort that have undergone a period of non-deposition.
- **Disconformity**: This is found in the sedimentary record when there have been periods of erosion or non-deposition between two layers of parallel sedimentary rock.
- **Nonconformity**: When sediments accumulate atop igneous or metamorphic layers that have undergone erosion or non-deposition, this nonconformity occurs.

Metamorphism

Metamorphic rocks form when any type of rock (igneous, sedimentary, or other metamorphic rock) is subjected to the intense heat and pressure of the Earth's depths. The differential relationship between heat and pressure results in rocks that have not melted yet are chemically and/or physically altered to become a suite of different rocks: metamorphic rocks. Metamorphic rocks are often identified by their **index minerals**, or minerals that form through the process of metamorphism. They are also identified by their textures. Remember, if it melted to form, it is not a metamorphic rock but an igneous rock.

There are two basic types of metamorphic rock and five basic textures:

- **Foliated**: Rocks that recrystallized into bands or layers due to differential pressure such as when two tectonic plates are pushed together. This banding or layering is usually visible to the naked eye. There are three textures of foliated metamorphic rock:
    - **Slaty**: Metamorphic rocks that have distinct flat sheets and crystals that are small, even microscopic in some cases. **Slate** is a predominant type, as is phyllite.

    - **Schistose**: Metamorphic rocks forming sheet-like layers often containing mica, or other visible minerals. **Schist** is the predominant example of this texture.
    - **Gneissose**: These metamorphic rocks are arranged in distinct compressed layers, often resembling a layer cake. **Gneiss** is the predominant type of this metamorphic rock.
- **Non-foliated**: Metamorphic rocks that do not exhibit foliation, or layering, were usually formed from a combination of intense even pressure, or intense heat. There are two textures of non-foliated metamorphic rocks:
    - **Granoblastic**: Metamorphic rocks that exhibit coarse grains. **Marble** and **quartzite** are often granoblastic.
    - **Hornfelsic**: Non-foliated metamorphic rocks that are extremely fine-grained, exhibiting tiny to microscopic crystals. **Hornfels** is a predominant hornfelsic rock, as is **skarn**.

Serpentinite, a metamorphic rock that forms in circulating hydrothermal waters deep underground, exhibits a wide variety of textures as well as unique mineralogy. Some varieties of serpentinite have been used as asbestos in the past due to the component mineral's fire-retardant properties. Recently, the use of asbestos has been widely banned because these fibrous minerals are linked to cases of cancer.

Mining and Mineral Resources

Since the beginning of human civilization Earth materials, essentially rocks and minerals, have been used for economic purposes. This includes building materials, chemicals for pigments, agriculture, ornamentation, money and trade, heating and light, toolmaking, and the making of weaponry. Since the advent of the industrial revolution in the 18th century, rocks and minerals have yielded ores for industrial use and energy production on a very large scale.

Rocks containing mineral **ores**, naturally occurring solids from which economically valuable metal or minerals can be extracted, as well as economically viable deposits of energy-producing Earth materials, must be located through an implicit understanding of the geology of a particular locale. This is often done through exploration or direct surveys. Exploration can also be done through the use of sensing and signaling equipment, which is part of the science of geophysics. Once the materials are found, they are extracted through some procedure such as mining or well production. Many of these materials are not suitable for use initially and must be altered physically and/or chemically to yield a product with viable use. For instance, metals often have to be separated from the rocks or sediments that contain them through some means of chemical separation. Oil that is extracted from underground deposits must undergo some form of a chemical process similar to distillation in order to yield the various energy-producing products that we use such as gasoline or jet fuel.

Mining for Materials

Mining is defined as the extraction of economically valuable materials, rocks and minerals, from the Earth. Many economic deposits are associated with sedimentary rocks. Mining is a nonrenewable activity; in other words, it depletes the resource to the point that resource cannot be replaced. Renewable activities can be replaced, put back into the Earth and the environment, but because of the physical processes and the geochemistry implicit in mining activities, mining in general has a tremendous negative impact on the environment and is not a sustainable practice. At the same time, for mining to be economically feasible the material of interest must be concentrated. This limits the amount of mining that can be done.

Metallic Ores

Most of the elements known to occur in the Earth are **metals**. Metals are generally lustrous, malleable, and opaque, exhibiting varying degrees of strength. Metals also conduct electricity and heat, making them valuable for the computer and heavy industries. Metals also have other properties that make them valuable to human civilization. Mined metals include copper, iron, lead, and tungsten.

Most metals are not found intact in the Earth and must be extracted from ores. Once mined, metal-bearing ores must usually be processed in order to obtain the mineral of interest. This is often done near the mine, although sometimes the ore is shipped elsewhere for processing. For example, it is uncommon to now find gold intact in **veins** within mines. A vein is a sheet-like body of some sort of crystalized material. Most metals of interest such as gold are usually dispersed throughout rocks these days and not found in veins. When extracted, these ores often processed with chemicals like cyanide as a dispersant to obtain the gold. The **tailings**, or waste products, then must be stored, or impounded, often on-site. If the storage impoundment is breached for some reason during rain or other precipitation events, these often-poisonous chemicals (e.g., cyanide) can enter local water systems and cause severe environmental degradation. Other impoundments may become exposed to weather and end up eroding, further degrading a local environment.

Also, when metals are processed, such as with **smelting**, heating metals beyond their melting point and then combining them with other metals or chemicals, air pollution is produced. Sometimes this air pollution is severe and must be regulated by a government agency. Steel production has often used coal to produce fire and heat, which not only pollutes the air but produces greenhouse gases, primarily CO_2. In addition, smelting and other forms of metal processing utilize enormous amounts of water for cooling, stressing local water systems and increasing the chance for water pollution.

Nonmetallic Ores

Nonmetallic substances are also mined, such as gems, gravel, limestone, phosphates, sand, and sulfur. Rocks such as gravel, marble, and slate are mined to be used in the building industry and as decorative stone. Mining these materials does not often cause the same kinds of environmental impacts as mining metal, but all mining has some sort of impact. Mining for gemstones often extracts a human toll, such as with exploitative mining practices in Angola and South Africa. Limestone is mined for pharmaceuticals, and phosphates are mined for fertilizer.

Energy resources are also mined. Although technically a metal, uranium ore is a mined material, which presents its own set of problems due to radioactivity. Coal, though neither a metal nor a mineral, is mined as well. It presents a host of problems in its extraction from **acid mine drainage** in the form of sulfuric acid produced from sulfur present in coal to pollution from its burning to greenhouse gas emissions. Acid mine drainage occurs when chemicals released from mining processes combine with ground or surface water to form acids due to the presence of free hydronium ions (H^+) in free waters. These excess acids degrade local water systems. Crude oil and natural gas are technically mined as well. Since they are carbon-based fossil fuels, again, all of the problems associated with fossil fuels come with them. Another liquid can be said to be mined, water. Water is often unsustainably extracted from groundwater, as it is not put back into the source; in fact, it cannot be put back into the ground at all. Groundwater is withdrawn in large amounts for use in large agricultural operations.

Mining Methods

There are several commonly used approaches to mining. The use of each of them depends on the type of ore being mined, the depth to surface from which the ores are extracted, and the type of material that must either be moved or drilled into.

- **Underground mining**: Mining that utilizes the digging of tunnels or shafts in the Earth in order to gain access to valuable rocks and minerals for extraction. This is often used to mine for metal ores, such as iron, zinc, copper, gold, uranium, and other metals. It is also used for coal mining.
- **Surface mining**: This is done through stripping layers of vegetation, soil, and even rock off of the Earth to gain access to valuable rocks and minerals for extraction. There are four types of surface mining operations:
    - **Open-pit mining**: An open pit is dug into the ground. Often used to mine for ores, minerals, and gems such as diamonds, phosphate, or gypsum. This can have devastating effects on local ecology and wildlife habitats due to disturbance of soil layers, and bedrock, as well as the removal of vegetation.
    - **Quarrying**: Very similar to open-pit mining except it refers to the removal of whole rock (usually for building materials) rather than ores; often used to mine building materials like granite, limestone, marble, or sandstone. Ecological and wildlife destruction similar to open-pit mining occurs with quarrying.
    - **Strip-mining**: Removing layers of soil and rock to access Earth materials; used to mine coal or clay such as kaolinite in the state of Georgia. Depending on the ore of interest, strip mining sometimes has less of an impact on ecology and wildlife than other methods.
    - **Mountaintop removal**: Removing the tops of mountains, usually through the use of explosives; used mainly to extract coal. This is possibly the most devastating type of mining; it even affects local human communities. The destruction from mountaintop removal increases the potential for dangerous floods and landslides. Disturbance of Earth layers leads to excessive acid mine drainage, which can potentially poison water supplies. In addition to this type of pollution, excessive coal dust in the air can cause illness in human populations as well. Local politicians often look the other way when mining companies cause this sort of damage because of the amount of money that is made by the mines.
- **Placer mining**: With this method of mining, miners sift through running water in riverbeds in areas where upland weathering and erosion yields mineral resources, often gems and precious metals, although other heavier elements of economic value are mined using the placer method. Ore types of elements, called **rare Earth elements (REE)**, are often placer mined. These elements are widely dispersed in the Earth and often used for consumer electronics such as cell phones. Placer mining often has a devastating effect on local riparian ecosystems.
- **Solution mining**: This mining method is used for salts, boron, bromine, copper, lithium, potash, and even uranium. Fluids are injected in deep deposits, and the material of interest is dissolved, then siphoned out. This method of mining disturbs less surface ecosystems, but there is always a possibility of hazardous materials leaking into local groundwater systems. Groundwater systems are not isolated from surface water systems.

Benefits and Impacts

Mining has benefited human civilization immensely. Our general global standard of living has been greatly enhanced by all of the products, all of the construction and infrastructure made possible with use of mined materials. From the

REE in cell phones and computers that we use to air condition buildings that keep us from being too overheated to the roadways that we pave for rapid transportation, mining has provided us all with enormous benefits.

At the same time, the costs and impacts of mining and mined materials cannot be ignored—for instance, the human rights violations that occur in heavily exploited mining areas of the world, in the extraction of REE, diamonds, and gold. It is also a human rights disaster since often individuals are conscripted into mining by military governments and other entities. This is another mining practice in which children are sometimes employed, adding to the human cost of mining in general. The extraction and processing of both metals non-metals has an impact an on human health because of pollution. Then there's the fact that mining is unsustainable by definition. Once a material is mined, whether it's iron, water, or uranium, it cannot be "put back" into the ground. So, what can we do, short of going back to living the hunter/gatherer lifestyle and shunning modern civilization completely?

Remediation Efforts and Recycling

One effort to **remediate** (reverse environmental damage) mining sites is that of **reclamation**, especially with sites than have been strip-mined or quarried. With reclamation there is an attempt to restore the mined site back to its previous state by removing mining machinery and structures, replacing overburden (topsoil that's "in the way" of the mined resource), replanting vegetation, and restoring ecosystems. It's nearly impossible to restore an ecosystem back to its original state, but it's worth at least getting started. Another problem with this effort is that mining companies often don't want to spend money on these efforts, and only government regulation will work in that regard. In less developed countries, there are often no regulations, and nothing is done.

Another way to ease the burden brought about by mining is through recycling. Since mining is non-sustainable, with overuse we run the risk of running out of certain materials to mine at some point in the future. The Earth is of finite volume; it can only take so much extraction. Recycling of metals is especially favorable since by their very nature they can be melted down to be reused again. Computer parts can also be repurposed, hopefully preventing overmining. This is especially true for valuable REE metals.

Finally, the fact that patterns of use and extraction change over time can ease the mining burden as well. As technology becomes developed, new technologies can make locating and extracting certain ores more feasible and less costly. This can also lead to new reserves being discovered, with the drawback being that this is often in remote sites within fragile, vulnerable ecosystems. Changing patterns of use can also be of benefit. The use of lead decreased dramatically during the last half of the 20th century, mostly as an additive to gasoline since it has been a target of concern by the **Environmental Protection Agency** (**EPA**). Also, the natural fibers from rocks such as serpentinite have been all but banned in recent times due to health concerns. Finally, in times of economic crises, which seem to occur periodically these days, people in general may start spending less on consumer electronics and the like.

Earth Hazards

Although life-giving and vital, the Earth itself can also be hazardous due to the nature of some of its features and processes. Danger to human life, limb, and property can often be avoided by being cognizant of the risks, although this is often ignored due to economic interest and other factors. Sometimes these hazards are just plain unavoidable and must be dealt with firsthand through emergency management procedures.

Mass Wasting

Mass wasting is the movement of rock, **regolith** (loose particles), and soil due to gravity and position. Mass wasting has a role in nature; it takes weathered material out of an upland area and moves it downslope. At the same time, **erosion**, a process similar to mass wasting, causes millions of dollars of damage to property annually. Unlike mass wasting, erosion is a somewhat lateral, as opposed to vertical, movement of weathered products across Earth's surface due to forces such as water and wind. Also, unlike mass wasting, erosion is mostly anthropogenic in nature.

The most active and potentially destructive forms of mass wasting occur in upland areas like mountains, which are geologically "young." **Landslides** are one of the most spectacular forms of mass wasting that often occurs in mountainous regions and areas near seashores that are inherently unstable. One way to avoid them concerning property is to simply be aware of the risks and do not build on or near such slopes. **Rockslides** are similar to landslides and occur in extremely rugged mountains. They can be unpredictable, so again awareness is key here, as are local emergency management protocols.

Debris flows can occur is unstable slopes under the influence of water. **Soil creep** and **solifluction**, which occurs in areas underlain by permafrost, can cause costly damage to roadways and other human structures, although these slow movements of soil are rarely directly dangerous to human life. Again, being aware of the risks is important, whether to life or property. Insurance is important concerning these types of hazards when cost and value are important.

Earthquakes

As we have discussed, earthquakes are powerful ground movements caused by the movements of faults and fractures in the Earth. Earthquakes occur nearly everywhere, but not all regions of the planet are subject to powerful ones. Most strong earthquakes of greater than around 4 on the **Richter scale**, which measures the energy generated by earthquakes, occur in areas of faulting between lithospheric plates, for instance the subduction zones of western South America and eastern Japan, and the San Andreas Fault, which runs through California.

There is nothing that we can do about earthquakes, and in fact they are somewhat unpredictable. The best that people can do is be aware of what to do in their area should one occur and build earthquake-safe structures. Most local and national governments have regulations and procedures concerning emergency management in areas prone to earthquakes. One should also be aware of after-affects from earthquakes, especially tsunamis.

Tsunamis

Tsunamis are not "tidal waves" as many have thought in the past. They have little to do with tides. A tsunami is a large, powerful wave generated from a landslide, a volcano, or an earthquake. A tsunami generated by an earthquake can be truly devastating.

Tsunamis can be generated by earthquakes when an overlying subducting plate "slips," displacing a large volume of water. When this occurs in open water the ensuing wave travels near the speed of sound. As the tsunami approaches the shore, the energy of that wave is redirected due to interference by the rising seafloor into an immense wave. Note: If you are at the shore and see the water recede from the beach rapidly, run immediately to high ground! A tsunami could be approaching.

Again, there is little we can do to prevent tsunamis other than be aware of the dangers and possibly try not to build structures on the shore in tsunami-prone areas, despite the fact that the economics of tropical tourism say otherwise. Otherwise, emergency management is vital.

On December 26, 2004, an earthquake off the coast of Sumatra triggered a massive tsunami that hit Indonesia, Thailand, Sri Lanka, India, and African countries. Around 228,000 people were killed, and 1–2 million people became displaced. Since then, early warning systems have been implemented throughout the Indian Ocean. Hopefully that will be enough to prevent such a tragedy from happening ever again.

Volcanoes

Volcanic eruptions are another Earth hazard that is impossible to control or even escape, other than simply not being around volcanoes in the first place. **Volcanoes** are geologic features, usually some mountains, that occur because of molten rock, hot gas, and ash that erupts through Earth's surface: lava. Due to the differential **viscosity** (resistance of liquids to flow) of lava, some volcanoes are not that dangerous to be around if one exercises caution. Lava from volcanoes on the island of Hawaii are basaltic and have low viscosity. On the other hand, continental volcanoes such as Mount St. Helens in Washington State, or Vesuvius in Italy erupt granitic and intermediate lavas. These lavas are highly viscous and explosive.

There are other hazards associated with volcanoes such as pyroclastic flows, ground-hugging clouds of ash, and gas and rock particles that travel faster than a human being can run. A pyroclastic flow was responsible for the burial of the Roman city of Pompeii in 79 CE. Lahars are superheated volcanic mudflows that originate when volcanoes erupt during heavy rains or when lava disturbs local groundwater. They too are extremely dangerous and travel faster than a human can run.

Volcanoes, like earthquakes, are generally associated with plate tectonic activity, so a large portion of Earth's population is not directly exposed to them. For people who live near volcanoes, they should be aware of safety measures to take should an eruption occur around them. Local and national governments generally have emergency management procedures for areas predisposed to volcanic eruptions in much the same manner as an earthquake-prone area. In fact, because of their association with tectonic plates, these areas are often prone to both earthquakes and volcanism.

Massive eruptions can have far-reaching impacts to the volume of ash and dust that is produced in such eruptions. Volcanic ash containing silica that is shot up into the atmosphere can damage aircraft engines to the point that they can become inoperable. Silica is harder than metal components in aircraft. During these eruptions, governments can declare no-fly zones for aircraft. Another effect of massive eruptions are aerosols, minute particles, that can circulate in the upper atmosphere globally. This can cause overall cooling of the climate for a certain amount of time because the particles reflect sunlight back out into space. When Mount Tambora, in Indonesia, erupted in 1815 it caused "the year without a summer" in 1816 due to the far-reaching effects of the volcanic ash and dust. The Mount Tambora eruption was the largest in recorded history.

Floods

Floods are overflows of water onto normally dry land that occur when channel capacity to carry water is exceeded in streams and the like, bodies of water are filled beyond capacity, or soil is saturated beyond its ability to allow water to infiltrate downward. Floods are a multiscale disciplinary problem: geographical, geological, and hydrological. Floods often occur because intense rainfall of high volume inundates land surfaces and water bodies, although floods can also occur because of dam failure. Floods are different from the other hazards discussed here because they are not always forces of nature that cannot be predicted. They are often affected and caused by human activity.

Humans create impermeable surfaces such as parking lots and roads, which can increase the severity of flooding due to the inability of water to properly drain into the soil or downslope into larger bodies of water. Dams can either fail due to accident or structural failure, causing catastrophic failure. Humans also alter riverbanks due to construction and development, which can increase the chance of flooding. Another aspect of flooding is similar to the other natural hazards discussed here in that people take risks in building within flood-prone areas. Insurance companies and management agencies such as the **United States Geological Survey** (**USGS**) implemented the **100-year flood** system. Simply put, this means that there is a 1 in 100 chance that there will be a catastrophic flood every year.

Engineered structures such as **canals**, **dikes**, and **levees**, which are built to either redirect or hold water, can be breached beyond capacity or fail due to unpredictable amounts of rainfall. In late August of 2005, Hurricane Katrina passed over New Orleans, Louisiana, in the United States, resulting in an enormous, unprecedented amount of rain. The system of canals and levees meant to redirect the Mississippi River for civil and commercial purposes could not withstand the floods that resulted from this downpour. The excess water flooded the city of New Orleans, resulting in one of the worst disasters in recent history. Ultimately, the toll it took on the city was tragic; close to 2,000 people lost their lives, and the damage exceeded $100 billion.

Conclusion

Despite the challenges of living on our planet, Earth is an amazing place; in fact, as far as we know it is one-of-a-kind. It's the only planet we know of that supports life. We should respect our home and its resources. The Earth is a complex system. Through knowledge and stewardship, we can continue to protect our home planet, and hopefully do more than our fair share of cleaning up the damage that we've done, even as we look to the cosmos to explore and hopefully respect other planets should opportunity arise.

Discussion Questions

Directions: Review the chapter in order to completely and correctly respond to the questions and prompts.

Keep this in the back (or maybe even front) of your mind: Science is system of methods that allows us to understand not only our planet, and ourselves, but of the entirety of the universe. Answering the following questions correctly will lead to a more robust understanding of how to apply science to ENVS problems.

1. Why is there currently no evidence of the existence of population III stars?
2. What is a possible reason for the fact that the inner planets in the solar system (Mercury, Venus, Earth, and Mars) are composed of small, dense, rocky, metallic solids and the outer planets are gas giants, large, less dense, and icy?
3. Why are some meteorites older than the Earth?
4. There was very little oxygen in the atmosphere during most of the Precambrian. What might the atmosphere have been composed of back then? Why?
5. Why do we know that the outer core of the Earth is liquid?
6. How might the position of the continents on the planet affect climate and weather?
7. Could heat from the magma deep in the Earth be used as an energy source? How?
8. What sorts of evidence might help to let us know that we are currently in a period of oceanic transgression?

9. Which type of mining would have the least impact on terrestrial ecosystems?
10. What type of geologic hazard, other than earthquakes, would cause tsunamis?
11. What is the 5-year flood?

Bibliography

Hatcher, R.D., 1995, *Structural Geology: Principles, Concepts, and Problems*: Oxford, Oxford, 3rd ed., 656 p.
Levin, H., and King, Jr., D., 2017, *The Earth Through Time*: Hoboken, Wiley, 11th ed., 608 p.
Rollinson, H., 2007, *Early Earth Systems. A Geochemical Approach*: Oxford, Blackwell, 8th ed., 285 p.
Winchester, S., 2001, *The Map That Changed the World: William Smith and the Birth of Modern Geology*: New York, Harper Collins, 368 p.

Credits

Fig. 4.1: White Sands Missile Range / Applied Physics Laboratory, "The First Photograph of Earth Taken from Space," https://commons.wikimedia.org/wiki/File:First_photograph_of_Earth_from_space_24_October_1946.jpg, 1946.
Fig. 4.2: Hardwigg, "The Geological Clock, a Projection of Earth's 4.54 Gy History onto a Clock," https://commons.wikimedia.org/wiki/File:Geologic_Clock_with_events_and_periods.svg, 2010.
Fig. 4.3: William Smith, "A Delineation of the Strata of England and Wales with Part of Scotland." https://commons.wikimedia.org/wiki/File:Geological_map_Britain_William_Smith_1815.jpg, 1815.
Fig. 4.4: Emilio Segrè, "Marie Tharp, Geologist," https://commons.wikimedia.org/wiki/File:Marie_Tharp_working_with_fathometer_record_%28cropped%29.jpg, 1968.
Fig. 4.5: Copyright © by Nealey Sims (CC BY-SA 3.0) at https://commons.wikimedia.org/wiki/File:Earth%27s_Inner_Layers_denoting_the_LAB.png.
Fig. 4.6: National Oceanic and Atmospheric Administration, "Map of Major Tectonic Plate Motion," https://commons.wikimedia.org/wiki/File:NOAA_tectonic_plate_all.jpg, 2018.
Fig. 4.7: United States Geological Survey, "The Three Types of Faults," https://commons.wikimedia.org/wiki/File:Fault_types.png, 2005.
Fig. 4.8: Copyright © by Awickert (CC BY 3.0) at https://commons.wikimedia.org/wiki/File:Rock_cycle.gif.
Fig. 4.9: Copyright © by Woudloper (CC BY-SA 1.0) at https://commons.wikimedia.org/wiki/File:Offlap_%26_onlap_EN.svg.

Chapter 5

Ecology, Populations, and Wildlife Biology

The Biotic and Abiotic World

Objectives

- Understand that life itself goes beyond that of the individual organism, ultimately forming Earth's entire biosphere.
- Organisms are composed of chemical materials and require energy to exist.
- Evolution, through natural selection, drives speciation and diversity in communities.
- In Earth's history of life, extinction occurs when a species cannot cope with a changing environment.
- Diverse natural communities make for healthy ecosystems.
- Species interactions are the backbone of communities.

KEY WORDS AND TERMS

Ecosystem (n.): An assemblage of communities within a distinct geographic area, in which the species interact with the abiotic components as well as with each other.

Community (n.): Groups of different species, living in the same area that interact with each other.

Population (n.): A group of a particular species that lives within a distinct area.

Consider This . . .

As stated before, Earth is the only planet, that we know of, that harbors life. Recently, researchers found evidence of life within the Isua Greenstone Belt in Greenland dating back some 3.7–3.8 Gya, 700 Myr (million years) after the formation of Earth. One would think that life itself is integral to Earth processes such as climate and weather, the hydrologic cycle, and rock formation. One would also be correct in making that assumption. Life is a complex and vital series of processes, fundamental to the very nature of our planet.

Introduction

Organisms are chemically and physically some of the most complex entities on Earth. The cell is the fundamental unit of life, from the smallest bacteria to largest blue whale (Figure 5.1). Even at the bacterial level, organisms show evidence of chemical and structural complexity. Beyond singular organisms are groups of organisms living across the surface of the Earth, under the soil, within the ocean, and even in the very air around us. All of these organisms comprise Earth's biosphere. This biosphere has existed since at least around 3.7–3.8 Gya and possibly even earlier. It has fundamentally affected Earth's other spheres: the atmosphere, the hydrosphere, and even the solid lithosphere. Biological processes have their own unique systems the guide the development and spread of life.

The basic tenets of living things, organisms, that live on Earth, are **the characteristics of life**. The characteristics of life are generally considered to consist of the following:

- All living things, organisms, are composed of at least one cell.
- Organisms need energy to function.
- Living things respond to their environment in some way.
- Organisms are able to form new living things: reproduce.

Through these processes, life has been able to inhabit literally every environment on Earth. Also, through reproduction life is also able to form into new configurations through successive generations. This has been taking place for billions of years.

Natural Selection

Within **populations** of organisms, evolution guides the development of organisms over time. Evolution is the genetic change in organisms. These changes may be due to the environment or due to **natural selection**. Natural selection is the process in which traits that ensure the survival of a species are passed down through generations. This is based on the **fitness** of organisms. Fitness, in evolutionary terms, is the ability of an organism to not only survive but also to pass down these genes to successive generations, at the expense of genes that do not ensure persistence.

Natural selection alters the genetic makeup of a population of species through successive generations. In this sense it is vital to understanding antibiotic and pesticide resistance in organisms with extremely short life spans. This has implications in agriculture and pharmaceutics. This is because organisms respond to their environment and change over time, through successive generations.

Darwin and Wallace

In 1859 the British scientist Charles Darwin published *On the Origin of Species* based on his work, as well as to a certain extent the work of Alfred Russel Wallace, another British scientist (Figure 5.1). Darwin and Wallace worked independently, yet supported each other through correspondence. *On the Origin of Species* brought to light ideas that many scientists had been aware of for years. At the crux of these ideas was the overarching theory of natural selection, which is integral to biological evolution.

Figure 5.1 The Darwin and Wallace medals of the Linnaean Society

The basic principles include the fact that, in essence, all organisms have to struggle to survive. Beyond even that, all organisms also face a challenge in order to reproduce. It may not seem, to us, like life is a struggle to survive and reproduce, but imagine what life would be like without electricity, powered transportation, running water, and even things like banks and mass-produced entertainment. Our relative luxuries are artifacts of civilization, of the struggles that human beings who came before us had to go through in order to achieve these societal and technological benefits.

Another aspect of Darwinian natural selection is the fact that all individuals of a species vary in their characteristics due to the environment, or environments, in which each individual is found. Environment drives evolution through an unspoken sort of trial-and-error testing. If a trait promotes survival it is passed on to the subsequent generation. Any trait that does not do so is terminal; the organism simply does not survive or continues to survive severely impaired. This is similar to an **adaptive trait**. Adaptive traits are traits that ensure reproductive success. Without these traits the lineage of an organism will die off. As you can see, some members of a species are better suited to survival within their given environment. This isn't enough though; an individual needs to have the ability to reproduce, ensuring that their offspring can reproduce as well. This is known as **fitness**.

At the time of the publication of *On the Origin of Species* it was not clear just how traits were passed on from generation to generation within a given species. At around the same time as Charles Darwin's journey into natural selection, a monk named Gregor Mendel experimented with different varieties of peas, which led to his discoveries in **heredity**, or the passing on of traits to successive generations. Mendel's work led to the formation of a branch of biology known as **genetics**, which is the study of genes and genetic inheritance in organisms. Later, the mechanisms behind genetics would be discovered, including DNA.

We Are All Mutants

Changes to successive generations occur on the molecular level, in DNA, and can be passed on. These changes are driven by **mutation**. Mutations are accidental changes to DNA in an organism that may occur due to changes in its environment, or even occur randomly. Mutations are either lethal or non-lethal. An example of a lethal mutation is cancer.

Non-lethal mutations are often unnoticeable and do not get passed on to successive generations. These types of mutations have happened to all of us, making all of mutants of a sort. Non-lethal mutations that provide beneficial genetic variation on the other hand may be passed on. This genetic variation, stemming from mutation, is a mechanism for natural selection, which can lead to evolution in organisms. In addition to mutation, sexual reproduction also drives the variation for natural selection.

Disruptive selection

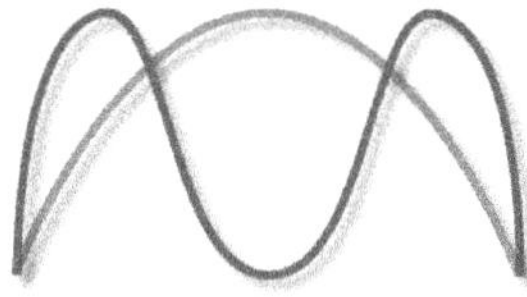

Stabilizing selection

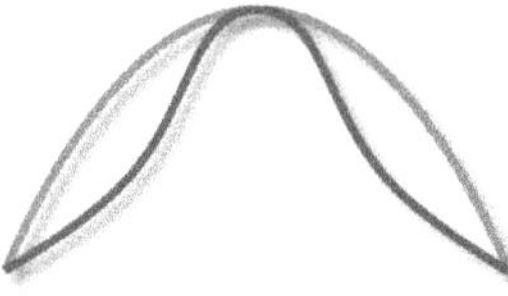

Directional selection

Figure 5.2 Disruptive, stabilizing, and directional selection

Variations in Selection

Natural selection, driven by environmental change, can take several forms. At the same time, human beings can actually be the agent of selection for other species.

Types of Selection

Natural selection due to environmental factors can take several forms depending on how organisms are affected by the environment (Figure 5.2). When a particular feature is favored with regard to survivability, it may be passed on to successive generations through **directional selection**, meaning that a trait moves toward favoring the most successful trait. For instance, rabbit ears tend to be longer in warmer climates because their ears act as temperature regulatory systems. Longer ears in rabbits allows for the dispersion of heat, leading to cooling. **Disruptive selection** occurs when a species lives across geographic environments warranting that members adapt to those differential conditions over time. For instance, a species of bird may live on two different islands. One island may be covered in flowers favoring the development of a long beak in order to feed on the nectar in the flowers. The other island may not be covered in flowers but may have many trees bearing seeds. The members of that bird species on that island may evolve strong beaks to crack open the seeds in order to eat. Finally, **stabilizing selection** preserves the status quo. Standard traits are favored, and the majority of a population features those traits because they are best for survivability. Thus, little change occurs over successive generations.

GROUND TRUTH

Artificial Selection

Natural processes are not the only means of selection within successive breeding populations; human beings also breed organisms through **artificial selection,** also known as selective breeding. In short, artificial selection is selection under human direction. This process is not new; human beings have been breeding organisms since prehistory.

Farmers, foresters, and animal breeders have been responsible for selective breeding throughout history. Animals that possess desirable traits are the only ones allowed to breed. Only seeds from plants that are particularly prized are used. Plant features that are not desired are cut away while desirable parts are spliced onto others. Through these simple methods, many of the animals that people have domesticated, and many of the plants that blanket entire landscapes, have been bred to order. These changes that people have made to other organisms have had a widescale impact on the biosphere itself. The list of organisms that people have bred is quite long. It includes dogs, rice, trees, and wheat.

A notable plant bred by humans is the wild cabbage, also known as *Brassica oleracea*. Wild cabbage also goes under the following names: broccoli, Brussels sprouts, cabbage, cauliflower, collard greens, gai lan, kale, and kohlrabi. These plants originated in Europe and can be historically traced to ancient Greece and the Roman Empire. All of the varieties of wild cabbage were bred over the years for particular traits related to the plants value as a food source. For instance, a preference for eating leaves that developed into tightly pack buds led to the breeding of cabbage. At another time, populations of humans favored eating immature buds, and this led to the development of broccoli and cauliflower. Keep in mind that all of these plants are genetically identical; they are all *Brassica oleracea*, with the only difference selective breeding by humans.

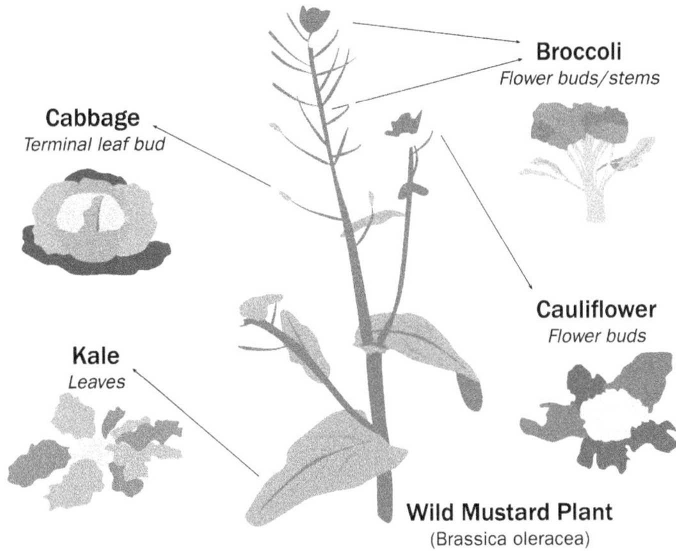

Figure 5.3 Artificial selection, or selective breeding, in the wild mustard plant (Brassica oleracea)

Since the beginning of the agricultural revolution, cattle have been bred by humans, not always for meat, milk, or hides, but as a draft animal for farm work. The widescale breeding of cattle for meat and dairy is fairly recent. This primarily occurred in Great Britain, during a period of time known as the British agricultural revolution between the 17th and 19th centuries. With increased industrialization came a geometric increase in population. In order to feed an expanding population, some of the first industrialized agricultural methods were introduced, including the widescale cross breeding of cattle. Different varieties of cattle were cross bred in order to increase the size of individuals for meat production. Breeding was also done to increase the volume of milk from dairy cows.

Biological Diversity

Biological diversity, or **biodiversity**, is important to biological sustainability. It is characterized by the diversity of organisms within specific areas, often on the **ecosystem** level. This includes diversity with populations of individual species, the genetic diversity within species, and the communities composed of these populations. Again, a **species** is defined as a group of organisms in which individuals share characteristics and can breed with one another, producing offspring that, in turn, can also potentially breed with other members of the same species. A group of the same species within a given area comprises a **population**. A group of species within an area is known as a **community** of organisms.

Speciation

Over time, new species are introduced to ecosystems through **speciation**. Speciation is the generation of new species from a singular species. There are two main modes of speciation that occur within natural settings: **allopatric** and **sympatric**.

Allopatric speciation occurs because of geographic, or even geologic, change. These changes include climate change, earthquakes, flooding events, glaciers, mountain-building events, or even volcanism. For instance, a population of some terrestrial animal may inhabit an ecosystem that includes mountains. A catastrophic flood, perhaps due to a river's course being altered by a powerful earthquake, floods the valley between the mountains, separating the species into various populations. Because these now isolated populations cannot contact each other, each population, over time, thousands to millions of years, will likely develop into new species. This is especially true if the environmental conditions on each mountain, now a series of islands, are vastly different. Allopatric speciation is the primary mode of speciation.

The second, and secondary, form of speciation is **sympatric speciation**. This form of speciation occurs when populations become split and reproductively isolated from each other within the same area. This may happen due to different feeding and mating habits within a population. It can also occur because of mutations or some other genetic phenomenon. As one can see, sympatric speciation is harder to discern than allopatric speciation.

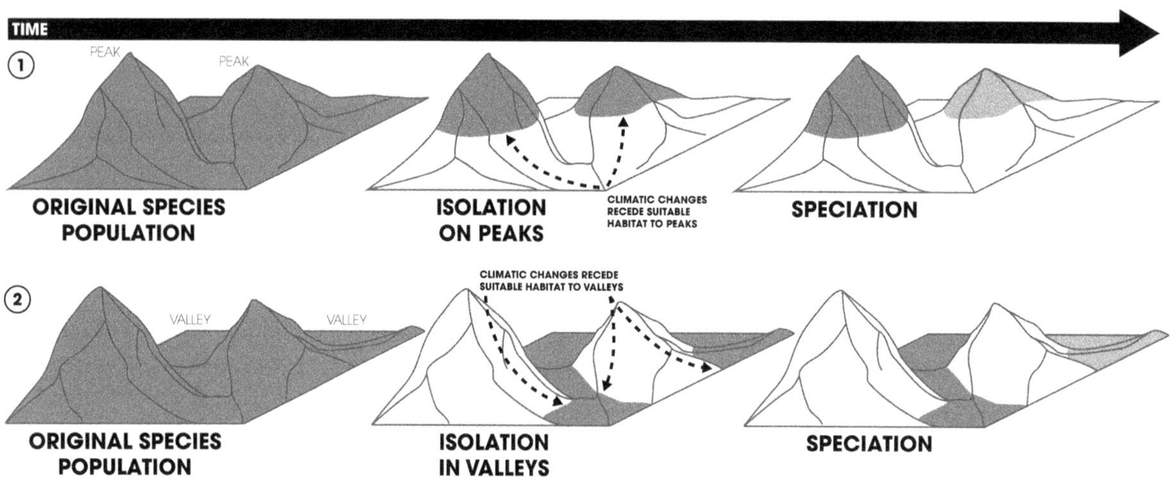

Figure 5.4 Allopatric speciation can result from elevational topography

Patterns of Speciation

Speciation generates ever-increasing patterns of diversity beyond the species level. These patterns can be analyzed on an evolutionary level and traced over time. This is often done by scientists through the creation of **cladograms**, also known as **phylogenetic trees**. Cladograms are charts that characterize a history of species divergence. These charts are generated using traits that evolved within species. These new traits are characteristic of speciation as well as evolutionary change in successive species through speciation. Cladograms can be relatively simple, such as the speciation history of one specific organism, to vastly complex, such as the entire pattern of evolution of all known organisms. As such, these cladograms are often a way to understand the evolution of current organisms from extinct organisms over vast spans of prehistoric time.

Challenges to Speciation

While speciation can lead to the development of new organisms from previous or existing organisms, there are conditions that can hinder this process or even end it altogether. Speciation often leads to the evolution of larger and more complex organisms from smaller and simpler ones, but it does not have to happen like that. Organisms can effectively "devolve" into new species due to a need, environmental or otherwise, for organisms to revert to simpler modes of biological development. There also exist organisms in which evolutionary progress is simply no longer required. Cockroaches and crocodiles are examples of these so-called evolutionary dead-ends. Neither animal has evolved new traits for millions of years. This is because each has reached an equilibrium point with regard to their respective environments. They are adaptable and efficient enough that there is no need to change to continue existing. In other cases, there are sufficient challenges for a species to continue existing at all.

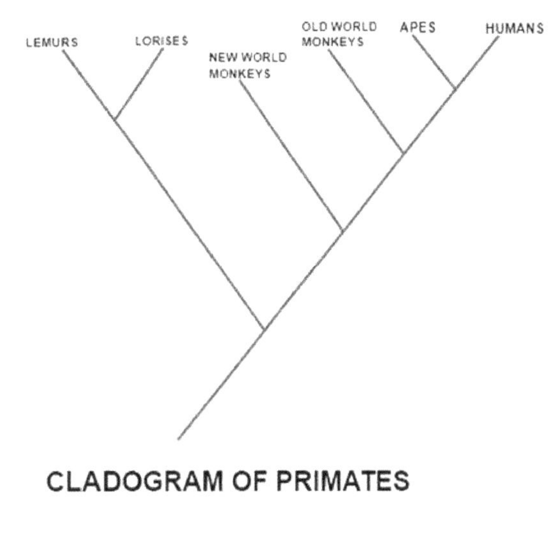

Cladogram of primates

Extinction and Extirpation

When a species faces severe challenges to its survivability it may undergo either extirpation or extinction. **Extirpation** occurs when a population of a species disappears from a particular geographic area but is still extant (exists elsewhere). Extirpation can occur because of any number of environmental factors, but human development and encroachment is currently a major factor in species extirpation. As an example, for a number of years in the late 20th century there were no gray wolves in Yellowstone National Park and surrounding areas. This was mainly due to ranchers hunting wolves in order to prevent their livestock being killed. Gray wolves had been extirpated from that area. In recent years gray wolves have been successfully reintroduced to the Yellowstone area due to the work of conservation organizations.

Environmental stress and other factors can go beyond pockets of extirpation to the total loss of a species from Earth itself. This is known as **extinction**. Extinction occurs when the last member of a species dies off. It is absolute. There is no going back from extinction; it is absolute and irreversible. Biodiversity is affected by extinction, although extinct species may be succeeded by new species through speciation. Extirpation can lead to extinction because of geographic constraints. There are quite a few causes for extinction.

Species Loss

Extinction, as well as extirpation, has many causes. Certain organisms are **endemic species**, species that inhabit very specific areas, or even an area, on the planet. The condition that endemic species find themselves in is called **endemism**. There are reasons that a species, or group of species, become endemic. One reason may be geographic isolation, for instance on islands. The emu is endemic to Australia; it is only found on the continent of Australia. On a related note, the emu used to inhabit nearby Kangaroo Island, King Island, and the Island of Tasmania but became extinct after human settlement on the late 18th century. This points to the fact that extirpation can easily lead to extinction as global biodiversity is reduced.

Some endemic species require very specific environmental conditions for survival. They may require a very specific set of moisture and temperature conditions, for example. These species may require very specific food as well. The now extinct golden toad of the Costa Rican cloud forests required very specific moisture and temperature conditions in the highlands where it lived. Once that changed, slightly, in the late 20th century, the golden toad ultimately went extinct. Back to Australia, the koala only eats eucalyptus leaves that are also endemic to Australia. It cannot survive in the wild elsewhere. Another example is the volcano snail, also known as the scaly-foot gastropod or the scaly-foot snail. It is specifically endemic to volcanic vents on certain islands in the Indian Ocean.

Cosmopolitan species are more or less the opposite of endemic species. Cosmopolitan species exist across broad areas of the planet, often within a wide range of ecosystems and climactic conditions. For instance, the aforementioned cockroaches exist worldwide in a vast variety of climates and environments. Norwegian rats are found all over the planet as well. Both cockroaches and Norwegian rats are hardy and can resist, and even thrive, on human encroachment. Flying birds are often cosmopolitan, especially migratory birds, because they are not often limited by geographic constraints. Cosmopolitan species are resistant to extirpation and extinction due to the nature of their existence. It takes global extinction crises to cause cosmopolitan species to go extinct. This has happened many times in the past. Severe changes in climate, and the environment in general, can lead to extirpation and extinction.

The encroachment of other species can cause extirpation and extinction. Human beings are a prime example of this. The extinction of the dodo and the passenger pigeon due to **anthropogenic extinction**, extinction directly caused by human beings, is well documented. The dodo, a flightless bird that was endemic to the island of Mauritius went extinct sometime in the late 17th century. It is believed this was due to overhunting by humans and invasive species. The last passenger pigeon, of North America, died on September 1, 1914, in the Cincinnati Zoo. The extinction of the passenger pigeon is directly related to human hunting as it was seen as a cheap source of meat in the 19th century.

The End

Extinction is the end of a species in the most absolute way possible. It is a fact of life, having originated at least 3.7 Gya with the first known living things. Nearly all organisms that have existed on Earth are now extinct. The **background extinction rate** was the normalized rate of extinction during Earth's past prior to increased human encroachment on the environment. The background extinction rate is different for different types of organisms. Since the beginning of human civilization, the background extinction rate has been altered due to human activity: climate change, development, population growth, and resource depletion, among others. It is now thought to be some 1,000 to 10,000 greater now than in the past.

Mass extinctions have occurred throughout Earth's history. Five of them are known to have happened in Earth's geological past:

- **End Ordovician extinction** (444 Mya)—Resulted in the extinction of many small marine organisms.
- **Late Devonian extinction** (360 Mya)—Many tropical marine species died out.
- **End Permian extinction** (250 Mya)—Also known as the Permo-Triassic extinction. The largest mass extinction event known, over 50% of many categories of species, including over 80% of all marine species, went extinct in less than 100,000 years.
- **End Triassic extinction** (200 Mya)—Many vertebrate species, other than the rising dinosaurs, went extinct.
- **End Cretaceous extinction** (65 Mya)—Also known as the K-T event, or the K-T extinction. The primary cause was a massive asteroid impact, the Chicxulub impact, off of Mexico's Yucatan Peninsula, although there were other factors (Figure 5.7). Up to 50% of many types of organisms went extinct, including the dinosaurs.

It is widely thought that we are currently within Earth's sixth mass extinction event. This is because of changes to the background extinction rate, as well the effects of climate change, human development, human population growth, and resource depletion.

GROUND TRUTH

The Pleistocene Megafauna

The Pleistocene epoch was a span of geologic time that lasted from about 2.58 million years ago to 11.7 thousand years ago. This was the time when human beings, *Homo sapiens sapiens*, and all of their ancestors evolved. The Pleistocene is marked by a series of global glaciation events, interspersed with periods of warming. The last of these glaciation events, known as the Ice Age, marked the end of the Pleistocene and the beginning of the current Holocene epoch 11,700 years ago. Pleistocene glaciation ice sheets, at their upper limit, extended down in southern Illinois in the United States. The Pleistocene, in general, was a much cooler and drier time than our current epoch.

There were a large number of remarkable, gigantic animals that lived during the Pleistocene known as the **Pleistocene megafauna** (Figure 5.8). Most of these animals were very large in size. They shared the planet with a lot of the species that we are familiar with; in fact, despite the periods of intense global cooling, the Pleistocene was not that much different from our Holocene epoch, geographically and geologically. The types of animals comprising the Pleistocene megafauna are probably very familiar to all of us thanks to recent movies and television shows about that period. These large animals include Diprotodon (a giant marsupial), dire wolves, giant ground sloths, Gigantopithecus (a primate that stood up to around three meters tall), Glyptodon (a giant armadillo), Irish elk, mastodons, saber-toothed cats, short-face bears, woolly mammoths, and woolly rhinos. Giant birds and even reptiles also existed during the Pleistocene, including Haast's eagle, the moa, and Meiolania (an extinct giant turtle). Toward the end of the Pleistocene, human beings shared parts of Europe and Asia with other human species, including the Neanderthals. In fact, a small portion of some Asian and European genomes contain genes originating in

Neanderthals, as well as Denisovans, another species of human that lived in Asia. The thing is, none of the gigantic and obviously formidable animals of the Pleistocene made it very far into the Holocene epoch. Neither did our own relatives, the Denisovans and the Neanderthals.

Figure 5.6 Some of the Pleistocene megafauna

The extinction of Pleistocene organisms is known as the **quaternary extinction event**. Scientists have speculated for quite some time as to why such large, hardy animals disappeared when the Ice Age ended. There is some evidence to indicate that one of the factors was human over-hunting, something which we would have done alongside our relatives the Denisovans and the Neanderthals. This is known as the **overkill hypothesis**. Comparisons can be made to recent cases of overhunting and population control, including the extinction of the passenger pigeon in the 19th century. In harsh colder climates, meat provides fats and proteins that deliver significant means for nutrition and, ultimately, survival. Animal hides and bones would also have been used for clothing as well as shelter.

There are other hypotheses concerning the Pleistocene extinction, including climate change and the spread of disease. Since evolution, and conversely extinction, is driven by changes in the environment, including changes in climate, the **climate change hypothesis** makes some sense. Many of the large mammals of the Pleistocene evolved under very specific climactic conditions. As the world became more humid and warmer at the beginning of the Holocene, this may have put too much stress of the survivability of these animals. With moister, warmer conditions come the spread of microorganisms. Another hypothesis as to why the Pleistocene megafauna went extinct is that higher humidity and temperatures brought diseases. Some of these diseases may have been indirectly spread to animal populations by nomadic humans. This is known as the **hyperdisease hypothesis**.

Currently, none of the hypotheses for late-Pleistocene extinction take precedence over any of the others. In fact, scientists have made arguments for and against all of them, so the investigation continues. In fact, all of the hypotheses may have had a hand in the quaternary extinction event to some extent. We may never truly know what exactly happened, but lessons in environmental change and survivability remain for us to consider.

Earth's Biosphere

Ecology and evolution are directly related. Speciation and the subsequent evolution of species results in structuring whole ecosystems. Ecosystems are formed through the use of soil, water, and biological matter, which even affect abiotic components. Species comprising populations are part of communities. Communities, along with the climatic and geologic bases for abiotic systems, are contained within ecosystems. On a grand scale, the patchwork of Earth's ecosystems comprises the sum total of all life on Earth, the **biosphere**.

Ecology and Ecosystems

The understanding of ecosystems, biotic communities within a setting of abiotic components, is informed by the science of ecology. Ecology's first two syllables, "eco," comes from the Greek word "oikos," meaning house. Interestingly enough, the term is also the root for "economy." The Cary Institute of Ecosystem Studies defines ecology as "the scientific study of the processes influencing the distribution and abundance of organisms, the interactions among organisms, and the interactions between organisms and the transformation and flux of energy and matter. The breadth of subject matters within ecology." In other words, ecology is the study of how **biotic** (living) entities (organisms) interact with each other as well as their interaction with the **abiotic** (nonliving) environment.

A Purpose Within an Ecosystem

Organisms within ecosystems find themselves in their **habitats**. A habitat is the environment that an organism lives in. Habitats include biotic and abiotic components that the individual organism interacts with and are scale dependent according to the activities of said organism. Organisms generally thrive in specific habitats, although some organisms may range across a variety of them. Many organisms select suitable habitats through habitat selection. One can sort of think of a habitat as an organism's home. The quality of a habitat, suited to an organism, is crucial to an organism's survival. Through development and other factors, humans constantly encroach on the habitats of other organisms.

On the other hand, an organism's **niche**, its role within a community, is dependent on habitat use. An organism's niche is also dependent on its role in energy and nutrient flow within an ecosystem, which is essentially what it eats and what eats it. Decomposition is another factor. An organism's niche, or role, is also dependent on the other organisms within an ecosystem with which it interacts.

Organisms have different sorts of niches. **Specialists** are organisms with very narrow niches, specific requirements. They are extremely good at fulfilling their roles but tend to be vulnerable to change. Tigers are an example of a specialist species. Tigers dominate their ecosystems but are vulnerable to change within their habitats, particularly human encroachment. Specialists tend to have a narrow range of habitats. **Generalists**, on the other hand, fulfill broader niches, can inhabit a variety of habitats, and utilize a wide array of resources. Rats are a prime example of generalists.

Population Dynamics

Populations of organisms change over time, whether within a state of stability or change. The study of populations is often quantitative and dependent on numerical calculation. Changes occur within a population's size, which is the number of organisms present at a particular time. These numbers may change, remain stable, or experience cyclical change over specific spans of time.

The **population density** of a species, on the other hand, is the population per unit area. Populations of high density increase the chance for members to find mates but also increase the rate of competition for mates, as well as resources. High-density populations are also subject to higher rates of disease, or predation; when these increase in occurrence the population can go into crisis, potentially leading to extirpation or worse. Lower density populations make it more difficult to find mates but increase the availability of resources particular to said species. Space is also more plentiful when population densities are lower. Population crises can occur in low-density populations due to lower rates of mating.

The spatial dispersal and distribution of a population within a unit area is also an important factor to consider. Unsurprisingly this is known as **population distribution**, or **population dispersal** (Figure 5.7). There are three basic kinds of population distribution in nature. **Random distribution**, or more appropriately dispersal, is haphazard with no discernable pattern. This is seen somewhat in the lack of patterns of trees within a forest due to the arbitrary dispersal of seeds by birds and the wind. Patterns are easily discernable in **uniform distribution**. This is when organisms are evenly spaced within an ecosystem, and it is often due to territoriality. Penguins show this sort of distribution. **Clumped distribution** occurs because of the availability of resources. It is the most common form of population distribution. Examples include species in the African savannah gathered around a watering hole, or schools of fish gathering around food.

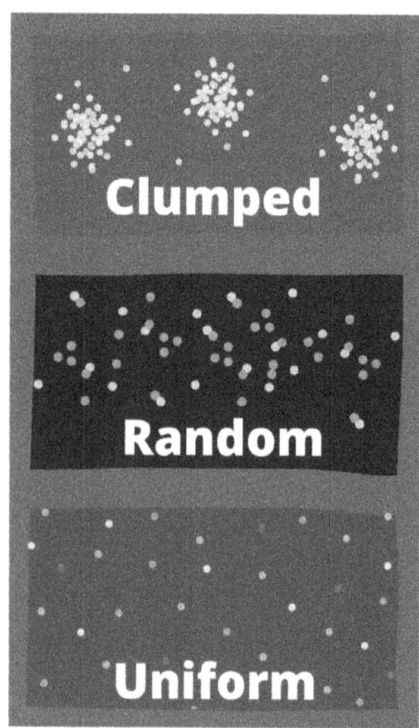

Figure 5.7 Clumped, random, and uniform population distributions

The sexual composition and age distribution also structures populations. In monogamous species, basically female–male mating pairs, healthy populations are generally 50/50 female/male, which maximizes population growth. When this ratio is skewed, it can alter the overall population and cause the numbers to lower. Age distribution can also affect population dynamics. **Age structure diagrams** are used to understand age distribution and its impact on future populations. Populations with a larger number of mating or pre-mating age individuals tend to increase in size over time. Stable populations are those with a more or less even number of individuals from pre-mating age to post-mating age. Declining populations have a higher number of post-mating–aged individuals up to the age limit.

The study of populations is often numerically constrained with regard to populations. **Crude birth/date rates** are simply understood as the inputs and outputs of a population per 1,000 individuals. This is impacted by the survivorship over time of an individual of a species, which is dependent on said species' particular biology and niche. **Survivorship curves** are a means to understand this phenomenon. There are three basic sorts of survivorship curves: types I through III. Type I species experience more deaths closer to the species life span, or end of life. Species that are type II experience births and deaths along a fairly even distribution. Type III species experience higher numbers of deaths closer to birth.

The essence of population studies can be pared down to four factors:

- **Natality**—Births within a given population
- **Mortality**—Deaths within a given population
- **Immigration**—Individuals entering a population from another area
- **Emigration**—Members of a population leaving the area

These four factors can then be placed within an analytical formula known as the **growth rate formula**, which is time dependent. The span of time of analysis is variable depending on the research.

(crude birth rate + immigration rate) - (crude death rate + emigration rate) = growth rate

Biotic Potential in Animals

Biotic potential is an organism's inherent ability to produce other organisms, their maximum reproductive capacity. There are essentially two sorts of organisms in that regard: **K-selected** and **r-selected**. These metrics are used to assess populations of animals.

K-selected species are animals with long gestation (childrearing) periods and relatively few offspring. They have **low biotic potential**, which is merely a measure of the maximum number of individuals a species can produce. These animals stabilize near carrying capacity and are good competitors. Thus, they are often more advanced species that structure the biota of ecosystems, although there are exceptions. The aforementioned tigers are an example of a K-selected species. ("K" stands for "carrying capacity.")

Animals that reproduce quickly and often in great numbers are r-selected species. Due to these greater numbers, they have high biotic potential. "r" stands for rate. Because of maximum numbers, they are said to have **high biotic potential**. These organisms show little parental care and are often less advanced than K-selected species, but again there are exceptions because octopuses are r-selected species. Octopuses are incredibly intelligent, more so than any other invertebrate. Many sorts of fish and amphibians are r-selected species.

Populations and the Environment

It is well proven with all organisms, including humans, that a steady rate of growth causes an exponential increase in a given population over time. This is because steady growth rates are due to highly favorable environmental conditions. With these conditions met, organisms are somewhat at a leisure to mate more frequently and productively. Of course, the overall analysis of exponential population growth is much more complicated than just that. The fixed percent factor aiding growth causes geometric increases that can be graphed. These graphs often look like j-shaped curves, although there are exceptions. Regardless, in unfavorable conditions, populations increase rapidly and exponentially. There are limits inherent within the environment though. Exponential growth is not sustainable over time. This sort of growth begins with smaller populations within ideal conditions. Once this population has grown beyond certain limits the environment will start becoming stressed, and conditions will no longer remain favorable for the population.

Because populations depend on the environment and ecosystems, there are environmental **limiting factors**. These factors restrain population growth and include available space, disease, food availability, predators, and water availability. All of the limiting factors together form **environmental resistance** to exponential population growth. The limiting factors contain a population's **carrying capacity**, which is the maximum population size that the environment will naturally sustain. This decreases and even stops exponential growth and is often graphically plotted as an s-shaped **logistic growth curve,** which represents steady to exponential then slowed growth over time. Limiting factors alter carrying capacity, but limiting factors can be manipulated. Human beings have been manipulating environmental limiting factors for millennia, increasing their carrying capacity, often at a cost to the environment itself.

Time is not the only factor in determining population dynamics within the environment; population density is important to consider as well. There are two modes: **population-dependent** factors and **population-independent factors**. Density-dependent factors are affected by population density. They increase the chance for both mating competition and predation. There is also an increased chance for disease transmission in high-density populations, as well as increased competition for food. Density-independent factors are mostly abiotic in nature and include floods, fires, landslides, volcanism, and weather and climate events. Development and pollution are also population-independent factors, thus human existence as a whole, with regard to other species, is a population-independent factor.

Human Impacts on Other Populations

Population dynamics are often differential within communities. When one population of species declines, other species may appear. In a somewhat related manner, human development has been displacing other species and threatening biodiversity for quite some time. Human economics and society have often informed human behavior in our relationships with other organisms. Historically, nature has been viewed as a hindrance to development and only as a source of resources for human growth. Thus, there have often been few checks to human development and resource extraction. Thus, biodiversity, which we are dependent on, has been under pressure for some time. In fact, biodiversity as a whole has become somewhat damaged.

Efforts are being made to reverse this historical trend through conservation and sustainable practices. Parks and protected areas aid in the overall effort to preserve biodiversity as well, despite often being underfunded. Where governmental agencies may falter due to various reasons, economic methods may provide a stimulus toward preserving biodiversity through things such as ecotourism as well as other environmentally sustainable practices. One can hope that all of these efforts are making a difference.

Species Interactions

The interactions within populations and between populations within communities structures entire ecosystems. There are two modes of interaction: **interspecies** and **intraspecies**. Interspecies interactions occur between different species; intraspecies interactions concern singular species. There are six basic community and species interactions:

- **Amensalism**—Species interactions in which one species is unaffected yet one or more other species is harmed
- **Commensalism**—Species interactions in which one species is unaffected while another one benefits
- **Competition**—Species interactions in which one or more species vie for the same resources, potentially harming one another
- **Exploitation**—Species interactions in which one or more species uses other species for their own gain; one species is always harmed in some manner
- **Mutualism**—Species interactions in which two or more species cooperate for mutual gain
- **Neutralism**—Species interactions in which two or more species gain or lose nothing

Amensalism

Amensalism is often unintentional. Since the organism that harms one or more other organisms gains nothing from the interaction, this makes sense. This type of interaction is often found when sessile organisms (living things that don't move around) interact with each other. For instance, several varieties of the black walnut tree are **allelopathic**, which means that they secrete toxins into the environment that harm other living things that may be competition, a type of amensalism. The only other living things affected by this toxin, juglone, are plants, and the walnut tree doesn't even directly secrete that chemical! Parts of a walnut tree, like roots and bark, contain hydrojuglone, which is very harmful to everything. Once the walnut tree sheds these parts, they decay and oxidize. Oxidation turns hydrojuglone into juglone, which is harmful to surrounding plants. There is no harm to walnuts themselves though, so keep eating them! They're healthy for us humans.

Commensalism

Due to its very nature, commensalism may also seem unintentional, more so than amensalism, actually. Commensalism, again, is a community interaction in which organisms benefit yet other others remain unaffected. Often, commensalism occurs between a larger host and a smaller **commensal**, the organism that benefits from the interaction. In nature a lot of insects and other smaller organisms ride on larger organisms to get a ride. In the ocean, a type of fish called a remora will latch onto larger sea creatures such as sharks to get a free ride. The larger organism is neither harmed nor helped; the remora gets to conserve its energy because of the ride. Birds making their nests in trees is another common example, especially with regard to smaller birds. The bird gains a reliable home, and the tree gains nothing. Trees providing shelter from the sunlight is another example of commensalism.

Competition

There is a chance for harm through competition in communities. Organisms that compete are seeking the same resources for survival: food, mates, shelter, water, and so on. Since energy and resources are expended during competition, all of the organisms involved in the competitive act are harmed in some way. The organisms that gain the benefit are the organisms that are harmed the least or have survived. Competition in some cases leads to the death of both organisms. There are two main types of competition in communities:

- **Interspecific competition**—Competition between members of two or more different species. This can have two basic outcomes. One outcome is **competitive exclusion,** in which one species completely excludes another one from gaining a resource. For instance, lions will often scavenge carcasses in the wild. If the kill was made by another animal such as a hyena, the lions will chase them off and prevent them from having the food by force. Sometimes a compromise is reached of sorts. This results in **species coexistence**, in which resources are shared. This can be due to low populations sizes, basically meaning that there's plenty to go around, or a lack of need to directly compete. For instance, during dry spells in the African savannah, all sorts of animals will share the watering holes that dot the landscape.
- **Intraspecific competition**—Competition between members of the same species. This is often for territory or mates, or both. It can also occur when populations become large, stressing the availability of resources within a given area. As an example of intraspecific competition, bighorn sheep are led by the largest, most powerful male ram. This ram will eventually be challenged by a younger male for leadership of the herd. The two rams will then compete by butting heads using their huge horns in order to see which of the two will be unable to continue. This competition is violent and brutal. Often both rams, regardless of the victor, are harmed. Sometimes one, or both, will be killed.

Competition and other interactions within communities are aspects of an organism's niche. Sometimes a species fulfills its entire role within a community, having no competition for all of its resources. An organism occupying that position fulfills a **fundamental niche**. In fulfilling a **realized niche,** an organism has a limited role that it must share with other members of the community. In this manner it may only use a subset of resources that it needs, while other organisms use other subsets.

Organisms need not engage in violent competition in order to compete. Through **resource partitioning** organisms can share certain resources through specialization. Members of a community may feed upon certain food groups at night rather than during the day or specialize in eating food groups that other members of a community do not. For instance, many species of birds, such as finches, evolve eating nuts of a particular kind of tree. Other members of that species may poke holes in that tree with their beaks in order to get insects that may live in that type of tree. With regard to this example, the finches probably evolved differential features through **character displacement**. Character displacement within a species may result in the evolution of certain features in order to realize a niche. For instance, the aforementioned finch that eats nuts may have evolved a large, strong beak to crack open the nuts. The other member of its species may have evolved a long, pointed beak in order to poke holes in trees to extract insects and worms.

Exploitation

Exploitation is a type of community interaction that usually involves feeding behavior in which energy is gained from the tissues of other organisms. This is a powerful form of community interaction that often forms the structure for entire communities. There are three main types of exploitation:

- **Herbivory**—A form of exploitation in which organisms, usually animals, feed on of the tissues of plants. This is a form of exploitation in which the exploited organism, the plant, does not necessarily perish, nor even seem to be discernably harmed. Think of when people pluck apples from apple trees. Many animals, especial insects, practice herbivory. Plants have evolved defense mechanisms to counter herbivory, such as toxic or distasteful chemicals, spines, or thorns. Some plants utilize herbivory to distribute their seeds through the feces of herbivore animals.
- **Parasitism**—A form of exploitation in which the exploiter, the **parasite**, uses the exploited, the **host**, for its own benefit, usually as nourishment. Some parasites live within their hosts, such as bacteria and roundworms. Others live outside of their host, making occasional contact, usually with multiple hosts. Examples of this are lampreys and ticks. Many parasites inadvertently transmit diseases through parasitism. With that in mind, parasites do not always outright kill their hosts; in fact, it may be against the parasites' best interest to do so. Hosts and parasites often generationally adapt to each other in an "evolutionary arms race," evolving adaptions against one another.
- **Predation**—Exploitation in which the exploiter, the **predator**, captures or ambushes, then outright kills the exploited, the **prey**. There is no gray area with this form of exploitation: One organism has to die for this to work. Predation is necessary for healthy ecosystems. It creates a sense of balance and structures feeding relationships, the transfer of energy and matter. Balance is achieved through the maintenance in the number of predators and prey, which in turn impacts herbivory and other community interactions. There is a cyclical response between predator and prey populations in the wild. Much as with parasites and hosts, predators and prey constantly evolve new means to deal with each other. Predators evolve better means to capture prey. Prey evolves better means to defend against or evade predators. These evolutionary adaptations actually lead to stronger organisms that live longer, healthier lives, better suited to taking care of their young.

Mutualism

Many organisms can benefit from mutualism, not just two. Pollination is a form of mutualism in bats, bees, birds, and other organisms that transfer **pollen**, fine powder used by plants for reproduction, from one plant to another. The animals usually feed off plant nectar, which usually does not harm the plant, and in turn the animals spread pollen, allowing plants to mate. This allows plants to produce fruit, which in turn feeds many sorts of animals, including humans. This is a type of community interaction that has far-reaching effects. **Symbiosis** is a form of mutualism in which organisms live in close contact or even within one another. Clownfish secrete a substance that makes them immune to the venom of sea anemones. Living in sea anemones keeps clownfish safe. On the other hand, clownfish attract other fish to the anemones, who eat them. For another type of symbiosis we need look no further. Human beings require bacteria in our intestines. The bacteria aid in digestion and regulate the gut environment. In turn, they feed off some of the food that humans consume. All other mammals have this symbiotic relationship.

Neutralism

There is often no way to prove that neutralism exists in biological communities. At the same time, neutralism is the most common form of community interaction. Neutralism is the species interaction in which members of a community have no impact on each other. One need look no further than one's yard to see examples of neutralism. Squirrels run across the yard and up trees, while ants build their mounds with the yard. Neither interacts with each other yet both inhabit the same community. In the desert, cactuses and tarantulas inhabit the same community yet never interact. Neutralism is hard to discern but is seemingly everywhere!

Exchanges of Energy and Matter

A biological **community** is an assemblage of populations, or different species, living together in the same area at the same time. Community members interact with each other. The interactions, based on the nature of the ecosystem, govern the form, structure, and species composition of the community. In order for communities to exist, there must be interchanges of matter and energy. These are fundamental biological interactions, and they manifest in feeding systems, which are ultimately exchanges of energy and matter.

Trophic Pyramid

A **trophic pyramid**, also called an ecological pyramid, an energy pyramid, a food pyramid, and an Eltonian pyramid is a graphical representation of energy flow in communities and ecosystems in the form of a pyramid shape. One of the most important species interactions is the one in which organisms gain energy, in the form of what organisms consume, how they produce nutritious substances to continue living, **food**. Matter and energy move through the biotic, and ultimately into the abiotic, sections of communities. At its base, the feeding levels within communities start with **autotrophic** organisms, self-feeders that utilize photosynthesis on the land and chemosynthesis in the ocean's depths. One land autotrophism is an open system, ultimately depending on energy from the Sun. **Heterotrophic** organisms feed on living tissue, whether it be that of autotrophs or other heterotrophs. **Decomposers** and **detrivores** consume nonliving matter.

Trophic Levels

There are four basic trophic levels along with a special component that interacts with all of the levels: the decomposers and detrivores. We will discuss terrestrial trophic systems, which are open systems, as opposed to deep ocean trophic systems, which are closed systems.

As energy is transferred upward in the pyramid, energy also dissipates; 90% of all energy is lost as it is transferred to the next level. Energy loss is due to **metabolic heat**, the energy used to burn through nutrients, as energy is transferred upward utilizing matter, in the form of carbohydrates.

In parallel to energy loss, matter is also consumed, lost, and transformed, which is concomitant to biomass in general. There is always a 90% loss as energy and matter is transferred upward, and only around 10% is retained for use, a quantitative relationship in feeding (Figure 5.8). Subsequently there are fewer organisms at higher levels and less available energy. The decomposers and detrivores are a special, and very complex, case that will be discussed briefly.

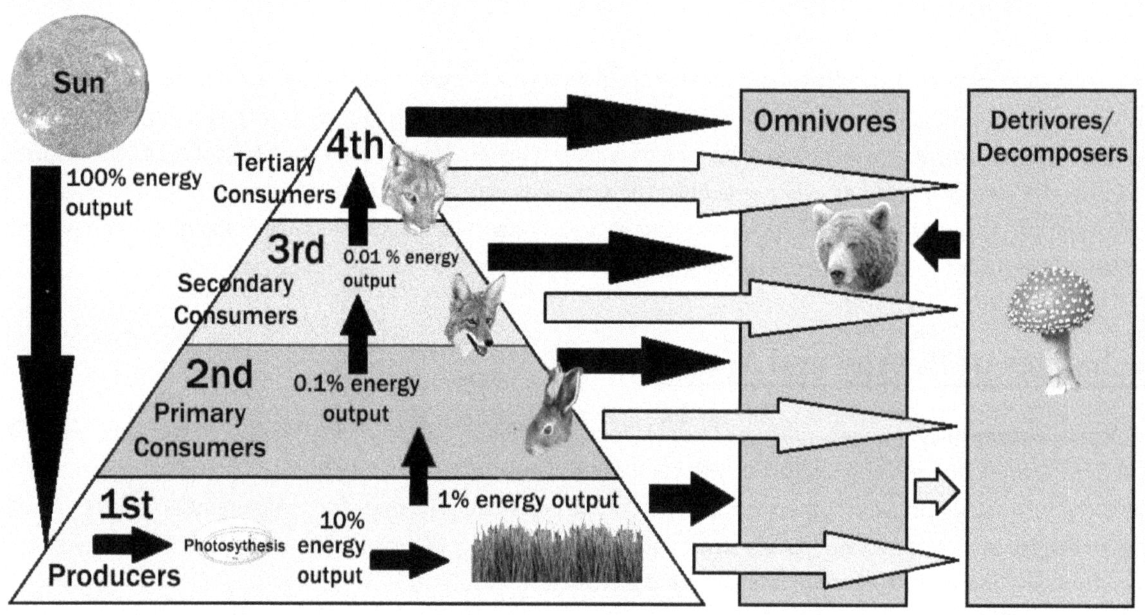

Figure 5.8 Modeled quantitative relationships in a trophic pyramid

Please note that there are organisms that consume organisms from multiple trophic levels. These are the **omnivores**. Omnivores consume both plants and animals, making use of the entire trophic pyramid, in theory. Omnivores include bears, chickens, crows, flies, pigs, and human beings.

- **First trophic level, producers**—Autotrophs, or self-feeders. Through photosynthesis these organisms are able to manufacture their own food. In doing that, producer organisms also only retain 10% of the energy they gain from the Sun for consumption. As that energy is consumed by other organisms, there is a further loss of 90% of the energy that the producers retained. Examples of producer organisms include algae, cyanobacteria, and green plants.

- **Second trophic level, primary consumers**—Organisms that consume the producers. These are the herbivores who consume plants. Examples are deer, grasshoppers, and rabbits.
- **Third trophic level, secondary consumers**—Organisms that prey upon the primary consumers. These are carnivores who consume meat. Examples are foxes and rodents.
- **Fourth trophic level, tertiary consumers**—Organisms that prey on the secondary, as well as the primary consumers. These are high-level carnivores who consume meat. Since there is a 10% loss upward from level to level in the trophic pyramid, the number of tertiary consumers is a lot less than organisms that occupy other levels. At the same time, tertiary consumers have an enormous impact on communities and ecosystems. Examples include hawks, sharks, and tigers.
- **Decomposers/detrivore**—Organisms that consume nonliving matter. Decomposers/detrivores interact with all levels of the trophic pyramid. These are the organisms that enrich soil through the processing of nonliving matter and break down the tissues of dead organisms. **Decomposers** break down nonliving matter, including leaf litter, and make it available to ecosystems. Examples of decomposers are bacteria, earthworms, and fungi. **Detrivores** scavenge dead bodies and waste products. These organisms hasten decay and important biological process. Examples include dung beetles, millipedes, and vultures.

Influential Species

Keystone species are species that have a wide impact on a given community or ecosystem, often beyond its actual abundance. Keystone species are often top predators. For instance, sea otters consume sea urchins, who in turn consume kelp that lives on the bottom of the ocean. If the sea otter were to disappear in that community, sea urchins would overgraze the kelp. With no kelp left, the barely mobile sea urchins would die off, leaving behind a somewhat barren community. When wolves were exterminated from the Yellowstone area, the deer population increased and overgrazed. This led to negative impacts on the forest structure. These are both examples of how the disappearance of predators at high tropic levels keeps intermediate trophic levels in check, indirectly affecting organisms at lower trophic levels.

Other organisms structure the nonliving components of ecosystems affecting the abiotic components. **Ecosystems engineers** are organisms that alter forested structures, rocks, soil, and waterways. On a very large scale these are human beings who basically have altered the planet across multiple ecosystems. Other ecosystems engineers include beavers, fungi, and prairie dogs.

GROUND TRUTH

Invasive Species

Invasive species are species of organisms transplanted from their native communities to communities elsewhere, often far away. These species may be beneficial to their new communities and ecosystems, for instance the European honeybee, but more often than not they bring harmful impacts to the new ecosystems in which they find themselves. These impacts manifest in various ways. For instance, disruption to trophic feeding systems can occur. Zebra mussels came to North America from Europe. These shellfish had no predators when they arrived in the Great Lakes, so they caused population crashes in other organisms due to overfeeding. Impacts on human activity, including economic impact, can occur as well. Kudzu is a non-native plant that was introduced to the United States in the late 19th century as ornamental. The American southeast was a very favorable habitat for kudzu, so it spread rapidly and widely. This effectively choked out other plant species, including trees harvested for timber, a major economy in the United States.

Human beings are predominantly the agent of transplantation for invasive species, whether inadvertently or not. For instance, aquatic organisms can "hitch a ride" in ballast water in ships that cross the ocean, which is what happened with the zebra mussel. By the way, ballast water is used to give ships stability while traveling. Ships intake ballast water from surrounding waters into ballast tanks. Organisms, particularly microorganisms and larvae, can be in intake water. When the water is discharged elsewhere on Earth, these aquatic organisms may be part of the discharge. Disease organisms can also be spread into a new environment from people. When Columbus landed on the Island of Hispaniola in the New World (in this case the Caribbean Sea) his crew inadvertently transmitted a long list of disease organisms into the Native population, including chicken pox, measles, and smallpox. This caused a horrific drop in the Native American population of the island, resulting in a drop of 80%–95% within the first 100 years of colonization, nearly decimating the Native population.

People also willfully introduce invasive species into new ecosystems. The aforementioned kudzu was introduced to the American southeast as an ornamental plant for the gardens and estates of well-to-do Americans. It is often impossible to contain non-native species that are acclimated to a new environment, which is exactly what happened with kudzu. Pythons are a more recent problem in the United States, particularly in Florida. Both rock pythons and the larger Burmese pythons are pets for people who enjoy reptiles and snakes. Unfortunately, in the 1980s pythons kept as pets in Florida escaped captivity. Since the climate and environment of Florida is very similar to places like Central Africa and Southeast Asia, the pythons did not die off and in fact thrived. Now there are hundreds of thousands of pythons in Florida, including a hybrid of the Burmese and rock pythons. The now Floridian pythons have wreaked havoc, even feeding upon alligators, which are native to Florida! Pythons are legal to hunt in Florida, with a permit, but that has done little to stop them from spreading. Efforts to eradicate the pythons have largely failed; they are now part of the environment in Florida.

Many efforts have been made to prevent the spread of invasive species across the globe with little success. Once a species thrives and has a significant population, it is almost impossible to eradicate. At that point, conservation measures are usually put into place, and it is accepted that the new species has essentially restructured a given community.

Communities in Crisis

Communities experience many types of exterior disturbances, such as the removal of keystone species, the spread of invasive species, and natural disturbances. Human impacts cause major changes in this manner. The organisms within communities exhibit varying responses to the crises that impact biological communities.

Some communities resist change and remain stable despite the disturbance. **Resistance** depends on the compositions of populations within communities as well as the nature of the disturbance. Some communities exhibit **resilience** and experience changes in populations in response to an exterior disturbance, later returning to their original state. These responses vary with the magnitude of a given crisis as well. Sometimes communities are nearly or completely wiped out.

Disturbances of great magnitude result in succession within biological communities. **Succession** is a series of changes that occurs once a disturbance has all but devastated a community. When a disturbance eliminates soil life and/or all vegetation, the community must undergo **primary succession**. This essentially occurs when there is nothing left and a community completely rebuilds itself. This happens after such disasters as the catastrophic loss of an entire body of water, the encroachment of glaciers, the ecosystem becoming flooded with lava, or enormous volcanic explosions. Immediately following such vast catastrophes, the ecosystem may seem to have become a barren wasteland but within disaster are the seeds of growth. Eventually hardy **pioneer species** emerge such as lichen and the community rebuilds itself. Recently, the ecosystem surrounded Mt. St. Helens, in Washington State experienced primary succession. Now, decades later, it is again a thriving ecosystem. Slowly but surely life finds a way.

Secondary succession occurs when a catastrophe alters but does not completely destroy a community. This happens in the wake of events such as widescale industrial farming or some other human development, wildfires, hurricanes, large mining operations, tornadoes, and clear-cut logging. The remaining vegetation and soil components form the basis for the return of a community back to a thriving original state. Ultimately a climax community will reemerge, the evidence being a diverse mix of grasses, trees, and undergrowth.

Humans, whether through economic or altruistic interest, undertake projects of **ecological restoration** from time to time. This is a common practice with large environmental enterprises such as mining operations. Ecologists known as restoration ecologists may have a large hand in planning these restoration projects. Because of the complexity and magnitude of ecological systems these efforts can be costly and time-consuming. When development is undertaken with natural ecosystems it is best to work with the existing natural system, sustainably, causing as little damage as possible, working in harmony with nature.

Biomes

Have you ever taken a trip that was that was thousands of miles long and noticed that the vegetation, even the land itself, became very different from that which is familiar to you? You may have crossed over one of Earth's largest features. On a grand scale ecosystems form enormous major regional complexes of similar communities that comprise the **biosphere**, the sum total of all living things on Earth. These are Earth's **biomes** (Figure 5.9).

Biomes are characterized on the plant life and vegetative structure that is draped across them, but ultimately the plant life is dependent on two key factors: climate and geology. Plants, especially larger plants like trees, exist in their current state due to both soil characteristics based on the bedrock that it formed from and the overall climate of a region. Since climate is a significant factor, similar biomes occupy similar latitude, the exception being arid biomes such as deserts. Aquatic systems have biomes as well, albeit with differences in temperature profile, plus the presence of currents, depth, dissolved nutrients, salinity, substrate type, and waves.

Transitional zones do occur, but biomes themselves are somewhat singular in their over-composition. Note that higher elevations within biomes, like mountains, tend to break up the characteristics of these biomes. Mountainous areas themselves contain their own biome systems, tending toward boreal forest and tundra conditions at high elevations.

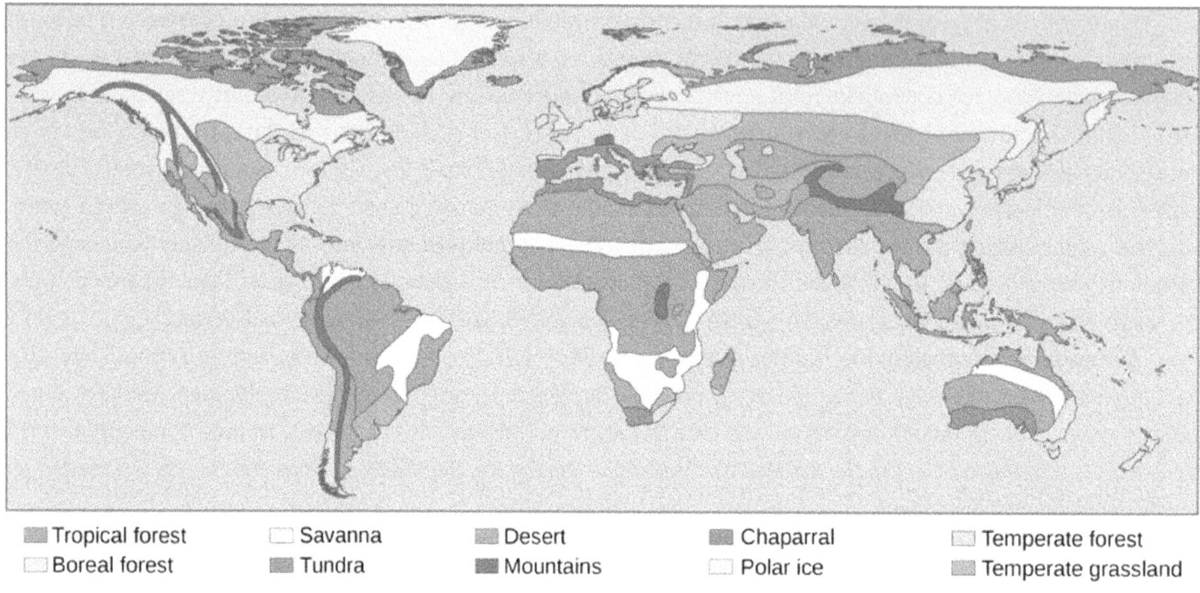

Figure 5.9 caption-Earth's major biomes, including polar ice and mountains

Boreal forest

- Also known as taiga, these are high latitude biomes at around 60°. They are characterized as having some variation in temperature in precipitation, although a significant portion of the year experiences temperatures that are below freezing. There are four seasons, although temperatures rarely get higher than 15° C in the summer. It snows throughout most of the year in the boreal forest. The magnitude of precipitation is not high though, rarely experiencing greater than 80 mm per month, with vast forests consisting of hardy evergreen trees. As such, significant logging takes place in boreal forests. Soils are poor and acidic, with snow and ice on the ground for at least half the year. Fauna like bears, moose, reindeer, and wolves thrive year-round in this biome. Boreal forests span places like Alaska, northern Canada, Scandinavia, and Russia.

Chaparral

- The chaparral is found in Southern Australia, Southern California, Chile, and the Mediterranean Sea. This biome tends to be found in middle latitudes, between 30° and 45°. The chaparral is characterized by relatively mild temperatures with little variation annually, hovering around 15° C consistently. Precipitation tends to be moderate as well, rarely getting more than 80 mm per month, with seasonable variability. This is a two-season system with a clear wet and dry season. Because of the mild precipitation, vegetation tends to consist of mostly shrubs and undergrowth that can tolerate long dry periods. Due to the dry season chaparral biomes experience frequent wildfires. Fauna consists of coyotes, lizards, jackrabbits, and the like.

Desert

- Deserts are extremely unique biomes for one reason: Water rarely falls from the sky. They are characterized by an almost complete lack of precipitation year-round. Deserts are found at all latitudes. Many deserts experience relatively high temperatures, but that does not have to always be the case. Temperatures can vary widely in the desert. The Gobi Desert of Asia is below freezing much of the year, and Antarctica is technically a desert. The soils, where they can be found, tend to be desiccated and saline. Some deserts, like the Sahara, are barren and draped in sand dunes. Other deserts, like the Sonoran Desert in the southwestern United States, are heavily vegetated with hardy plants adapted to extreme aridity. Animals that live in deserts tend to be nocturnal or nomadic. These include camels, lizards, snakes, and scorpions.

Savanna

- Grasslands characterize the savanna. Savannas are found in the tropics between the 15° and 30° latitudes. Temperatures are warm but consistent, rarely going below 15° C and rarely going above 25° C. Precipitation is seasonal in the savanna with a pronounced wet and dry season: a two-season system. It may barely rain during the dry season, yet the wet season can experience significant rainfall, up to 200 mm per month at times. Vegetation, again, is mostly grassland interspersed with copses of trees adapted to the savanna. This biome spans Africa, parts of Australia, India, and parts of South America. Fauna consist of elephants, gazelles, giraffes, hyenas, lions, and zebras.

Temperate deciduous forest

- Deciduous forests contain deciduous trees, trees that lose their leaves each fall, go dormant throughout the winter, and grow new leaves in the spring. Temperate biomes are found in the middle latitudes between 30° and 60°. These are four season systems: winter, spring, summer, and fall. Temperatures are variable but generally mild, rarely staying below freezing in the winter and rarely staying above 20° C in the summer. Precipitation is consistent with some variation. It rarely rains more than 100 mm per month. Soils are fertile and tend toward acidity. There is a wide variety of plants, including hardwood trees such as beech, oak, and maple, interspersed with evergreen trees. The temperate deciduous forest spans places like Eastern China, Western Europe, and Eastern North America. Animals include black bears, cardinals, hawks, owls, squirrels, and many varieties of spider.

Temperate grassland

- Temperate grasslands are also known as prairies or steppes. These are four season systems that experience less relative precipitation annually compared to other temperate biomes, rarely more than 60 mm per month. As a temperate region, temperate grasslands are found in the middle latitudes between 30° and 60°.Temperatures tend to be more extreme seasonally than other temperate zones, below freezing in winter to above 20° C in summer. Lack of rain in summer, combined with high temperatures, can lead to minor dry seasons. Vegetation is mostly grassland interspersed with adapted trees. Soils are extremely fertile. Temperate grasslands are found in the Great Plains of North America, as well as Central Asia. Fauna include badgers, grasshoppers, and prairie dogs.

Temperate rainforest

- Temperate rainforests are, again, found between the 30° and 60° latitudes. These are four season systems that experience mild temperatures, rarely below freezing in the winter and rarely above 20° C in the summer. The hallmark of temperate rainforests is a lot of rain! Rainfall is usually between 80 and 200 mm year-round, with some months seeing over 300 mm. There is an abundance of adapted vegetation, including vast conifer forests. As such, significant logging and lumber operations often take place in these biomes. Soils are fertile in these biomes yet prone to erosion and landslides due to extreme rainfall. Temperate rainforests are found in the Pacific Northwest of the United States and Canada and the main islands of Japan. Flora and fauna include fir trees, elk, bears, and moisture-loving animals like the banana slug.

Tropical dry forest

- The tropical latitudes are technically at or below the 30° latitude line. Temperatures in tropical dry forests, like any other latitude close to the equator, tend to be relatively high compared to other biomes with little variation, tending to hover around 25° C annually. What significantly characterizes the tropical dry forest is extreme seasonality; it's a two-season system. Half of the year is extremely dry and arid. The other half of the year experiences extreme rainfall, close to 400 mm per months at times! This makes for soils prone to erosion and flooding very likely, which is frequent during the wet season. Plant life consists of hardy scrubs, undergrowth, and adapted trees, which are often evergreen. Animals of the tropical dry forest include large cats, monkeys, and lizards.

Tropical rainforest

- The tropical rainforest is the most biologically diverse terrestrial biome. These are found below the 30° latitude line, and in fact are usually found near the equator. There is little variation in temperature in these biomes, with temperatures staying at around 25° C constantly. There are no real seasons in this biome. Tropical rainforests experience consistently high rainfall as well, around 300 mm per month, with some months seeing up to 500 mm. There is a high density of lush, tropical vegetation in the tropical rainforest, including very tall trees that create a canopy above the forest floor. Soils of the tropical rainforest tend to be poor and very acidic due to the extreme amount of drainage. Plant life is supported by a large amount of organic material from decaying matter in the soil. Tropical rainforests are found in West Africa, Central America, Southeast Asia, and South America. Fauna include anacondas, jaguars, gorillas, toucans, and a huge variety of insects and spiders.

Tundra

- All tundras are found in either the Arctic or Antarctic Circles, the placement of which varies with time but are generally above, or below, 60° latitude. In truth, most tundras are found very close to 90° C latitude, the North and South Poles. Tundras are characterized by extremely cold temperatures. Temperatures rarely climb over freezing in summer and tend to average between -10° and -20° C in winter. There is some annual seasonality with a summer of no more than 10° C and moderate rainfall, no more than 40 mm per month. Regardless, it is often as dry as a desert in the tundra. Winters are long and extremely cold. Tundras are also characterized by a layer or permafrost beneath the surface of the soil that never thaws out. There are almost no trees in the tundra; most of the vegetation is low-lying and hardy. Lichen abounds. The tundra can be found in Alaska, Canada, Russia, Scandinavia, and extreme southern South America. Animals include the arctic fox, lemmings, musk ox, and polar bears.

Conclusion

The patchwork of ecosystems that span the breadth of the Earth comprise our home. This home has function and purpose on an effective biological level. These systems, while robust, are also fragile. As such, understanding Earth's ecosystems is of vital importance as the planet itself undergoes changes, both naturally and anthropogenically. With the ensuing advent of climate change, these vital ecosystems and all of the communities that exist within them are currently seeing challenges. We must learn to live in harmony beyond our own human development, in accordance and concordance with our amazing yet vulnerable biosphere!

Discussion Questions

Directions: Review the chapter in order to completely and correctly respond to the questions and prompts.

Keep this in the back (or maybe even front) of your mind: Science is system of methods that allows us to understand not only our planet, and ourselves, but of the entirety of the universe. Answering the following questions correctly will lead to a more robust understanding of how to apply science to ENVS problems.

1. What are four characteristics of life? List them, and explain them.
2. What is meant by fitness with regard to evolution?
3. Explain directional, disruptive, and stabilizing selection. Find some examples of each.
4. Why is allopatric speciation the predominant form of speciation? Explain.
5. Take your favorite animal (or plant, or fungus, or any other organism) and draw a cladogram leading to its evolution. Research this. Use the example in this book, and logic, as well as your imagination to create the cladogram.
6. How and why are endemic species prone to becoming extinct?
7. Many scientists think that we are in the midst of the sixth geological mass extinction event. Do you agree? Explain.
8. Name an organism with low biotic potential and one with high biotic potential, to the best of your ability. Explain your choices.
9. What is a type of community interaction in the area in which you live? Explain why and what organisms are involved.
10. A patch of grass receives an equivalent of 10,000 units of energy, let's say calories. A rabbit eats the grass, then is caught by a fox. How many calories, or units of energy, does the fox get from eating the rabbit?
11. What biome do you live in? Why do you think so?

Bibliography

Amstutz, L.J., 2018, *Invasive Species. Ecological Disasters*: Minneapolis, MN, Essential Library, 112 p.

Di Febbraro, M., Carotenuto, F., Castiglione, S., Russo, D., Loy, A., Maiorano, L., and Raia, P., 2017, *Does the jack of all trades fare best? Survival and niche width in Late Pleistocene megafauna: Journal of Biogeography*, vol. 44, no. 1, p. 2828–2838.

Odum, E., and Barrett, G.W., 2004, *Fundamentals of Ecology: Farmington Hills*, MI, Cengage Learning, 5th ed., 624 p.

Credits

Fig. 5.1: Popular Science Monthly, vol. 73, 1908.
Fig. 5.2: Copyright © by Mliu92 (CC BY-SA 4.0) at https://commons.wikimedia.org/wiki/File:Selectiontypes-no_images_(vector).svg.
Fig. 5.3: Copyright © by Liwnoc (CC BY-SA 4.0) at https://commons.wikimedia.org/wiki/File:Wild_Mustard_Plant_Selective_Breeding.svg.
Fig. 5.4: Copyright © by Andrew Z. Colvin (CC BY-SA 4.0) at https://commons.wikimedia.org/wiki/File:Allopatric_speciation_caused_by_topography.svg.
Fig. 5.5: Biologymama53, "Cladogram of Primates," https://commons.wikimedia.org/wiki/File:CLADOGRAM.jpg, 2022.
Fig. 5.6: Copyright © by Mauricio Antón (CC BY 2.5) at https://commons.wikimedia.org/wiki/File:Ice_age_fauna_of_northern_Spain_-_Mauricio_Ant%C3%B3n.jpg.
Fig. 5.7: Copyright © by Projectoer (CC BY-SA 4.0) at https://commons.wikimedia.org/wiki/File:Patterns_of_Population_Distribution.png.
Fig. 5.8a: NASA/SDO (AIA), "Sun," https://commons.wikimedia.org/wiki/File:The_Sun_by_the_Atmospheric_Imaging_Assembly_of_NASA%27s_Solar_Dynamics_Observatory_-_20100819.jpg, 2010.
Fig. 5.8b: Copyright © by Phyllis (CC BY 2.0) at https://commons.wikimedia.org/wiki/File:Cougar_(4275815216).jpg.

Fig. 5.8c: Copyright © by Paul Asman and Jill Lenoble (CC BY 2.0) at https://commons.wikimedia.org/wiki/File:Coyote_Canis_latrans_(22020336743).jpg.

Fig. 5.8d: Copyright © by Åsa Berndtsson (CC BY 2.0) at https://commons.wikimedia.org/wiki/File:Hare_Hare_%2820300967221%29.jpg.

Fig. 5.8e: Copyright © by Nevit Dilmen (CC BY-SA 3.0) at https://commons.wikimedia.org/wiki/File:Grass_dsc08672-nevit.jpg.

Fig. 5.8f: Copyright © by Robert F. Tobler (CC BY-SA 4.0) at https://commons.wikimedia.org/wiki/File:Kamchatka_Brown_Bear_near_Dvuhyurtochnoe_on_2015-07-23.jpg.

Fig. 5.8g: Copyright © by Timothy A. Gonsalves (CC BY-SA 4.0) at https://commons.wikimedia.org/wiki/File:Fly_agaric_mushroom_Ooty_Jul17_DSC04142.jpg.

Fig. 5.8h: Copyright © by Kārlis Kalviškis (CC BY-SA 4.0) at https://commons.wikimedia.org/wiki/File:Chloroplast-drawing.svg.

Fig. 5.9: Copyright © by OpenStax (CC BY-SA 4.0) at https://commons.wikimedia.org/wiki/File:Major_terrestrial_biomes_in_the_world.png.

Chapter 6

Ecosystems and Conservation Biology

Energy and Matter in Ecosystems

Objectives

- Understand that Earth is a vastly complex system and series of subsystems.
- Understand that the components of systems, including Earth's systems, must work in accordance with each other in order for the system to function.
- Come to know that there are some singular cycles of matter and energy in the Earth system that are vital to life.

KEY WORDS AND TERMS

Ecosystem (n.): An assemblage of communities within a distinct geographic area, in which the species interact with the abiotic components as well as with each other.

Biodiversity (n.): A measure of the sum total of all organisms within a given area.

System (n.): A network of interactions between components, elements, or parts that relate with each other in determinable manner.

Consider This . . .

Life is a process that is subject to the same physical and chemical conditions as any other natural process. One of the thermodynamic precepts governing overall energy use is the law of conservation of energy. Concordantly, there is the law of conservation of matter, another principle that we covered in a previous chapter. These two principles apply to singular organisms and **ecosystems** as well.

Introduction

Ecosystems that are biologically diverse are robust and are functionally self-sustaining. **Biodiversity** is also a function of scale. The sustainable functionality or "health" of any population, community, or ecosystem is directly related to the amount of biodiversity at all of these scales. It also well known that anthropogenic encroachment within natural ecosystems often disrupts the amount of biodiversity on all levels. Within these levels of biodiversity, energy and matter are exchanged between individual members of populations and communities and across whole ecosystems.

Biodiversity: The Big Picture

Biodiversity, or biological diversity, is a measure of the sum total of all organisms within an area. The higher the number of organisms, the greater the biodiversity. There are three different scales or levels of biodiversity: genetic diversity, species diversity, and ecosystem diversity (Figure 6.1).

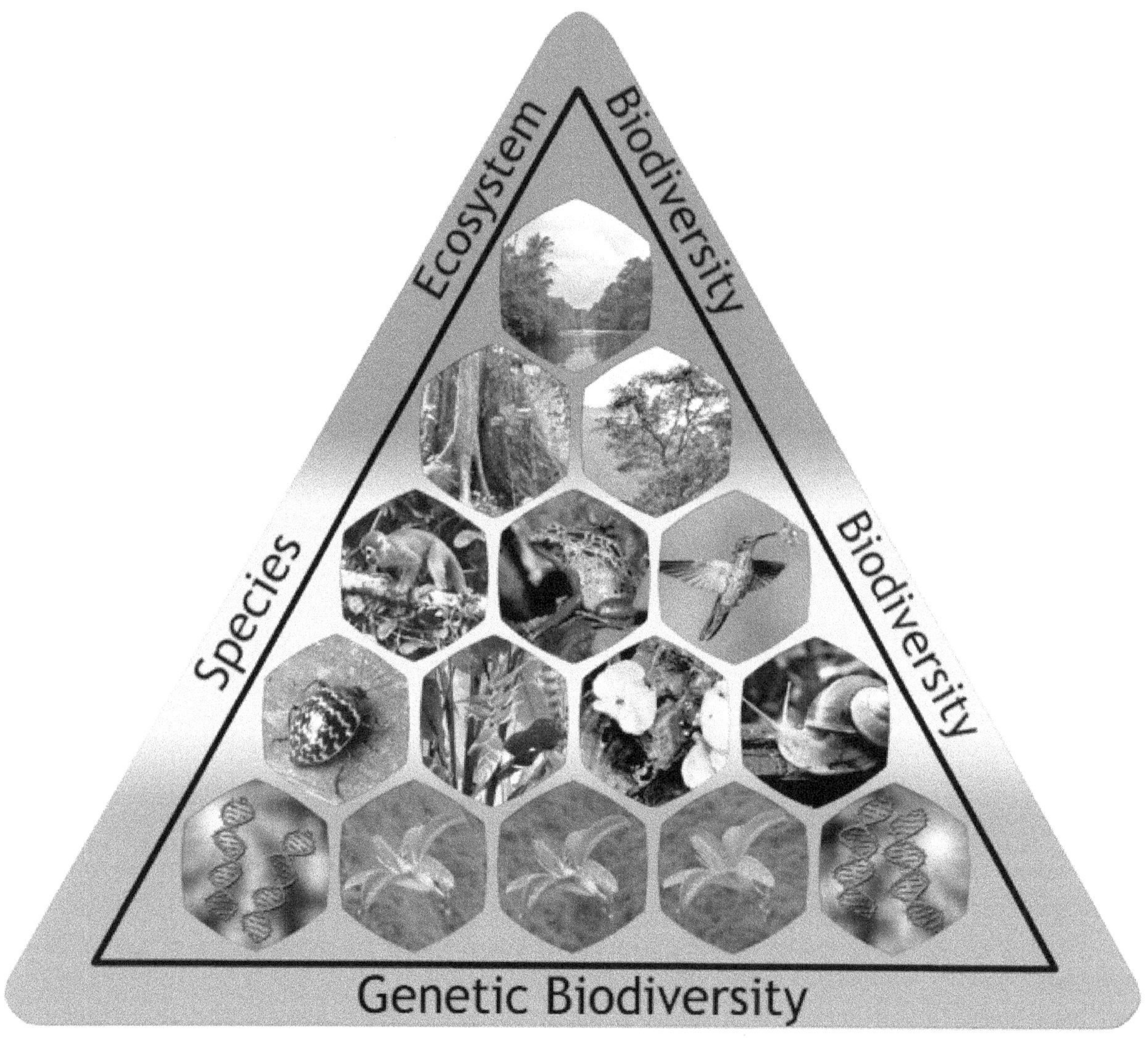

Figure 6.1 Three types of biodiversity

Genetic Diversity

Genetic diversity refers to the number of differences in DNA within individual species or populations. Biological fitness, at the most fundamental level, is dependent on adaptations, or mutations, within a given organism's genes. Species that exhibit genetic differences that improve the fitness of individuals within given populations are more likely to pass successful genetic material down through successive generations. Populations with high genetic diversity are more likely to cope with environmental change because they can adapt to a variety of conditions.

Populations that exhibit low genetic diversity are vulnerable to changing conditions within ecosystems. Low genetic diversity is depleted of the raw materials that are necessary for adaption to take place. These conditions can take the form of changes within climate or weather patterns, geologic hazards, disease, or even human development.

Low genetic diversity within populations can also lead to **inbreeding**. Inbreeding is a crisis within populations that stems from the actual members of it; it is not an exterior threat but an interior one. Inbreeding occurs when genetically similar members of a species mate. It occurs within successive generations, in other words, families. Fitness is a function of genetic diversity, so inbreeding is a loss of fitness and can lead to genetic abnormalities and diseases.

Species Diversity

Species diversity is driven by changes to successive generations that occur on the molecular level, in DNA, that can be passed on. These changes are driven by mutation. Mutations are accidental changes to DNA in an organism that may occur due to changes in its environment or even random changes. Mutations are either lethal or non-lethal. An example of a lethal mutation is cancer.

Non-lethal mutations are often unnoticeable and do not get passed on to successive generations. These types of mutations have happened to all of us, making all of us mutants of a sort. Non-lethal mutations that provide beneficial genetic variation, on the other hand, can be passed on. This genetic variation, stemming from mutation, is a mechanism for natural selection, which can lead to evolution in organisms. In addition to mutation, sexual reproduction also drives the variation for natural selection.

Ecosystem Diversity

The breadth of diverse ecosystems across the Earth, or within biomes, is **ecosystem diversity**. Within these ecosystems are a vast array of communities and habitats which also characterizes ecosystem diversity. Diverse ecosystems within a given geographic area, as well as a diversity of changing land- and lifeforms within an ecosystem, promote advanced levels of ecosystem diversity.

GROUND TRUTH

The Predominance of Insects

When we consider which group of organisms has come to dominate the Earth, it's easy for us to say human beings, or even vertebrates. This is, of course, a biased viewpoint since, after all, we are human beings, a type of vertebrate. From an evolutionary and even an ecosystems engineering standpoint there may be some truth to the argument that human beings are the predominant species on the planet, but when considering adaptation on an organic level, the occupation of the widest number of niches possible, that argument may not hold true. In this regard, insects may be the most successful and predominant group of organisms on Earth (Figure 6.2).

The number of species of any given group of organisms on Earth is not evenly distributed. Groups of organisms acquire species through adaptive radiation, allopatric speciation, and low rates of extinction. Given their overall success at adapting to a wide variety of environments, speciation through adapting to changing environments, and overall resilience insects are again more successful in comparison to all other groups of organisms. For instance, consider the cockroach. It has become something of a cliché that the cockroach has not needed to evolve for hundreds of millions of years. While this may be true for some species of cockroach, members of their order Blattaria have undergone intense speciation and adapted to a large number of environments. At this point in time, there are over 4,600 species of cockroach on Earth, and the cockroach is only one type of insect!

As of recently, out of around 1,494,000 species of known animals and animal-like lifeforms on Earth, including protists, there are about 1,250,000 arthropods, animals of the phylum Arthropoda to which insects belong (Figure 6.2). That is close to 84% of all animal species on Earth. Of all known species of life on Earth animals, not counting protists, comprise around 76% and of all those animals 73% of them are insects. This means that 55% of all known species of life are insects! Clearly insects, generally classified as animals within the class Insecta, are quite the success story both evolutionarily and functionally.

Figure 6.2 A very small number of the vast number of insects on Earth

Quantifying Biodiversity

There is an upward estimate of possibly 8.7 million species of organisms on Earth, but only around 1.2 of those species have been classified, most of those being insects. That means that around 86% of all organisms on Earth remain a mystery. There are a few reasons this is the case.

There are still many places on Earth that have not been fully explored. Many ecosystems remain inaccessible to us, for the most part, cut off from us without the most extreme effort due to rugged terrain, hostile climates, and remote locations. This, of course, includes the ocean's abyss, a place that is incredibly difficult to visit even with advanced technology. Biodiversity also includes microscopic species, which are difficult to discover and characterize due to the very nature of their size. In addition to all of this, many species of organisms are very similar in appearance to other species, requiring intensive investigation to differentiate them from similar species.

Latitudinal Gradient

Organisms are also distributed across the surface of Earth unevenly due to geographic factors. Biodiversity tends to be greater overall in lower latitudes. This is known as **latitudinal gradient**; species richness increases as one approaches the equator, the lowest latitude. There are many reasons behind this condition.

Lower latitudes tend to have more favorable climates for organisms, and of course lower temperatures. Most organisms on Earth tend to thrive in places that have steady, warmer temperatures as well as higher moisture levels, with the availability of water. There are more diverse habitats at lower latitudes, especially within tropical biomes. These biomes support more species, species diversity, and species evenness. Lower latitude biomes contain more diverse habitats, which, in turn, tend to increase species diversity. Also, lower latitudes contain greater numbers of human beings, which, counterintuitively, tends to promote habitat diversity through human development. This is a tendency though and somewhat conditional, mostly occurring on the local level.

Absolute Biodiversity Loss

Absolute biodiversity loss occurs both locally and globally. When a population of a singular species dies off within a particular area or ecosystem but not globally it is known as **extirpation**. **Extinction** is when a species disappears completely on a global level, all members of that species die off, irreversibly. Extirpation can lead to extinction as species disappear from all of their habitats on Earth.

It is estimated that over 99% of all species that have ever existed on Earth are extinct. Of all of the extinct organisms that paleontologists study, countless others are lost in time, having left behind nothing, to our knowledge, in the fossil record. Extinction is a constant condition of life. The current **background extinction rate**, natural rate of extinction over time, is one species per million species per year. That is not the current extinction rate. The current rate has more than doubled, and it is known that humanity's encroachment on wild habitats is the main cause.

Since the beginning of the Cambrian period there have been five mass extinction events that have eliminated over 50% of all known species. As a result, life responded by adapting to the changes that brought about and resulted from mass extinction, flourishing and radiating after every extinction event.

Increasing numbers of species are currently going extinct, something that began occurring during the ongoing Holocene epoch, which coincides with the advent of human civilization. It is well known that humans have been the direct agents of extinction, for instance in the case of the dodo and the passenger pigeon. Undoubtedly, human development and encroachment has been the indirect cause of extinction for many other species, as 18%–88% of large mammals and birds have gone extinct since the later Pleistocene epoch, some 160,000 years ago over all of the current continents and large land masses.

The International Union of Conservation for Nature (IUCN) manages the **Red List of Threatened Species**, which monitors species that are in danger of going extinct on a global level. The overall result of this monitoring **system** is clear: Biodiversity is in decline and has continued to do so for quite some time. There are currently more than 150,000 species on the Red List, with over 42,000 directly threatened with extinction. This includes amphibians, birds, conifer trees, mammals, rays, reef-building corals, and sharks. Since 1970 the following has been documented: the extinction of 58 species of fish, nine species of birds, and one mammal species. That one species of mammal is the Caspian Tiger. In actuality, several other mammal species have since gone extinct, including another subspecies of tiger, the Javan tiger.

These losses in biodiversity through extinction are accompanied by losses in ecosystem and genetic diversity. The geographic range of species is also resultant from these biodiversity losses, likely a result of decreased habitat ranges due to development. The World Wildlife Fund (WWF) and the Zoological Society of London (ZSL) maintain the **Living Planet Index**, which utilizes data to measure biodiversity across the planet. There are current increases to biodiversity, particularly with amphibians.

While it may seem easy to immediately blame overall human development for biodiversity loss, the actual causes may be complicated to disseminate. For one thing, causes for biodiversity loss are often **synergistic**, meaning that multiple causes combine with increasing complexity. As it stands, there are five main reasons for biodiversity loss: habitat alteration, invasive species, overharvesting, and pollution. The fifth cause is global climate change.

Habitat Alteration

Habitat alteration is the reduction of a given species, or group of species, geographic range due to development. There are many aspects to this, including farming, which artificially truncates the composition of natural communities. In a related manner, grazing reduces grassland composition and structure, limiting habitats through the reduction of vegetative structure. Clear cutting of forests for timber eliminates resources for communities, thus reducing habitats. Dams built for not only agriculture but for hydroelectric power flood the land upstream, destroying habitats. Urbanization, as well as commercial and industrial development, strip the land of vegetation, pave it over, and eliminate entire natural communities. Conversely, there are populations that actually thrive on human development, such as the aforementioned cockroaches and rats. Habitat alteration is the greatest current driver of biodiversity loss.

Invasive Species

Human activity is the main cause of invasive species, which is another driver of biodiversity loss. Humans have been introducing new species to non-native environments for millennia, some of them accidentally, such as pythons in Florida, and some deliberately, such as the introduction of new crops to an area. Once introduced, many of these invasive species have no natural competitors, parasites, or predators, at least initially. This often results in the devastation of communities, as well as billions of dollars in economic loss. Isolated geographic habitats such as islands are particularly vulnerable due to the limits on populations and resources.

Overharvesting

The overharvesting of organisms, usually vertebrates, for various reasons directly results in biodiversity loss. This was the cause of extinction for many organisms in the past few centuries, often through hunting, such as the Caspian tiger and the passenger pigeon. Many other species have been hunted to the brink of extinction, such as gorillas, sharks, whales, and wolves. Currently the Siberian tiger is also threatened. Many of these animals are hunted for folk medicine or for sport. Often there are few rules and regulations regarding the hunting of these animals in a given country.

Pollution

Pollution is the introduction of harmful substances into the environment. Pollution impacts biodiversity both directly and indirectly. Air pollution results in the introduction of harmful chemical species to all known ecosystems, because of prevailing winds and air currents. Pollutants have been found thousands of miles from their source in places like Antarctica. Agricultural runoff harms ecosystems through the introduction of excess nitrogen and phosphorus to natural systems, especially aquatic systems that may experience algal blooms as a result. Acid mine drainage, often from coal mining, poisons entire aquatic systems, including groundwater. Chemical and oil spills devastate natural communities and ecosystems. Finally, pollutants may persist in the environment, such as plastics, causing damage to wildlife for decades.

Climate Change

Over the past 100 years, climate change has emerged as a major mechanism behind biodiversity loss. While, on a geological scale, Earth's climate constantly goes through periods of warming and cooling, the current increase in warming is apparently occurring at a much faster rate than Earth's geological past. As a result, we have seen an increase in more intense weather events, with greater impact on human life, limb, and property. These weather events have also caused drops in biodiversity, wherever they occur. All of the hottest years on record, going back over a hundred years, have occurred within the last decade. When temperature and rainfall patterns become altered, entire populations become stressed. Migratory patterns become altered; in fact, since winters have become much shorter in many parts of the world the migratory patterns of birds have been markedly altered. The volume of Earth's atmosphere is finite, and since the beginning of the industrial revolution humanity has been steadily pumping greenhouse gases into it, especially carbon dioxide (CO_2) from the combustion of fossil fuels. It is more than likely that this has been a major driver behind the increased rate of overall global warming as averaged across the surface of the Earth annually.

Benefits of Biodiversity

The benefits of biodiversity include ecosystems services. These include energy resources, food, and shelter. Through mainly the activity of microbial communities, biodiversity provides the purification of air and water. These microscopic lifeforms also act to break down waste products. Biodiversity helps to stabilize climate and weather patterns, as well as provide a buffer for floods. Biodiversity is directly responsible for nutrient cycling and soil fertility, as well as disease control and plant pollination. Biodiversity automatically enhances genetic diversity. Human beings benefit from biodiversity on an aesthetic and cultural level as well. We also benefit from the trillions of dollars of value brought about by biodiversity-driven ecosystems services annually.

The flexibility and stability of an ecosystem are directly and positively affected by increased biodiversity. Reduced biodiversity has a negative impact on an ecosystem's total functioning. This stability can further be reduced by the loss of keystone species in a community. Ecosystems are often complex; damage and loss of the components of an ecosystem can have a profound effect on the entire ecosystem. In addition to ecosystem diversity, genetic diversity is linked to biodiversity. This can actually have a positive benefit for human beings. Genetic diversity in agricultural crops results in more disease and pest resistant crops. This genetic diversity saves agricultural production millions of dollars per year. Many wild species are valuable as resources for food such as wild-caught fish or pharmaceuticals in the form of fungi and plant species.

Biodiversity clearly provides us economic benefits. In addition to the aforementioned animal and crop species, the development of parks provides people with a host of economic benefits, including revenue on the local, state, and federal level. The local residents of economically challenged populations may see great benefit from biodiversity, especially people who live near or in parks. As a counterpoint to this, parks that have a large number of visitors annually may experience negative impacts from the visitors.

Despite our engineered ecosystems, human beings naturally seek a connection with raw, untamed nature. It is literally where we come from. We are animals despite our unique adaptions. There is also no escape from being in an ecosystem. Even the engineered ecosystems of rooms and buildings and cities are still just that, ecosystems. We love parks, keep pets, and place high economic value on spaces within wildlands. Studies have been undertaken that indicate that the emotional and mental problems of so many people, often leading to crime, result from being disconnected from nature, from living in the artificial "concrete jungle." This has often been called nature deficit disorder, a disconnection from nature itself.

Conservation Biology

Although we human beings need to extract resources in order to keep functioning within the civilization that we have created, mismanaging those resources has obvious negative impacts. People can make ethical and informed choices resulting in the sustainability of resources without undue damage to the environment. This includes species management done in an ethical manner. **Conservation biology** is the branch of the life sciences that is concerned with the conservation of nature and of Earth's biodiversity through the protection of species and ecosystems. Conservation biologists utilize their knowledge of evolution and extinction to solve habitat- and ecosystem-level problems. One of the goals of conservation biology is lessening or even eliminating the impacts of human development.

One aspect of conservation is understanding population sizes within communities and ecosystems. Populations approach extirpation or even extinction level when the number of members drop significantly. The **minimum viable population** is smallest number of individuals possible before lesser numbers indicate a problem with survivability. A network of individual subpopulations or metapopulations over an area, be it ecosystems or biome level, prevents outright extinction, but these smaller individual populations may still run the risk of extirpation. Extirpation can indicate inherent problems of conservation regarding a particular species.

Environmental stress and other factors can go beyond pockets of extirpation to the total loss of a species from Earth itself. This is known as **extinction**. Extinction occurs when the last member of a species dies off. It is absolute. There is no going back from extinction; it is absolute and irreversible. Biodiversity is affected by extinction, although extinct species may be succeeded by new species through speciation. Extirpation can lead to extinction because of geographic constraints. There are quite a few causes for extinction.

There is a model, called the **equilibrium theory of island biogeography**, that acts as an approximation for all subpopulations (Figure 6.3). It utilizes the response that populations of, say, birds exhibit as they inhabit islands off of some imaginary coast. This can be applied to habitat islands particular to species within an overall ecosystem, the sea. Species richness is modeled through an island's size and distance from the mainland. Extirpation and migration have to be in equilibrium. Essentially, (1) there will be fewer species that colonize islands far from the mainland, which is a proxy for the length of the toll that a migration would have on population size; (2) immigration rates to large islands would be higher, resulting in larger populations, which explains the size of populations in ecosystems of relatively large size; and (3) the extinction and extirpation rates on larger islands are lower because species of metapopulations inhabiting larger areas have more habitats to choose.

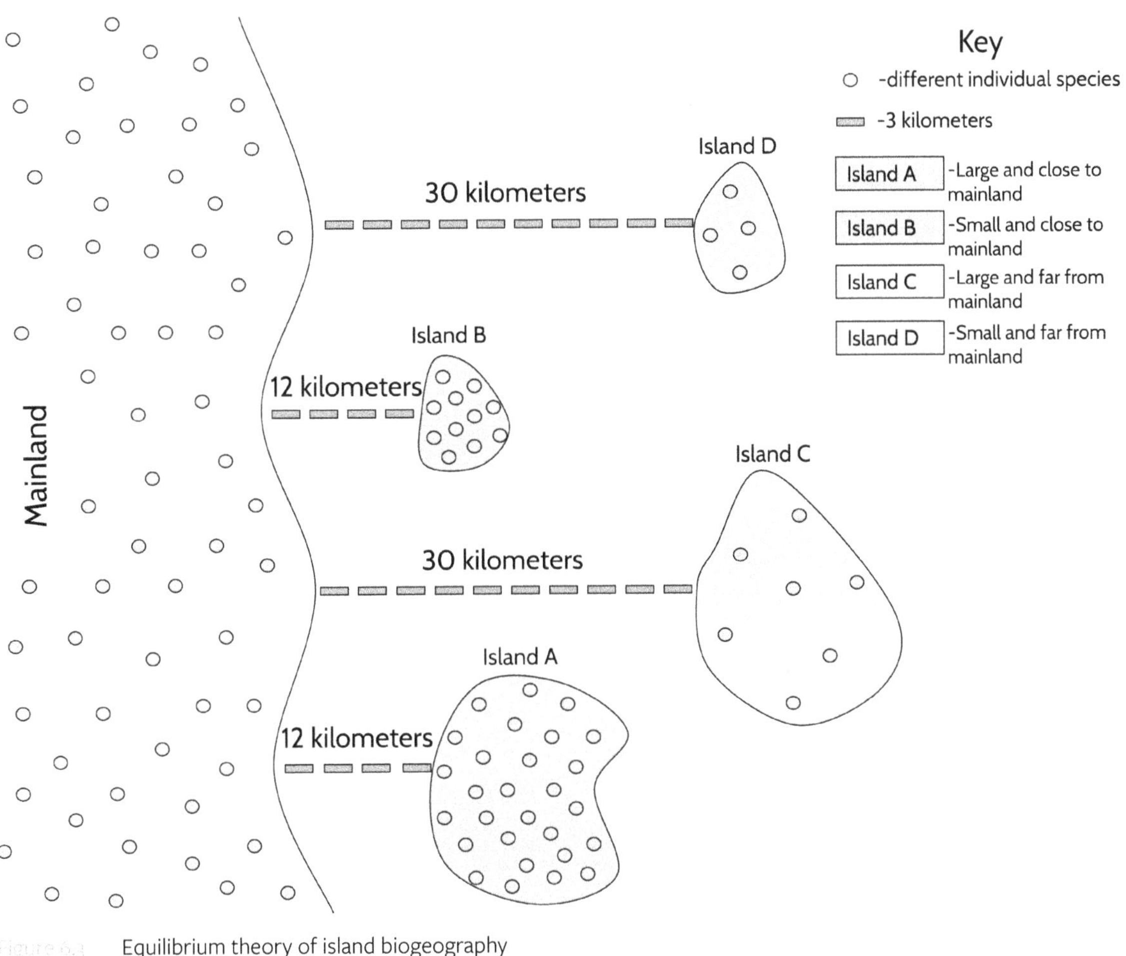

Figure 6.3 Equilibrium theory of island biogeography

The model of an island also applies to deforestation. Forests are habitats to many organisms that rely on them. Through deforestation such as development, logging, and road building, forests are fragmented and become habitat "islands" for woodland species. As deforestation continues, islands increase, while the general size of islands decreases. This leads to losses in diversity. This fragmentation of forests is one of the biggest threats to biodiversity, and it is ongoing since many temperate forests have been cleared.

Methods of Conservation

Since the 19th century concerted efforts have been made in conservation. In truth, human beings have always made these efforts, but it took until the industrial revolution to witness the severity of the threats that development poses to habitats. Extinction is also a relatively recent concept as far as science is concerned.

One of the landmark conservation efforts came through legislation in the United States. The **Endangered Species Act** (**ESA**) of 1973 enacted very specific laws forbidding individuals and the government from harming endangered species and their habitats. The goals are to prevent extinction and ensure population stability as well as recovery. There are currently over 1,600 threatened species on the ESA's list.

The ESA, as well as other conservation efforts from the United States and other governments, has met its share of opposition. Certain corporate and business interests see the ESA as a threat to economic development. Despite this, the ESA has had its successes; for instance, it helped the American bald eagle come back from the brink of extinction. This is ironic in that the bald eagle is literally the national symbol for the United States. Other species have had their populations stabilized through these conservation efforts. As with many conservation organizations, governmental or not, the U.S. Fish and Wildlife Service as well as the National Marine Fisheries Service experience varying levels of underfunding because of mismanagement due to political conflicts.

Captive breeding programs are efforts made by conservation organizations to raise individual species in captivity with the intent of reintroducing them back into the wild. The captive species are not domesticated but prepared to survive in their native habitats. Wolves were successfully reintroduced back to Yellowstone National Park through a captive breeding program. At the same time, this is met with opposition from ranchers who thought that livestock would again be threatened by the wolves. In that regard some habitats are so fragmented that reintroduction is all but impossible. This is probably not the case with Yellowstone since Yellowstone is such a vast geographic area. Once a species is reintroduced, large areas of habitat must be maintained and protected.

Regions that are prioritized as globally important for biological diversity are known as **biodiversity hotspots** (Figure 6.4). These places are ecologically sensitive areas that support great species richness. Many endemic species, species found nowhere else in the world, live within biodiversity hotspots. There are some stipulations for determining a biodiversity hotspot. The area must 0.5% of the entire planet's plants species, and 70% of its habitats must have been lost due to human impacts. There are currently 36 global biodiversity hotspots maintained through the United Nations Council for Biological Diversity and partner organizations. These hotspots account for a little over 2.5% of Earth's land surface, containing around 50% of all plant species and around 42% of all vertebrate species.

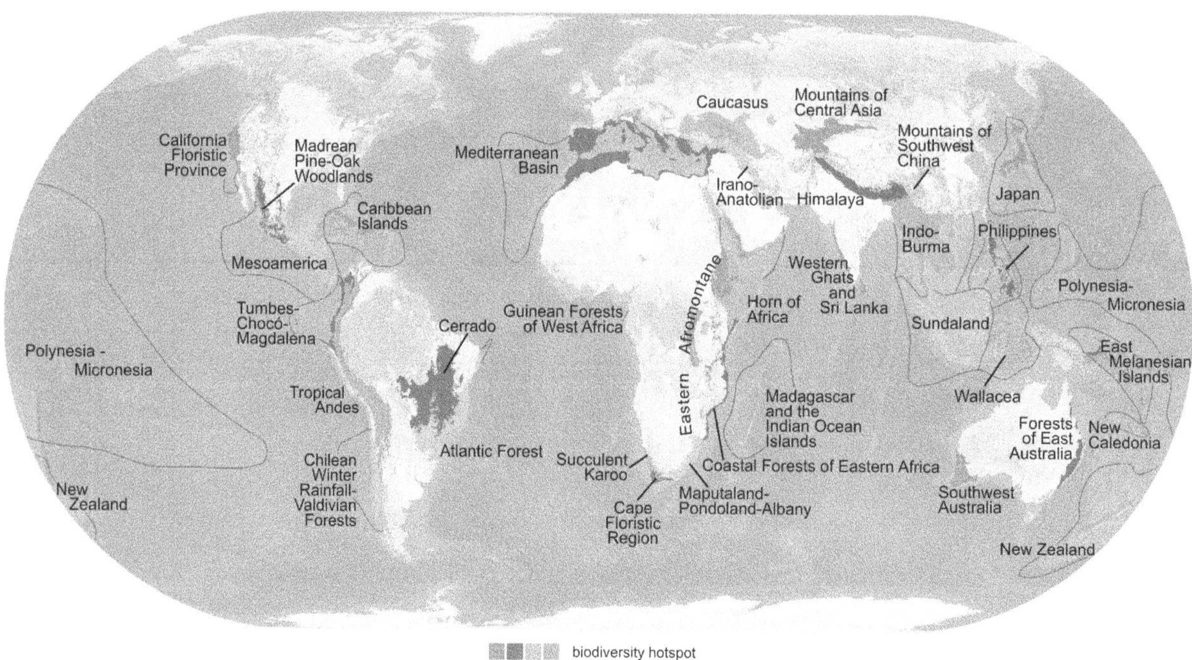

Conservation International (conservation.org) defines 35 biodiversity hotspots — extraordinary places that harbor vast numbers of plant and animal species found nowhere else. All are heavily threatened by habitat loss and degradation, making their conservation crucial to protecting nature for the benefit of all life on Earth.

Figure 6.4 Biodiversity hotspots

Finally, local conservation efforts can make significant impacts in conservation. It can be as small as a local park or bird sanctuary, or a business or farm that sets aside some of its land to allow for wildlife. The Beltway in Atlanta, Georgia, is designed to not only be a walkable path for human beings but a nature sanctuary as well. It provides corridors for animal movement as well as the elimination of dangerous motorized vehicles from habitats.

Ecosystems and Environmental Systems

The planet Earth as whole is composed of an almost unimaginable series of systems and subsystems, that all somehow interact with each other to form a vast functioning whole. A **system** is a network of components, elements, or parts that interact with each other in a determinable manner. Regardless of the complexity, a system is ultimately understandable through the integrated characterization of its components. In this regard, a system may interact with other outside systems, which would be part of that system's operations.

Systemic components exchange energy, matter, or even information within said system. These subsystems receive inputs of energy, matter, and information. These are then processed before the subsystem output, which feeds into another, or a series of other, components. As one may ascertain, systems are often extremely complex, especially vast systems like Earth, but ultimately these systems, their components, and the interactions therein are quantifiable in that one is capable of characterizing the whole and the parts. Of course, what is meant by "one" is the scientific and engineering community at large! For instance, groundwater contamination is understood through the interaction of physical groundwater transmission underground, the groundwater connection or interaction with surface water, and the transport or exchange of water that contains contaminated biological matter, another subsystem.

Feedback Loops

It is important to understand **feedback loops** when one regards systems. A feedback loop is a system, or subsystem, output that also serves as input to that system or subsystem. Feedback loops are circular processes that in turn may also regulate the output of a system or subsystem into a separate system or subsystem. There are two basic types of feedback loops: negative and positive.

Negative feedback loops are outputs resulting in systemic behaviors that move in one direction, acting as inputs that cause the system to move in the opposite direction. A prime example of this is the human body's temperature regulating system. The optimal human temperature is 98.6° Fahrenheit (F). When outside factors cause a person's temperature to exceed 98.6° F, output, that person may seek shade or sweat, input. Thus, a negative feedback loop causes the system to maintain stability, which is the function of feedback loops. A person may also enter a cold environment where their body temperature may start to drop, output. That person may put on extra clothing or shiver, input, maintaining the standard temperature of 98.6° F. Negative feedback loops are the most common feedback loops in nature. They help maintain systemic stability regardless of the type of system.

When a feedback loop does not input back into the system in an oppositional manner, it is a **positive feedback loop**. Positive feedback loops are continuous, feeding back into a system in only one direction, driving it forward toward one extreme or another. Positive feedback loops are rare in nature. Examples are exponential population growth, the spread of disease, cancer, and erosion. Erosion is a positive feedback loop that is more than likely familiar to us all (Figure 6.5). For instance, a small body of water runs through a field that has been stripped of vegetation due to an abandoned construction project. As it rains, over time, runoff will carry particles of soil off of the stream bank into the stream and out of the system. Eventually the entire stream will become forever changed, a vast system of gullies and dangerous soil slumps. While rare in nature, positive feedback loops are common where there are human impacts, which is global.

It is important to get past the counterintuitive terms "negative" and "positive" feedback. It should be obvious that negative feedback loops usually bring about positive effects since they stabilize systems. Positive feedback loops often have negative effects since the alter, even destroy, systems.

Figure 6.5 Erosion, a positive feedback loop

Dynamic Systems

Systems, especially natural systems like ecosystems, are dynamic, with interactions often being time dependent. We now know that negative feedback is essential to systems stability. **Dynamic equilibrium** is a state in which the component subsystems of a system are in constant oppositional direction at equivalent rates, which leads to balance and stability, which is vital for any system. Dynamic equilibrium leads to **homeostasis**, which is a state in which a system maintains constant or stable internal conditions. Key to the maintenance of a system is the balanced interactions and functioning of the **emergent properties**, which are the subtle characteristics of the systems components or subsystems that are not apparent without minute examination. To add to the complexity, systems are connected to other systems. Furthermore, sharp boundaries between system components, subsystems, and other systems often are not apparent. This is especially true in natural systems. It is the task of engineering to understand human-made systems, although these systems are often highly complex in their own right.

Rivers are complex systems that have multiple components, smaller streams that feed into the main river. The main river also accepts input from other systems such as soil, which is carried away by runoff into the stream system. **Riparian**, or river systems, are complex natural systems wherein all components must be understood. With this in consideration, the output of the river must also be understood, which is ultimately the mouth of the river feeding into the ocean, output into input. One such complex system is the Mississippi River.

Dynamic Systems of the Mississippi River

The Mississippi River watershed is one of the largest in the world. It drains 3.2 million square kilometers, and its discharge is 16,792 cubic meters of water a second into the Gulf of Mexico, the 15th largest river in the world in terms of discharge. The Mississippi River is important to 32 states in the United States and is a significant source of drinking water and irrigation for a large part of the Great Plains, one of the major agricultural regions in the world.

Every year from March through May, depending on the latitude, planting season begins for the enormous farms and agricultural operations of the Great Plains. These farms are essentially industrial operations utilizing enormous amounts of fossil fuels to power gigantic machinery. Large amounts of artificial fertilizer are also used on these farms. The demand for food increases every year, so the demand for fertilizer increases as well. These fertilizers contain mostly nitrogen (N) as well as, in some cases, phosphorus (P). Both elements are vital for plant growth, especially N.

Meanwhile, every year in the Gulf of Mexico at around the same time as planting season in the Great Plains something happens to the water. Along a swathe of the Gulf Coast from the coast of Texas to as far as the coasts of Mississippi or Alabama there is a significant drop in dissolved oxygen (DO) levels. Fish and other aquatic animals need DO to breathe just like terrestrial animals. They do not "breathe water" but instead breathe DO. Oxygen levels in the water often drop to below 1.5 PPM, a level at most aquatic organisms die or leave that volume of water. This is exactly what happens in the Gulf of Mexico's annual "dead zone," costing the fishing and shrimping industry millions of dollars per year. This also puts a stress on local family fishing operations, people who depend on the Gulf for their livelihood. The size of the Gulf dead zone grows every year. Its size, in 2021, was 6,334 square miles in area, around the area of New Hampshire. The Gulf dead zone has even been given a name by the United States National Oceanographic and Atmospheric Administration (NOAA), the Gulf of Mexico Dead Zone, or sometimes the Gulf of Mexico Hypoxic Zone. Hypoxic refers to **hypoxia**. Hypoxia is a state in which oxygen is not available for living things to survive.

The reason for the Gulf of Mexico Dead Zone is excess N and P from the fertilizer used upstream, mostly in the Great Plains, which drives a positive feedback loop downstream. Enormous amounts, as in millions of metric tons, of fertilizer are used every year by the large farming operations of the Great Plains. This excess fertilizer, including N and P chemical species, from fertilizer is transported downstream into the Mississippi River via runoff from irrigation and rainfall. The fertilizer is within water discharged into to the Gulf Mexico, and it is used by phytoplankton, microscopic plant-like organisms. Run-away growth occurs within these populations of phytoplankton. Bacteria in the water eat dead phytoplankton and their waste. These enormous populations of bacteria deplete the DO in the water since they undergo respiration. In turn, the drop in DO eliminates many species of aquatic organisms. This over-enrichment of nutrients with a subsequent drop in DO is known as **eutrophication** (Figure 6.6).

Figure 6.6 Eutrophication

Efforts are being made to decrease the size of the Gulf Dead Zone by nonprofit organizations like the Nature Conservancy as well the United States government's Hypoxia Task Force. This includes government funding of wetlands at the mouth of the Mississippi, as well as another nearby river, the Atchafalaya. The Harmful Algal Bloom and Hypoxia Research and Control Act was passed in 1998. Under its directive, there is an initiative for an assessment of hypoxia in the Gulf of Mexico. Some of the possible solutions outlined by the act include the reduction of nitrogen fertilizer use in Midwestern and Great Plains farms, changing the timing of fertilizer applications to minimize transport via runoff, the use of alternative crops that require less artificial fertilizers, restoring wetlands as buffers and creating artificial ones, and improving sewage treatment, another source of nitrogen. There are provisions for evaluating these methods.

So far, these efforts have been met with varying results. One of the foremost solutions would be to limit the amount of fertilizer entering waterways within the Mississippi River watershed, something that would likely require more legislation at the federal level.

GROUND TRUTH

Parts per Million

The term "parts per million" (ppm) is used to describe a small amount of some substance, a solute, dissolved in large amount of another substance, the solvent, usually a liquid and often water in the natural sciences (Figure 6.7). The amount of dissolved solute is measured against the solvent using equivalent units.

Operationally one part per million is the same as 1 milligram (mg) of some substance dissolved in 1 liter (L) of water or some other solvent: 1 mg/L. Equivalent mass can be used as well rather than volume, so 1 mg of something can be dissolved in 1 kilogram (kg) of something else: 1 mg/kg.

Figure 6.7 Visualization of ppm on the largest block, including at the top from left to right, parts per ten, parts per thousand, and parts per ten thousand

Although a ppm might seem like an incredibly small amount, dissolved chemical species in water can have an enormous impact on an even smaller level. Parts per million is useful for measuring pathogens and toxins in natural waters such as bacteria and heavy metals like cadmium (Cd) and mercury (Hg), but some of these species are not only present in smaller amounts, but even in those tiny amounts may still present a danger to natural waters. These minute potent chemical species and organisms have to be measured in parts per billion (ppb) or even parts per trillion (ppt).

Ecosystems

An **ecosystem** is defined as the interaction of all biotic entities to each other, as well as the interaction of all biotic entities to abiotic entities, and vice versa, within a particular area during some period of time. These interactions, on a large scale, involve the energy flows, matter cycling, and even information exchanges within a given ecosystem as well as between different ecosystems. The relationships are complex and systemic and involve feedback loops.

On the ecosystems level, even on up to the level of the entire Earth system, energy originates from the Sun with regard to Earth's surface and near surface. Since this is an exchange of energy from one system into another, this is an **open system**. This energy from the Sun, in the form electromagnetic radiation, is processed and transformed, flowing through ecosystems, utilized to transform matter by the photosynthesized organisms. On the other hand, matter on Earth is a **closed system**, for the most part discounting the small amount of meteorites and cosmic debris that makes it to Earth, as well as the small amount of gases that escape every year into space. For the most part all matter on the surface of the Earth is transformed and moved through systems on Earth. In terms of nutrients this would be in feeding systems in which food is passed upward through the trophic pyramid. In this system the inevitable dead and decaying matter is consumed by detrivores and decomposers, then sent back into the system, transformed into basic components, or even into other systems such as soil.

Production and Productivity

Throughout ecosystems, energy is converted into biomass through the actions of the producer organisms or autotrophs, photosynthetic organisms on and near the surface of the Earth. Initially solar energy is converted to chemical energy by the autotrophs. This is known as **primary production**. This initial assimilation of energy by autotrophs is specifically called **gross primary production (GPP)**. After autotrophs have used GPP for their own energy through respiration, what is left is **net primary production (NPP)**, which generates biomass that is available for heterotrophs, consumer organisms. In turn, biomass generated by heterotrophs that can be used by organisms in higher trophic levels is called **secondary production**. Throughout these processes the rates of production are important, as **productivity** can also be measured in terms of gross primary productivity and net primary productivity.

Some areas in the world generate higher amounts of net primary productivity than other areas. This is usually constrained by latitude. For instance, tundras generate orders of magnitude less net primary productivity than tropical rainforest. This is a function of vegetated growth, so deserts also generate very small amounts of net primary productivity than more humid and warmer areas. As a rule, terrestrial biomes depend on precipitation and temperature to support autotrophic life, and aquatic systems depend on light filtering through the water and nutrients.

Nutrients are important for organisms on both land and in the ocean. **Nutrients** are compounds and elements that are required for survival and consumed by organisms. Nutrients originate with primary production. There are three **macronutrients** essential for life: carbon, nitrogen, and phosphorus. Carbon is essential for life in that all organisms are carbon-based lifeforms. Carbon is present in essential compounds such as carbohydrates. Nitrogen and phosphorus are essential to all living things and required by plants, and other autotrophs, for growth. There is a host of **micronutrients** essential for living things, such as calcium (Ca), potassium (K), and sodium (Na). Micronutrients are needed in smaller amounts. The need for various micronutrients varies from species to species.

Spatial Considerations

Ecosystems differ greatly in size. An ecosystem is usually thought to be of somewhat moderate geographic extent and self-contained. In truth, all systems on Earth are interconnected, including individual ecosystems. Ecosystems adjacent to each other interact, exchanging energy and matter. Organisms may also migrate between ecosystems. The zones between ecosystems are called **ecotones**. Ecotones are transitional systems in which the characteristics of adjacent ecosystems combine. For instance, two ecosystems, a forest and a swamp, may be next to each other. In between both will be an ecotone that combines both characteristics. This zone may be forested but feature little of the standing water that is present in swamps. Organisms that occupy both forest and swamp may be present in this ecotone.

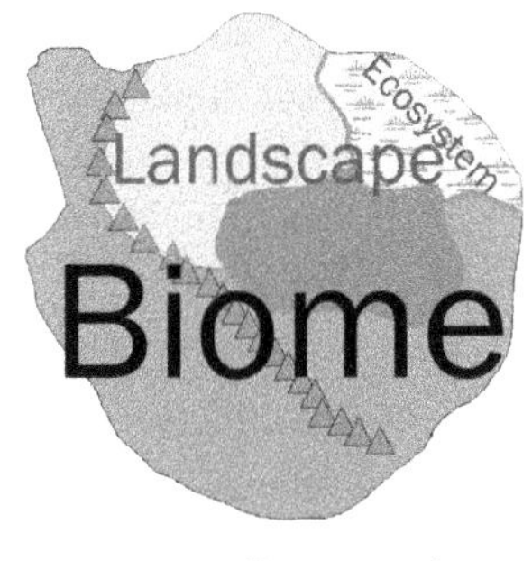

Figure 6.8 Ecosystem in a landscape in a biome

There are other categories of natural systems that are larger than ecosystems (Figure 6.8). A **landscape** is a group of ecosystems that may all share some of the same characteristics and occupy similar topographies. As discussed in the previous chapter, **biomes** are large-scale systems, composed of landscapes, that occupy vast areas of the Earth dependent on underlying geology, climate, and latitude. These in turn support specific types of vegetation characteristic of these biomes. So, biomes contain landscapes, and landscapes contain ecosystems.

GROUND TRUTH

Remote Sensing and Geographic Information Systems

Remote sensing is a group of methods and technologies that allow scientists and engineers to create a complete picture of Earth's multitude of landscapes, geographies, and systems. Geographic information systems (GIS) are remote sensing technologies that utilize computational systems. Geographic information systems include computer software that is used in nearly all of the environmental sciences such as ecology, geology, geography, climatology, and forestry. These systems can help people analyze how elements within Earth's systems are arranged to help in planning and land use decisions

Biogeochemical Cycles

Biogeochemical cycles, also called **nutrient cycles,** are movements of nutrients (the building blocks of matter used by organisms) through ecosystems. These cycles, on a large scale, move throughout Earth's three spheres: the atmosphere, the biosphere, and the hydrosphere and the lithosphere, on a continuous basis. The organisms that exist within these three spheres make up the biosphere.

The **atmosphere** is the gas that surrounds the Earth and to a certain extent penetrates the soil. The soil is the outer layer of the **lithosphere**, the layer of rocks and soil that we live on down to the upper mantle where we cease to find organisms. The **hydrosphere** is the water that permeates both the atmosphere and the lithosphere. Water is literally everywhere on Earth and in the universe. The **biosphere** is everywhere that living organisms are found.

Nutrients often reside in **reservoirs**, or **pools,** for varying amounts of time. Matter and energy remain in those reservoirs for certain amounts of time depending on the state of **flux**. Flux is the movement of nutrients between reservoirs. Flux changes over time depending on changes to the environment. Human beings influence the flux of nutrients as well, such as in the use of artificial fertilizers and mining. The amount of time that each reservoir of nutrients remains in each reservoir is their **residence time**. Reservoirs can release more nutrients than they accept, making them **sources**. **Sinks** are reservoirs that accept more nutrients than they release.

Four Primary Cycles

There are four cycles vital for life on Earth. There are three macronutrient cycles: carbon, nitrogen, and phosphorus. The fourth cycle that will be discussed is the hydrologic cycle, water. Water is important in that it transforms and transports nutrients. Through its presence in all living things it binds together all life. Other cycles of nutrients and chemical species exist, including the sulfur cycle, the copper cycle, and even the gold cycle, but these cycles are fairly minor in scope compared to the carbon, nitrogen, phosphorus, and hydrologic cycles.

The Carbon Cycle

The **carbon cycle** is characterized by the routes that carbon (C) atoms take through the environment. Photosynthesis transports carbon from the air to organisms, and in turn, respiration returns carbon to the air and oceans. When organisms die, decomposition returns carbon to sediment and ultimately to rock, which is the largest reservoir of carbon on Earth. This carbon may ultimately be extracted from the Earth by human beings and converted into fossil fuels. These fossil fuels undergo combustion, releasing carbon back into the atmosphere at rates higher than natural rates. The world's oceans are the second largest reservoir of carbon. Excess carbon, in the form of carbon dioxide, dissociates into carbon and carbonate ions. As the ocean absorbs more carbon dioxide, it becomes more acidic.

Human beings profoundly affect the carbon cycle. Burning fossil fuels transfers carbon from the ground into the air, contributing to pollution and climate change. In this regard, cutting forests and burning fields transfers carbon from organisms into the air. The current atmospheric carbon dioxide reservoir is the largest in the past 650,000 years. This undoubtably drives climate change.

The Nitrogen Cycle

The **nitrogen cycle** is characterized by the routes that nitrogen (N) atoms take through the environment. Nitrogen is present in proteins and nucleic acids. Nitrogen comprises 78% of Earth's atmosphere, but that nitrogen is inert. Nitrogen atoms naturally form nitrogen molecules (N_2), which are triple covalently bonded, meaning that three outer valence shell electrons strongly bind N_2 molecules.

Nitrogen is made available to plants by nitrogen-fixing bacteria. In the nitrogen-fixing process, nitrogen gas is combined, or fixed, with hydrogen by nitrogen-fixing bacteria to become ammonium, which can be used by plants. Another type of bacteria converts ammonium ions into nitrite ions, then into nitrate ions. Plants can then uptake these nitrates. Animals also need nitrogen and obtain it by eating plants and other animals. Nitrates in soil and water are converted back into gaseous nitrogen by **denitrifying bacteria**.

Human beings have profoundly affected the nitrogen cycle. The **Haber–Bosch process** was developed by two German chemists in the early 20th century, Fritz Haber and Carl Bosch, in order to synthetically produce fertilizers through the combination of nitrogen and hydrogen in order to synthesize ammonia. While the Haber–Bosch process ensured plentiful amounts of food for many people, it also changed the nitrogen cycle. Currently, humans are fixing as much nitrogen as is produced through nitrogen fixation. In addition to that, calcium and potassium in soil is washed out of the system by nitrogen-containing fertilizers.

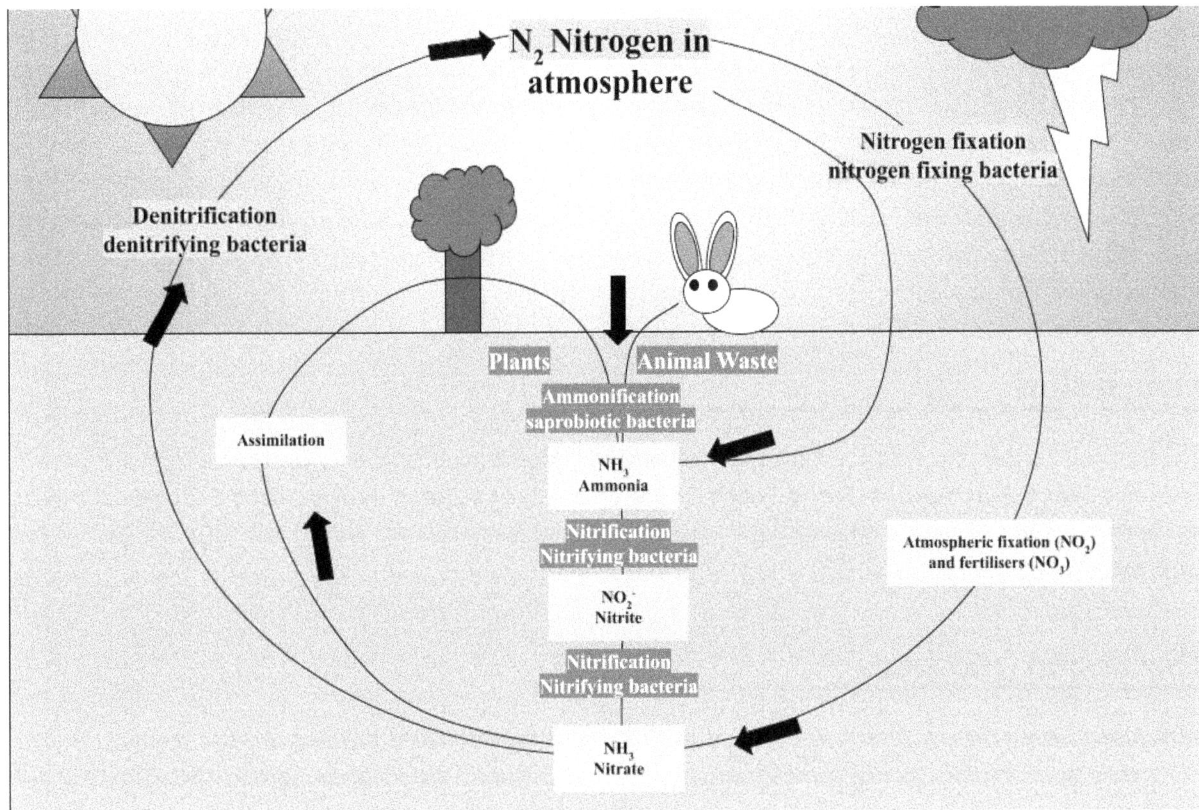

Figure 6.9 Simple diagram of the nitrogen cycle

Nitrogen is also present in fossil fuels, leading to the emission of nitrogen-containing greenhouse gases as well as nitrogen-containing pollutants. Excess nitrogen also leads to soil acidification. Excess nitrogen produced by humans leads to drops in biodiversity and has an impact on fisheries.

The Phosphorus Cycle

The **phosphorus cycle** is characterized by the routes that phosphorus (P) atoms take through the environment. Phosphorus is important to living things because it is a key component of cell membranes. It is also present in nucleic acids, such as DNA and RNA, and their components. Phosphorus, along with nitrogen, is important for plant growth.

It is of note that there is no significant amounts of atmospheric phosphorus other than the relatively minute amounts in ocean spray. Phosphorus is mostly found in rocks and is released from those rocks through weathering, often entering waterways. Phosphorus is a limiting factor for plant growth, so when humans use artificial fertilizer containing phosphorus, environmental imbalances are likely to occur.

Human beings obtain most phosphorus used in agriculture and industry through mining, which often results in the release of phosphorus into water systems. Wastewater that is discharged into natural waters also releases phosphorus. Phosphorus in natural waters leads to eutrophication in bodies of water by spiking the amount of available phosphorus in these waters, making it available to aquatic autotrophs.

The Hydrologic Cycle

The **hydrologic cycle** is characterized by the routes that water (H_2O) molecules take through the environment.

Water is vital for biochemical reactions and is involved in nearly all environmental exchanges. Water defines large-scale systems from ecosystems to entire biomes. The ocean is the main reservoir of water.

Water cycles through the environment in two primary ways: **precipitation** and **evapotranspiration**. Precipitation is the transfer of water from the atmosphere, through rainfall, snow, sleet, or some other process. Evapotranspiration transfers water from biosphere, hydrosphere, or lithosphere into the atmosphere. Evapotranspiration is a combination of **evaporation,** which is the spontaneous phase change of liquid water to water vapor, and **transpiration,** which is the outgassing of water by plants, primarily trees. The hydrologic cycle will be discussed at length in a future chapter of this textbook.

Conclusion

The Earth is a complex system that contains almost countless systems and subsystems. Understanding that these systems are all interconnected is vital for our understanding of this planet and all of the life it harbors. It is also vital for us to understand these systems in order to live in harmony with nature and to solve many of the current environmental problems of our times.

Discussion Questions

Directions: Review the chapter in order to completely and correctly respond to the questions and prompts.

Keep this in the back (or maybe even front) of your mind: Science is system of methods that allows us to understand not only our planet, and ourselves, but of the entirety of the universe. Answering the following questions correctly will lead to a more robust understanding of how to apply science to ENVS problems.

1. Why is low genetic diversity a problem? What are some of the problems that it causes?
2. What is the most predominant group of organisms on Earth? Why is that the case?
3. Look up a place on Earth that has high biodiversity due to latitudinal gradient, then look up a place on Earth that has low biodiversity due to latitudinal gradient. Compare and contrast the two using what you've learned.
4. What leads to extinction?
5. Why is it possible for extirpation to lead to extinction?
6. List the five main causes for biodiversity loss and why they lead to biodiversity loss.
7. Explain the equilibrium theory of island biogeography.
8. Find an example of both negative and positive feedback loops and explain them.
9. What is eutrophication, and why is it a problem?
10. Explain ecotones.
11. Explain biogeochemical cycles.

Bibliography

Amstutz, L.J., 2018, *Invasive Species. Ecological Disasters*: Minneapolis, MN, Essential Library, 112 p.
Jolivet, P., and Verma, K.K., 2005. *Fascinating Insects: Some Aspects of Insect Life*: Sofia: Pensoft, 310 p.
Odum, E., and Barrett, G.W., 2004, *Fundamentals of Ecology*: Farmington Hills, MI, Cengage Learning, 5th ed., 624 p.

Credits

Fig. 6.1: Copyright © by Fährtenleser (CC BY-SA 4.0) at https://commons.wikimedia.org/wiki/File:Biodiversity_Pyramid_%28English%29.png.
Fig. 6.2: Copyright © by Anaxibia (CC BY-SA 3.0) at https://commons.wikimedia.org/wiki/File:Insects_collage.jpg.
Fig. 6.3: Copyright © by Lindsayt18 (CC BY-SA 4.0) at https://commons.wikimedia.org/wiki/File:APES_Final.svg.
Fig. 6.4: Copyright © by Conservation International (CC BY-SA 4.0) at https://commons.wikimedia.org/wiki/File:Biodiversity_Hotspots_Map.jpg.
Fig. 6.5: Copyright © by Jim Barton (CC BY-SA 2.0) at https://commons.wikimedia.org/wiki/File:Bank_erosion,_Hedderwick_Burn_-_geograph.org.uk_-_2084984.jpg.
Fig. 6.6: Copyright © by N Chadwick (CC BY-SA 2.0) at https://commons.wikimedia.org/wiki/File:Eutrophication_-_geograph.org.uk_-_3784065.jpg.
Fig. 6.7: Copyright © by Cmglee (CC BY-SA 4.0) at https://commons.wikimedia.org/wiki/File:Visualisation_parts_per.svg.
Fig. 6.9: Copyright © by EssauMejia (CC BY-SA 4.0) at https://commons.wikimedia.org/wiki/File:The_Nitrogen_Cycle_(1).png.

Chapter 7

Public Health and Toxicology

Environmental Impacts on Human Well-Being

Objectives

- Understand that one's environment has an impact on their overall health and well-being.
- Learn about the various environmental hazards present on our planet.
- Learn about how communicable diseases get spread even on a global level.
- Know that with everything in the environment there comes some form of risk.

KEY WORDS AND TERMS

Pathogen (n.): A microorganism or organic components, such as bacteria, protists, or viruses, that cause disease.

Toxicant (n.): A toxic substance that enters the environment.

Toxin (n.): A substance of some kind, such as a chemical species, that produces a negative effect in organisms.

Virulence (n.): The intensity or severity of harmfulness in diseases, poisons, or toxins in general.

Consider This . . .

There is an old adage that we will return to, "the dose makes the poison." This is very true with all of the chemical species that we come in contact with but can easily be applied to contact with everything in our environment. The overall health of an organism is in some way due to its resistance to harm from the environment. On the other end of the spectrum, our environments are where we live and provide us the benefits of living on planet Earth. With abundance must also come the risk of overabundance, as well as toxicity from elements that resist the vigorous life we might enjoy.

Introduction

Human beings are ultimately interested in their own well-being with regard to Earth's environment. While our planet provides us with a home and a multitude of resources, there are hazards implicit in our existence on planet Earth. Factors that influence human health and quality of life are known as **environmental health** (Figure 7.1). In this regard, our current discussion will be mostly focused on human beings and human survival.

Figure 7.1 Environmental health is often linked to self-sufficiency

Environmental Hazards to Health

There are a multitude of hazards present within human populations, both anthropogenic and natural. These threats reflect modern conditions on Earth, in some part due to development brought about by human civilization. All of the factors currently present on our planet are based on conditions that we face daily, yet these conditions are also climatically, geographically, and geologically dependent.

Biological Hazards

Biological hazards, or **biohazards**, are present in the environment due to either human activity, natural conditions, or even a combination of both. This is because we are all still part of some sort of ecosystem, regardless of the engineered environments that we create for ourselves like living within buildings. Biohazards result from interactions within ecosystems, which are the interactions between organisms (in this case human beings) and the environment as well as with other organisms. These "other organisms" include other human beings.

People are constantly succumbing to communicable diseases within the human population. An infectious disease comes from a species of organism, often microscopic, parasitizing a human being. While hazardous to us, these parasites are only fulfilling their ecological roles within a given ecosystem. These parasitic organisms take the form of bacteria, protists, and viruses. Larger, metazoan organisms such as lizards, mosquitos, worms, and even other humans are often the **vectors** for these microscopic parasites. A vector is an organism that transmits an infectious agent or parasite to a host. Organisms like Gila monsters, or mosquitos, contain infectious organisms in their mouth parts and transmit these organisms upon biting or stinging a person. Human beings usually act as vectors inadvertently, such as when a person infected with the common cold spreads it by not covering their mouth when they cough.

Biohazards often come from the natural world, but human activity can be the root cause of infectious diseases as well. The fact that human beings gather in large numbers in small spaces, whether it be buildings or even cities, increases the chances for exposure to disease. Rapid travel by airline, boat, motorized vehicles, or rail allows for infected individuals, or animals, to move from one place quickly and easily in our modern times. As human development encroaches upon land that was previously wilderness, the chance for exposure to once dormant or new infectious diseases also increases. As the planet warms, conditions are made more favorable for disease organisms to spread since they, like all living things, favor warmer climates.

Types of biohazards include the following:

- **Bioactive substances**—Compounds that can intrinsically affect cells, organisms, and tissues. Examples include antibiotics, enzymes, and vitamins. **Prions** are another type of very dangerous bioactive substance. Prions are misfolded proteins (enzymes are also proteins) that transmit the misfolding trait to other proteins. A type of prion disease in the news in recent years is "mad cow disease." Indeed, one can get a prion infection from eating contaminated meat. Prion disease can be transmitted through contaminated medical equipment, and some forms are hereditary. Thankfully prion disease is very rare.
- **Fungi**—Organisms that belong to the kingdom *fungi*. Most fungi are not harmful. Toxic fungi include mold and poisonous mushrooms. These fungi can cause illness from being eaten, or in the case of mold, inhaled. Some toxic fungi secrete chemicals that can be absorbed through the skin. There are quite a few wild mushrooms that can be harmful to human beings and cause hallucinations, illness, or death. It's best to avoid picking wild mushrooms unless one has detailed knowledge of them.
- **Microorganisms**—These are typically bacteria but also include pathogenic protists, as well as macroscopic parasites. Most bacteria in the environment are harmless. Some types of bacteria are beneficial, such as the bacterium that inhabit mammalian digestive systems, including humans. Less than 1% of bacteria are harmful, including *E. coli* that can contaminate water, as well as *Listeria* and *Salmonella,* which can both contaminate food. It's best to pay attention to whether food is spoiled or not. Also, cook meats and wash fruits and vegetables thoroughly. The bubonic plague pandemic of the 14th century was caused by bacteria. Most protists, single-celled eukaryotes, in the environment are harmless, but a small number of them can be harmful to humans. *Giardia* is an infectious protist that is transmitted through feces. It can be found in places that are unsanitary as well as in contaminated water, and it causes a severe gastrointestinal sickness called giardiasis. Several types of amoeba are dangerous and can cause amoebic dysentery and other illnesses. Another type of amoeba, *Naegleria fowleri*, can cause a dangerous inflammation of the brain known as encephalitis. It lives in soil and water in tropical to subtropical climates, including the southern United States. To prevent the possibility of infection, avoid swimming in stagnant warm water since the amoeba often enters the body through the nasal tract. Malaria is caused by protists. Mosquitoes are the vector for malaria. Macroscopic parasites include fleas, ticks, and worms. These parasites invade the human body from contaminated food or the natural environment. Macroscopic parasites can, in turn, infect a host with bacteria, protists, or viruses. Organisms and viruses can enter the human body through breathing, the mouth or nose, and from skin contact.
- **Spores**—These include reproductive units from fungi, plants, and protists and include bacterial spores, which are a dormant form of some bacteria adopted when conditions are adverse. Spores are often inhaled,

such as mold spores, causing respiratory problems. Spores, like mold itself, accumulate in damp, older buildings with poor ventilation. Diseases caused by bacterial spores include anthrax, botulism, and tetanus. Infection from these comes from contaminated meat, and in the case of tetanus, infected flesh wounds.

- **Poison or venom**—Certain organisms produce toxic substances for either defense or predation. Poison is a biochemical substance that is present in the tissues of an organism that causes an adverse reaction upon contact or ingestion. For instance, many plants are toxic to animals if ingested, like the deadly nightshade plant, which is harmful to humans and causes illness, hallucinations, and even death. Certain species of frogs that live in Central and South America are poisonous upon contact. Animals are sometimes venomous. Venom is a **toxin** that is directly injected into another organism (Figure 7.2). For instance, poisonous jellyfish and snakes have features, such as fangs in snakes, that are used to take down prey. A human attacked by such an animal may succumb to its venom, falling ill or dying.
- **Viruses**—Submicroscopic packets of lipids, nucleic acids, and proteins that only reproduce within the bodies of organisms. Viruses are as ancient as life itself, although most biologists do not consider viruses to be true life forms since they do not share all of the characteristics of life. Any type of organism can be infected by a virus from bacteria to protists to all multicellular life forms. Viruses have caused some of the worst epidemics and pandemics in history, including the Spanish flu of the early 20th century, the avian flu of 1997, and the recent COVID pandemic. Viruses infect organisms, including human beings, through multiple pathways and are literally everywhere on Earth. Most viral infections are not life-threatening though; think the common cold. Regardless, prevention is the best strategy for avoiding viral infection.

Chemical Hazards

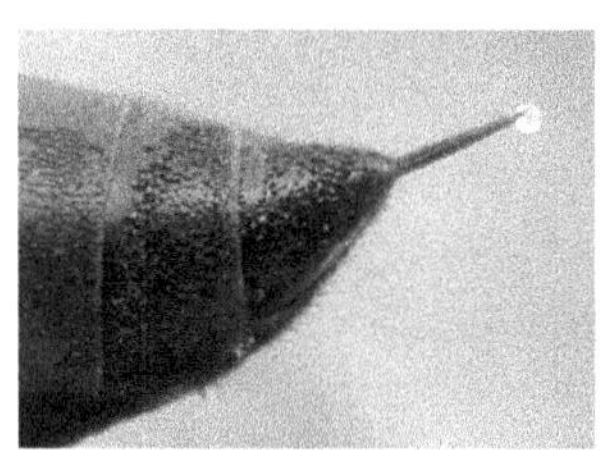

Figure 7.2 A close-up of a wasp's stinger and venom

Chemicals that exist within the environment, with the potential to harm humans, constitute the chemical hazards (Figure 7.3). Chemical hazards can be either anthropogenic or natural depending on the source. Exposure to anthropogenic chemicals, toxins, can occur in commercial, industrial, or residential settings. Many industrial chemicals are hazardous or toxic. Workplace management policies are critical with regard to many commercial and industrial settings. The Occupational Safety and Health Administration (OSHA) sets standards for workplace handling of potentially hazardous chemicals on the job. One must also keep in mind that many chemicals, both human-made and natural, are very beneficial to human beings and have given us a high standard of living.

Although some of the most dangerous chemicals can be found in industrial settings, toxins can be found in places of business and in our homes. For instance, many dry-cleaning businesses still use perchloroethylene (PERC) as a solvent. This chemical is carcinogenic (cancer causing), and unless properly stored and disposed of, may end up in local water supplies. You probably already store some potentially dangerous chemicals in your home such as ammonia, bleach, gasoline, herbicides, many other cleaning supplies, pesticides, and a host of others. By the way, never mix ammonia and bleach!

Chemical hazards include the following:

- **Asphyxiants**—These are chemicals that deprive the body of oxygen, cutting off the body's ability to use it. They are commonly found in gaseous form. Examples of asphyxiants include acetylene, carbon monoxide (from automobile exhaust), and methane.

- **Carcinogens**—Cancer-causing chemicals that are extremely dangerous for their long-term effects. These chemicals can cause health hazards many years after exposure. There are currently over 200 known carcinogens that pose a risk to human beings. Examples of carcinogens include benzene, formaldehyde, and vinyl chloride.
- **Corrosives**—These chemicals cause damage to substances, and living tissue directly, often severe damage. Chemical corrosives cause visible and/or irreversible changes to the composition of a material due to direct contact. Since many corrosives are acids or bases, they can cause adverse effects through mixing with other substances or when they are concentrated. Examples of corrosives include ammonia, hydrogen peroxide, sulfuric acid (battery acid), and sodium hydroxide (in household lye).
- **Endocrine disruptors**—Compounds that disrupt the body's hormonal (endocrine) system's developmental, immune, and reproductive effects. Examples include arsenic, PCBs, many pesticides, and many plastic products.
- **Flammables**—These are combustible chemicals, causing a rapid flame and heat-producing reaction in the presence of an oxidizing agent, usually oxygen. There are many chemicals found in many places that can ignite and undergo combustion, some more dangerous than others. Examples of flammable chemicals and substances include alcohol, charcoal, and gasoline.
- **Irritants**—These chemicals that can harm the eyes, skin, or respiratory tract of a person. These chemicals are highly, moderately, or slightly water soluble. Irritants can cause bleeding, coughing, inflammation, or rashes. Harm caused by irritants is often short-term, but long-term effects are possible, since some irritants are persistent. Some irritants cause allergic reactions in some people, and some of these chemicals can cause long-term, life-threatening harm. Examples include formaldehyde, nickel chloride, and chromic acid.
- **Mutagens**—These chemicals cause genetic change to a cell's DNA and RNA, potentially causing harm to a person. These genetic changes can initiate cancer, prevent normal biological functioning, or may even result in the malfunction of certain organs. Examples of mutagens include benzene, hydrogen peroxide, and radioactive substances.
- **Reactives**—These are chemicals that can explode when mixed or combined with other chemical chemicals, when in contact with air or water, or upon impact. Reactive chemicals include benzoyl peroxide, nitroglycerine, and silane.

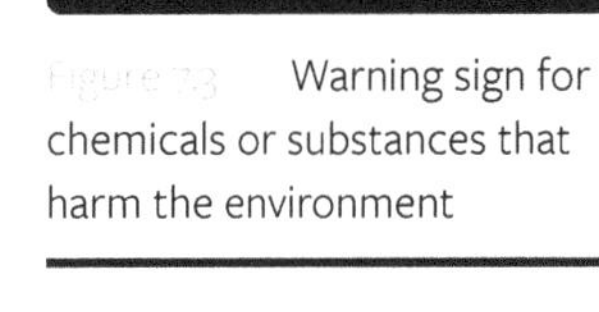

Figure 7.3 Warning sign for chemicals or substances that harm the environment

- **Sensitizers**—These chemicals are also known as allergens and can cause allergic reactions in people who are sensitive to them or who are exposed to them over a long period of time. The response to sensitizers varies from individual to individual, with some individuals being essentially immune to certain allergens while others becoming ill. Responses to these chemicals may be acute or chronic, mild, or even life-threatening. When exposed to these chemicals, irritation or swelling of respiratory tissue may occur, ramping up to the possibility of life-threatening reactions. Some people who are overexposed to chemicals may come down with asthma and other conditions. Allergens include some alkalis, chemicals found in certain foods, chemicals from insect bites, and certain medicines, such as penicillin.
- **Teratogens**—These chemicals can harm or hinder the development of human embryos or fetuses, causing birth defects or disruption to pregnancies. These sorts of substances should be completely avoided by pregnant women. Examples of teratogens include alcohol, polychlorinated biphenyl (PCB), organic mercury compounds, and thalidomide.

Cultural Hazards

Cultural environmental hazards, also known as social hazards, are behaviorally based habits and practices that may endanger human beings. These hazards come about from personal behavioral choices, a person's occupation, socioeconomic status, and the workplace. The hazards that result are crime; drug use, including alcohol; irresponsible physical behavior such as reckless driving; poor diet and nutrition; and smoking. Even excessive sunbathing is a cultural hazard because tans are often considered beautiful by cultural standards. In almost all cases cultural, or behavioral, hazards can be eliminated, prevented, or at least reduced. In other words, many of these cultural hazards are choices that can be made. For instance, one can choose to drive safely, to not do drugs or smoke, and to follow a healthy diet. These choices are made for a number of psychological and sociological reasons, which is why safer choices can be made as well. This is often through simply educating one's self. At the same time many cultural or social hazards are not only habit forming, they are addictive, something that is very hard to stop (Figure 7.4).

Changes in behavior can come about from incentives. For instance, one may turn to drugs and alcohol, or overeating, to cope with stress and trauma in their personal life, resulting in poor physical health and even pain. The incentive to no longer be in pain from the effects of obesity, or dangerous alcohol and drug withdrawal, may cause a person to seek medical assistance. A person who has a habit of driving fast and recklessly may get into a few minor accidents or get a few speeding tickets. This causes their insurance premiums to become more expensive, in addition to the expense of automobile repair. Not wanting to spend more money on auto repair and insurance, that person may will themselves to start driving carefully and within the speed limit.

Some cultural hazards are due to social connections. Many people take up drinking alcohol, drug use, or smoking cigarettes from spending time with other people who do so in social situations. The same can be said for doing crime in some situations, although crime can also be driven by poverty, which is ultimately the failure and irresponsibility of business, corporations, governments, and society as a whole. The need for human beings, social animals, to belong to a group of some kind is strong. Regardless, self-awareness and willpower are two powerful aspects of the human psyche as well. Bad habits can be broken by education, socially seeking out people who lead healthier lifestyles, and through the sheer desire to not cause harm to one's self as well as other people. Safer, healthier environments also lead to better choices such as agencies for the homeless and poor, drug and alcohol rehabilitation programs, good public schools, safe public parks, and well-maintained public works. Ultimately, though, cultural environmental hazards exist because of human behavior and society. As such, these problems are inherently complex, often requiring complex solutions implemented through cooperative efforts.

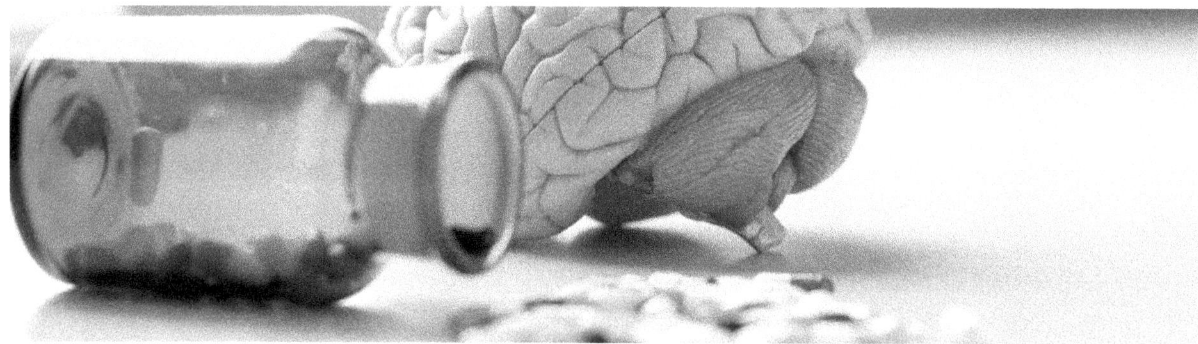

Figure 7.4 Drug addiction is only one cultural hazard

Physical Hazards

Physical environmental hazards include both human environments and natural environments. There is often little we can do to eliminate the potential for these environments to do harm, especially in the case of natural hazards. Hazards in the human environment include those we might find in the home, in the workplace, in industry, or at large within public and commercial spaces. Physical environmental hazards often manifest in clusters, for instance an earthquake may cause buildings to collapse, fires, and floods if dams are damaged or destroyed. Physical hazards are those that involve mass, speed, and other physical forces, in other words energy.

Types of physical environmental hazards include the following:

- **Building and construction**—These are hazards that come about from the human-made spaces we live in, work in, and do business in. There are a multitude of hazards present in our environment, from the very buildings that we inhabit. Often buildings collapse when subject to the stress of other hazardous events. For instance, earthquakes will topple buildings. The construction industry is potentially a hazardous one, but, in general, safety measures as well as training prevent many accidents. At the same time, construction accidents occur. Poorly constructed shelving in stores may even pose a hazard. In economically challenged parts of the world there may be no building codes for the construction of homes and workspaces. Errors in construction occasionally cause collapse, sometimes with catastrophic results. For instance, on June 24, 2021, part of the Champlain Towers South, a condominium outside of Miami, Florida, collapsed, killing 98 people. These types of tragedies are thankfully rare and can be prevented with stringent building codes and construction best practices. Many older buildings contain asbestos and lead pipes, both of which are potentially harmful to human beings. Asbestos is a group of naturally occurring silicate fibers that have been used in the past as building materials, mostly because of their fire-retardant properties. In the 20th century it was discovered that asbestos fibers, that become dislodged and airborne, can be inhaled, leading a number of serious illnesses such as asbestosis, lung cancer, and mesothelioma. Since then, large-scale efforts have been made to remove asbestos from buildings. If you think that there is asbestos in a building, ask around, and don't spend too much time there. Lead, a metallic element, has been used as a building material since ancient times. In the 19th century it was discovered that lead poisoning occurs in people exposed to lead. Lead poisoning can lead to serious illnesses, such as organ damage, neurological diseases, and even death. During the 20th century a major effort was made to eliminate lead in gasoline, paint, and water pipes. Radon, another element, is a toxic, radioactive gas that is colorless and odorless. It is emitted by certain rocks such as granite, gneiss, phosphate rocks, schist, shales, and uranium ores. It can even be found in limestone. Radon can build up in basements and get into water pipes. If you suspect the possibility of radon in your home or workplace, most local fire departments offer free radon testing kits.
- **Earthquakes**—This is the release of energy that causes shaking of the Earth when layers of rock beneath or near the surface of the Earth experience structural failure. Earthquake-prone areas vary. These places tend to be near the juncture of two or more tectonic plates. Tectonic lithospheric plates move constantly, slipping across each other, but often friction causes them stick, slowing or stopping movement. Potential energy builds up when this happen and is released when friction is overcome, causing an earthquake. Certain areas of the world are prone to earthquakes while others are not. For instance, the entire east coast of Asia and the west coasts of North and South America are within "the ring of fire," the rim of the Pacific Ocean, which is basically one large continuous zone of tectonic plate activity. Meanwhile, Africa's west coast and the east coast of North and South America are generally far from tectonic plate junctures, so

earthquakes in these areas are infrequent. Since an earthquake is simply a shaking of the Earth, other events may cause them, such as explosions, hydrofracking (the economic practice of breaking apart Earth materials to obtain fossil fuels), landslides, and volcanoes. Generally, there is little that we can do to prevent earthquakes since they are a natural phenomenon. Living away from these zones is basically out the question at this point in time, so earthquake-safe buildings and education are best when living in these places.

- **Electricity and lightning**—Electricity is the property inherent in electrical charge, including current and voltage. Lightning is a natural discharge of electricity that usually occurs in thunderstorms but can also occur during other natural events such as catastrophic volcanic eruptions. The power of electricity is current times voltage, so large amounts of current and/or voltage is potentially dangerous. Lightning discharges on average around 30 kiloamperes. Many household appliances average around 3 amps, which can cause a painful shock. This is a difference of 10 million orders of magnitude! When in a powerful thunderstorm it is best to stay in indoors or low to the ground to avoid a lightning strike. Lightning strikes on humans are comparatively rare, but they do happen. Electricity in general is what powers our entire civilization. This means that electricity is always around us. Wiring that is safe and well-designed electrical systems are vital in all settings: commercial, home, and industrial. Loose wiring and unsafe systems can shock people in contact with them or cause electrical fires. The use of power in our daily lives also generates **electromagnetic fields** (**EMF**), zones of moving electrical charges. There is some research to indicate that these fields can have long-term adverse health effects on people. It is best to occasionally walk away from electricity, walk away from the screen, into a more natural environment to engage in physical activity. Even a daily walk in nature makes a difference.
- **Extreme weather and temperatures**—These are meteorological events that are powerful in intensity. Examples include dust and sandstorms, droughts, extreme cold, hail/ice storms, high heat/heat waves, high winds, hurricanes, snowstorms, thunderstorms, and tornadoes. Experiencing these weather events depends greatly on where one lives on the face of the Earth. For instance, someone living in Libya, in the Sahara Desert, will never experience extreme cold or a snowstorm, but they will experience high degrees of heat and possibly sandstorms. Extreme weather and temperatures are nearly impossible to escape without actually going to another part of the planet. Records indicate that the intensity of certain storms, like hurricanes, and the global average temperature have been increasing in recent times. According to NOAA, global average temperatures have been increasing at a rate of around 0.18 °C per decade since 1981. Heating of the ocean in equatorial zones leads to the formation of hurricanes and other strong storms due to convectional heating. These storms are costly to people living on the coast in terms of human life and property. Powerful oceanic storms can cause storm surges, a local rising of the sea surface that can flood the coast far inland. Adapting to local conditions and educating one's self are the best measures taken against extreme weather events, as is the willingness to evacuate the area during hurricanes.
- **Fires and wildfires**—Fire is a rapid exothermic (releases heat), oxidation chemical process. During fires, energy is released as heat and light. Flame is the visible portion of fire, consisting of superheated gases. Fires are inherently dangerous since they emit high heat and cause the burning destruction of matter, both living and nonliving. The essential conditions for fire consist of the "fire triangle": fuel, heat, and oxygen. In the human-made environment there are many ways that fires can ignite fuels given the right conditions, including cigarettes, outdoor grilling, and unshielded/unsafe electrical wiring. Once started, the right amount of fuel and oxygen is all that it takes to keep a fire going: fabrics, oil and gas, wood, and so forth. Uncontrolled fires are potentially deadly for a variety of reasons. Local fire departments exist to put out

fires in our homes, businesses, and in industry. Wildfires are uncontrolled fires that burn through large areas of wilderness. Lately, with shifting weather and climate patterns, wildfires have been starting at increasing rates. These sometimes deadly wildfires have been extremely hazardous in drier areas of the world such as Greece, Southern California, and Eastern Australia. Again, safe management is the best practice with respect to fires in wilderness areas. Forest and park services around the world are responsible for fighting and managing wildfires.

- **Floods**—Floods are uncontrolled overflows of water onto land that is not usually submerged. Floods are a natural function of rivers and streams, which form flood plains, flanking each side of a stream. Extreme flooding is caused by exceptionally heavy rainfall, at times of long duration. Flooding can also be caused by coastal storms, meltwater from mountain glaciers, and melting snow and ice. Dam failure is a human-made source of potential catastrophic floods. In fact, engineers have, in the past, underestimated the input of water concerning human efforts to control waterways. For instance, even the high levees on the Mississippi River in Louisiana could not handle flooding caused by Hurricane Katrina in 2005. This resulted in the complete inundation of the City of New Orleans, a costly disaster in terms of human life and property. Certain areas of the world are more prone to natural flooding than others due to being incised by a multitude of streams and rivers. Examples include the Eastern United States, Southeast Asia, and Western Europe, although any lowered land surface is prone to flooding. Education is important in understanding the dangers posed by floods, as is a healthy respect for the power of water.
- **Human conflict and war**—Whether it's crime or regional conflict, human aggression takes its toll of life, limb, and property through violence. As such, crime and war are both cultural and physical environmental hazards. Human beings have always been ambitious and competitive, which has led to conflict throughout history. On a local level, crime is a constant condition of human society regardless of where one lives. Whether it's because of economic disparity or psychological abnormalities such as personality disorders resulting in poor impulse control, crime damages human lives in many ways. The physical hazard that crime poses is in the form of physical violence, especially violence that involves firearms. While it is the role of local law enforcement to control crime through police forces, communication, education, and peaceful conflict resolution can lessen the effect of crime with local human environments. Protecting one's home and loved ones with security technology and even weaponry of some sort are options as well. Unfortunately, some regions of the planet are not well policed and are subject to increased amounts of gang violence. War is another form of violent hazard. It is a condition of humanity, in both history and, unfortunately, in modern times. Idealistic, political, and religious differences abound between groups of people, as does economic disparity. Often wars begin out of ignorance and misunderstanding. Sometimes war becomes inevitable as nations defend themselves against undue aggression, such as during World War II. The physical hazard brought about by war is similar to that of crime, physical violence through the use of weaponry, but on a much greater scale. Humanity has possessed weapons that could destroy civilization globally, killing billions, for over 50 years. As it stands today, regional conflict is responsible for millions of deaths annually. One would think that with humanity's current level of intellectual, scientific, and technological development we would be beyond the need for war, but that does not appear to be the case.
- **Landslides and slope failure**—This is the downslope movement of Earth materials, usually in hilly and mountainous areas. Hazardous landslides and other associated phenomenon are not ubiquitous across the Earth's surface, only in specific areas. High mountainous regions are subject to landslide by their very nature. Rocks making up the surfaces of high mountains become weathered and maybe experiencing catastrophic failure. These rocks can be very heavy and at high velocities may pose a hazard into

mountainous regions. These landslides can also block roadways in high mountains. This is of concern in areas like the Alps, the Himalayan Mountains, and the Rocky Mountains. Hilly regions draped in unstable, clay-heavy soil may experience slope failure of some sort. Slope failure is responsible for property damage and even the occasional loss of life in parts of the world. These are places like Northern California and the Norwegian shore. The best method of dealing with this hazard is to not build on unstable slopes.

- **Radioactivity**—This is the spontaneous decay of unstable isotopic elements into more stable daughter products, also known as radioactive decay. Radioactive decay occurs constantly in nature, for instance elements within igneous rocks such as potassium and even uranium decay. Even bananas contain a certain amount of isotopic potassium that undergoes radioactive decay, in a very small amount. A high level of radioactive decay from isotopes is potentially lethal to human beings. Low levels of continuous radioactive decay cause cancer. Thankfully, these levels of radiation are normally only found in military installations, scientific laboratories, and nuclear power plants. Nuclear power plants are operated globally to provide extra sources of energy for regional electric grids. While generally a safe and clean source of electricity (no air pollution or greenhouse gases are produced), there are two problems with nuclear power plants: disposal of nuclear waste and catastrophic plant failure. While nuclear waste disposal is generally handled well, catastrophic plant failure is the physical hazard in question. Throughout recent history several nuclear power plants have experienced total failure, the most severe being Chernobyl in 1986 in what was then the Ukrainian SSR of the Soviet Union. It took encasing the entire power plant in a concrete and steel "sarcophagus" to halt the spread of the damage. Before it was all over, clouds of radioactive elements had spread throughout much of Europe and Western Asia, ultimately causing an increase in cases of cancer throughout that vast area. Maintenance and safety measures of nuclear power plants is necessary to prevent catastrophic failure. Security is also necessary in areas of conflict since nuclear power plants are subject to sabotage. All of these efforts are expensive.
- **Solar radiation**—The Sun itself poses a physical hazard. Ultraviolet radiation from the Sun causes sunburns in humans, which can lead to increased cases of skin cancer. The Sun, as a black body, emits electromagnetic radiation across the entire spectrum of light, from gamma rays to visible light to radio waves. As such, the Sun can be quite dangerous for those at higher altitudes or people who fly in vehicles within the upper atmosphere or even space. Most of us do not travel to those heights, though, so the concern for most of us is with ultraviolet light as well as the heat effect of the Sun. During periods of high temperatures, especially coupled with high humidity, levels of **insolation** (solar heating) can become unsafe for many people. Heat stroke can be potentially deadly, especially to the elderly, so it is often best to go indoors to cooler, well-ventilated spaces or at least seek shade during extremely hot, sunny days. Unfortunately, many populations, especially in equatorial regions, are often subject to dangerous levels of insolation. These populations are often impoverished with few adequate solutions for excessive solar heat.
- **Transportation and vehicular collisions**—These are hazards posed by the use of motorized vehicles on land, in the water, and in the air. The danger posed by motorized vehicles is somewhat obvious: They are large containers, often composed of metal, containing human beings that are propelled across the surface of the Earth or in the air at high rates of speed. Modern roadways are hazardous by nature, especially in highly populated areas. The National Highway Traffic Safety Administration estimates that, in 2022, 42,795 people died in motor vehicle accidents (mostly cars and trucks) in the United States alone. Unfortunately, many people are not adequately educated or trained to operate a motor vehicle. Accidents occur due to carelessness, conflict, exhaustion, inebriation, or some other form of impairment. Often people find enjoyment in driving a car, but this may be somewhat mistaken. Operating an automobile or truck is a very

serious thing and should always be taken seriously. Boating is potentially hazardous on waterways, with incidences of accidents rising in recreational settings. In such a setting people operating boats are sometimes inebriated, which only heightens the hazard level. Most boats, especially large ships, are operated by trained professionals, which, in theory, reduces the incidence of accidents occurring. The same can be said for piloting airplanes. In most nations, piloting an aircraft is strictly regulated, and only the most well-trained individuals are allowed to pilot aircraft. Unfortunately, as with all transportation, human error does occur and with that, accidents.

- **Volcanoes**—These are ruptures in the Earth's surface allowing for the expulsion of lava, solid materials like ash, and gases, originating from a magma chamber beneath the surface of the Earth. As a physical environmental hazard, volcanoes pose varying degrees of threat due to the nature of the lava eruption and their location. Many areas of the planet simply have no nearby volcanoes, while other locations, like within the Pacific ring of fire, are replete with volcanoes. Volcanic eruptions involving low-viscosity basaltic lava are generally not explosive in nature and pose a limited threat to human life. Excessive eruptions of this kind can be the catalyst for fires, wild or otherwise, when the lava contacts combustible materials like buildings or trees. Basaltic lava is what occurs on the island of Hawaii. Volcanic eruptions involving granitic or intermediate lava pose greater threats. These types of eruptions expel lava containing larger amounts of silica, which make the products more viscous. As such, explosivity is a real possibility, as is the expulsion of dangerous volcanic ash. Volcanic ash contains silicates, which are harder than the metallic components of aircraft, resulting in the possibility of aircraft accidents due to engine damage from volcanic ash. Volcanic ash can also be entrained within volcanic gas and debris, resulting in dangerous, high-speed pyroclastic flows. Pyroclastic flows from nearby Mount Vesuvius resulted in the city of Pompeii becoming encased in volcanic ash, including its inhabitants, in 79 CE. Volcanic explosions due to high-silicate lava can produce large fragments like highly dangerous volcanic bombs, which can go soaring through the air. Volcanic dust from large eruptions can be expelled into the upper atmosphere, creating an aerosol effect, blocking sunlight across the surface of the Earth for months or even several years. The year 1816 is known as "the year without a summer" because when Mount Tambora in Indonesia exploded it sent up an enormous amount of volcanic dust, cooling the entire Earth for at least a year. The "volcanic winter" will likely occur when the gigantic supervolcano (large, powerful volcano) under Yellowstone National Park in the United States finally erupts. Like most natural environmental hazards, there is little we can do to prevent volcanic eruptions. Having respect for their power and having a willingness to evacuate the area during dangerous eruptions is all that we can do concerning this hazard.

Figure 7.5 Mount Pinatubo, a hazardous volcano, erupting

GROUND TRUTH

SARS-CoV-2

In 2020, nearly everyone on Earth had their lives changed, temporarily and unfortunately in a lot cases permanently. In December 2019, a new and severe respiratory illness, COVID-19, emerged in Wuhan, China. It was caused by what was soon identified as the SARS-CoV-2 virus. The illness caused by this virus spreads between human beings through respiratory droplets that contain the virus. It was quickly discovered that this virus is highly contagious, making every human being a potential vector. COVID-19 quickly spread throughout China. That country went under quarantine quickly, but it was not fast enough; COVID-19 had already spread to other countries. It appeared in the United States in January 2020, and by February the entire world was under lockdown.

While the origin of COVID-19 is a contentious subject, its spread did make it clear that new **pathogens** are capable of emerging and quickly spreading at any time. It also tested the public's resilience, as well as its lack of resilience. While an inordinate amount of people were dying of COVID-19, many other people resisted lockdown measures, spreading disinformation and misinformation about the virus. COVID-19 became politicized, and this likely led to more people dying. Within less than a year a vaccine had been synthesized, a testament to the ingenuity and work ethic of scientists and medical personnel. Tests for COVID-19 were formulated, and by May 11, 2023, the COVID-19 pandemic was declared to be officially over, but the hard lessons it brought us will linger for some time (Figure 7.6).

Figure 7.6 Testing centers were a common sight during the COVID-19 pandemic

Human Disease

Despite all of the aforementioned environmental hazards, disease is what kills most human beings. Of these diseases, the most prevalent are cardiovascular and infectious diseases. Environmental and genetic conditions are often the root cause of diseases in humans. Past adequate family planning, there is little one can do about genetic propensity, but one's response to the environment can be somewhat controlled. By altering one's relationship with their environment, one can put measures in place that can stave off dangerous diseases. For instance, regardless of where a person lives they can practice good hygiene or alter their eating habits. Cancer, diabetes, heart disease, respiratory and digestive disorders, and pathogenetic diseases can all be somewhat prevented or at least remediated through best practices, education, and self-control. Unfortunately, poverty also results in an increased incidence of disease. This is a public health issue.

Infectious Diseases

In 2019, infectious diseases killed 13.9 million people (keep in mind that this was prior to COVID). Nearly half of those deaths occured in developing countries that do not have the same level of health care, money, and sanitation as developed countries. Infectious disease tends to spread quickly in highly populated urban areas, such as many of the cities in equatorial developing countries. Multiple, uncounted vectors spread these diseases quickly, often unaware that they are infected themselves.

Certain diseases that were once thought wiped out have reemerged. These include cholera, hantavirus, malaria, and tuberculosis. The reasons for this are somewhat complex. One of the factors is global mobility. Due to the ability of people to get on an airplane and travel anywhere, disease vectors can spread rapidly. It took little time in January and February of 2020 for COVID to spread globally. Bacterial diseases are evolving resistances to antibiotics because antibiotics are often overused, exposing bacterium enough to build immunity over generations. When we talk about generations with bacteria we talk of, on average, between 12 and 24 hours, producing millions of individuals from thousands.

As climates change across the Earth, areas on the planet are experiencing pronounced warming. Disease-causing organisms and viruses often thrive in warm, humid conditions, leading to more severe outbreaks. The highest concentration of people on Earth occurs near the equator, especially in Southeast Asia and sub-Saharan Africa. These warmer, highly populated areas tend to see frequent outbreaks of very virulent diseases.

As infectious disease outbreaks occur, experts at agencies like the Centers for Disease Control (CDC), the National Institutes of Health (NIH), and the World Health Organization (WHO) investigate, predict, and study infectious diseases, implementing measures to halt their spread. This work takes coordination, expertise, effort, and time. The nature of this work is complicated and nuanced but ultimately helps save lives.

Toxicology

Toxicology is the discipline of investigating and studying poisonous and toxic agents and their effect on human health as well as on the health of other organisms. **Toxicity** is the degree of harm a **toxicant** can have. "The dose makes the poison," often credited to the 15th- and 16th-century Swiss physician Paracelsus is very apropos. Some toxicants, like cyanogen chloride, a type of cyanide, is very toxic in low doses, whereas other substances, like water itself, require extremely large amounts consumed in a small period of time to be toxic. This can also apply to biological toxins.

Environmental Toxins

Environmental toxicology is concerned with the discharge of toxic substances into the environment and is a cornerstone of the field known as **public health**. Public health is the scientific field of preventing disease, prolonging life spans, and promoting general health through communicating to communities and individuals, both public and private, as well as through public outreach and educational efforts to organizations within society. Although the setting for public health is within ecosystems, the emphasis is on human health, with animals used as test subjects. While unfortunate, using these test subjects has saved human lives, and studies using animals can act as indicators for future threats to public health.

Toxins Around Us

Many toxic chemical substances, natural and synthetic, are dispersed throughout the environment and can potentially pose health risks. For that matter, nearly all human beings carry traces of these chemicals within their bodies. Many natural waters also contain a multitude of dispersed toxic chemicals.

There are over 100,000 synthetic chemicals on the market today. Unfortunately, many of them have not been adequately tested, so the only way to find out whether a given chemical poses a threat to public health is when, and if, that chemical causes adverse effects.

Dichlorodiphenyltrichloroethane, otherwise known as DDT, is an insecticide that saw widespread use in the 1950s and 1960s. It was generally thought to be harmless to human beings. Pictures even exist of children playing in clouds of DDT. In 1962 the American biologist Rachel Carson published a book called *Silent Spring*. In the book she wrote about the potential harm that the widespread use of pesticides could have on the environment and human beings. As a result, many of these chemicals were investigated, including DDT. It was found that DDT had harmful and persistent effects in the environment and that it was a carcinogen. The chemical was banned in many places. Today DDT is still in use but only in limited applications.

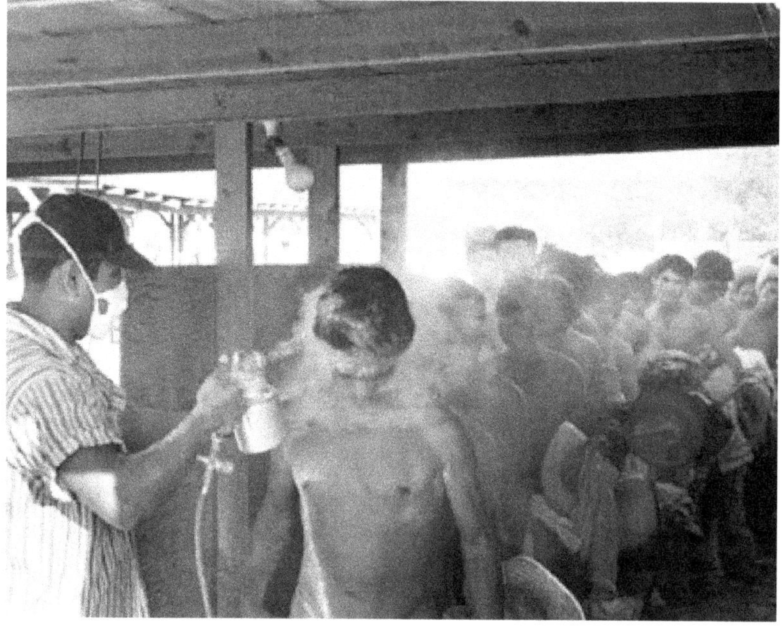

Figure 7.7 Laborers being fumigated with DDT in 1956

Airborne Toxins

Many chemicals used in industry and on the market are applied as sprays and fine mists, especially fungicides, herbicides, and pesticides. When applied outdoors, winds can carry these chemicals far from the area of application. Sometimes these chemicals are applied in such large amounts, and are so minute, that they can be transported in the upper atmosphere across the entire globe. **Pesticide drift** refers to the airborne transport of pesticides away from the site of application. The effect is so pronounced that pesticides have been found in bears in the Arctic, penguins in the Antarctic, and in humans who live in Greenland!

Soil Toxins and Contaminants

Chemicals often reside in soils for some amount of time before being transported elsewhere via runoff. Since soil systems are connected to ground- and surface-water systems, these chemicals leave the soil to either become concentrated or dispersed elsewhere. Some chemicals persist in soil, though, such as heavy metal elements and compounds. These metals have long been used in agriculture, industry, and transportation. The most prevalent metal contaminants are arsenic (As), cadmium (Cd), chromium (Cr), copper (Cu), lead (Pb), mercury (Hg), nickel (Ni), and zinc (Zn). Soil metal contaminants are often immobile, meaning that they are not readily transported out of the soil by runoff or other processes, nor do they easily react with other substances forming more mobile chemical compounds. As such, these heavy metals may cause health problems for human beings long after they are initially deposited.

Waterborne Toxins

Runoff transports chemical pollutants across the land into channels such as streams. Within these channels chemicals may concentrate, becoming more toxic. Chemicals can also become leachate in the soil, transported through infiltration deeper into the ground into the water table. Since groundwater is connected to surface water, this contaminated water may find its way into streams where organisms either absorb or drink it. These contaminants may also find their way into human drinking water. Often, high concentrations of waterborne contaminants will become apparent when aquatic organisms become sick or die. Aquatic organisms are sensitive to chemical changes in the water and can act as indicators of contamination.

Persistence and Breakdown Products

Toxins, like other chemicals, can degrade within a matter of time, becoming harmless. In other cases, they can persist, unaltered for decades or longer. These degradation rates depend on the individual properties of the chemical in question, including its reaction to light, moisture, and temperature.

Some chemicals in the environment are persistent, such as the aforementioned heavy metals like lead and mercury. These chemicals have great potential for ongoing harm since they are not easily eliminated from the environment, and subsequently from the potential to cause harm.

Other toxic chemicals form **breakdown products**, degrading into simpler compounds. This does not always mean that the resultant products will be less harmful. For instance, DDT can break down into DDE.

Bioaccumulation and Biomagnification

Often, toxicants in the environment accumulate in living tissue. Many of these toxicants are fat soluble and become stored in fatty tissues. As more toxicants are consumed in food, these toxicants accumulate and concentrate. This process is known as **bioaccumulation**. Meanwhile, as consumer organisms higher in the food pyramid eat these organisms that have been subjected to bioaccumulation, toxicants become concentrated in the tissues of upper level predators. This is a process known as **biomagnification**. Biomagnification can affect human beings who eat these organisms. For instance, mercury tends to accumulate and biomagnify in the tissues of commercially available tuna. Biomagnification has resulted in certain marine predatory birds, like pelicans, experiencing sharp population declines.

Investigating Toxins

Epidemiology is the public health field dedicated to the study of, and investigations into, the distribution of diseases within a given human population. Since testing on human subjects is both unethical and illegal, scientists often study the case histories of sick individuals as well as the autopsy reports of individuals who have succumbed to illnesses. These studies are often inadequate in understanding and combating the spread of diseases, especially with regard to future risks. Epidemiologists, in contrast, study comparisons between populations on a large scale. These studies of exposed and unexposed populations can last for years but often yield accurate predictions concerning populations at risk for diseases. Studies between exposed and unexposed people last for years.

Animals are also used as test subjects. While this saves human lives, there are ethical objections to using animals for test subjects. Thus, new forms of investigation are being attempted, such as using human cell cultures and bacteria to replace animal testing.

Dose-Response Analysis

Dose-response analysis is a technique for determining the toxicity of biological and chemical agents in a laboratory setting. It employs measuring how much of an effect a toxicant produces in different doses within populations. Unfortunately, on an ethical level, organisms, usually animal test subjects, need to be used for this technique. While organisms need not be killed to produce in some cases, they usually are in order to study particularly virulent toxicants.

The testing is done on discrete members of a population in incremental doses. A dose is a given amount of toxicant that the test animal receives. After the experiment the dose is plotted on the x-axis of a graph. The **response** is the effect of the incremental dose, usually the death of individuals, plotted on the y-axis, the functional axis. Thus, a **dose-response curve**, a graph of the dose-response analysis, is generated from the experimental data in order to understand the **virulence** of certain concentrations of a dose. The **threshold** is the amount of dose at which effects can be observed. The **LD-50** on a dose-response curve is the lethal dose for 50% of the test subjects. Many dose-response curves exhibit a sigmoidal-, or s-, shaped curve, although this is an artifact of the data. Like with all scientific analyses, it is the factual communication of data and results that matter.

Scientists can use the data from dose-response analysis and extrapolate it into possible results in human populations. Since animals in general, especially mammals (think laboratory mice), are not terribly different from human beings, this is reasonable for estimations. This sort of testing is often well regulated by governmental agencies.

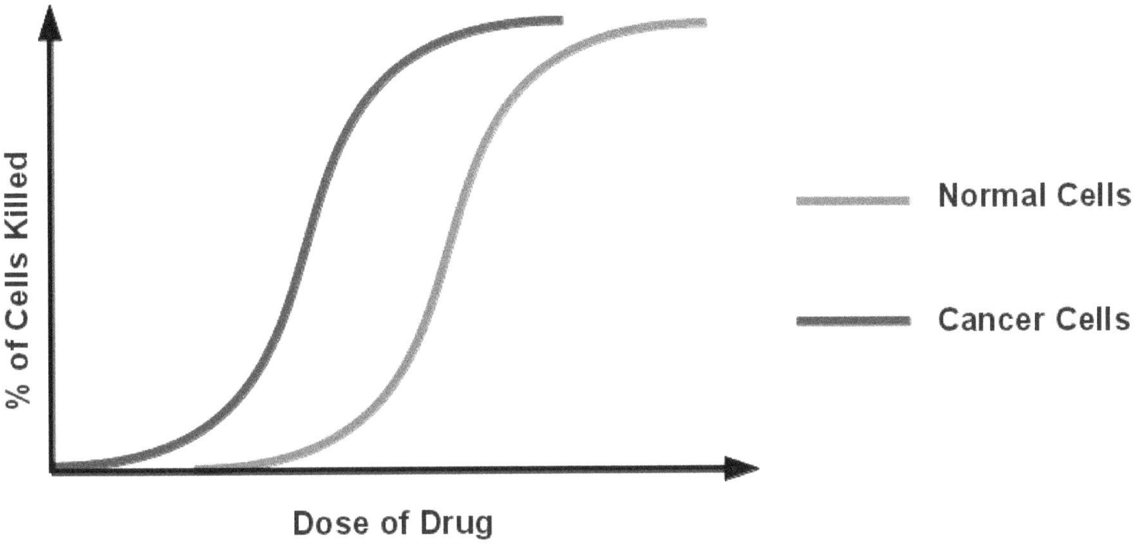

Figure 7.8 Dose response curves comparing the response of normal cells to cancerous cells undergoing chemotherapy, a treatment for cancer

Individual Response

Individuals within a population respond differently to certain environmental hazards, including biohazards and chemical toxicants. There are many factors within a given population, such differences in genetics and living environments, that may determine overall dose-response for individuals. People who already have predetermined health conditions can be more sensitive to certain toxicants. Sensitivity can also vary with regard to age, sex, and weight, all of which can influence an individual's response to toxicants. Extremely young and extremely old people tend to be more sensitive to toxicants than other people. Standards for toxicants in the environment in the United States are set by the EPA, although these standards are often not low enough to protect individuals, especially babies and infants.

Synergistic Effects

Nature is complex, and toxicants that enter the natural environment become part of that complexity. It is hard to predict the effects of two or more toxicants becoming mixed in the natural environment. Sometimes simply mixing something with water results in a new toxicant. For example, hydrogen sulfide that becomes airborne in water drops undergoes a chemical reaction to form sulfuric acid.

Other chemicals mixing in the environment may undergo **synergistic effects**. Synergistic effects occur when the impacts of mixed toxicants become amplified or even vastly different from the sum of their constituents. Mixed toxicants can even cancel out each other's effects. Since these mixtures are often unexpected, it is hard to predict the impact of mixed toxicants in the environment.

Risks and Regulation

Risk is the statistical probability of some harmful outcome resulting from some action. This essentially means that the introduction of some toxicant into the environment does not automatically produce a threat to public health; rather, it produces a probability that some negative effect will occur. Public health risks include these factors: the type of potential threat, the strength of the potential threat, the chance and frequency that an individual or population will encounter the potential threat, the amount of exposure to the potential threat, and an individual's sensitivity to the potential threat.

Risk management is the assessment of perceived risks, in combination with determinations and strategies to combat the threat from a perceived risk. Scientific assessments are made in combination with the needs and values of economic, political, and social entities. Developed nations as well as multinational groups have agencies that determine policy to manage public health risks.

Conclusion

A favorable environment leads to healthier inhabitants of that environment, including humans. There are many hazards in our environment that cannot be avoided. It would seem that more strident measures could be taken, across the globe, to make for a more livable environment for all organisms as free from contamination as possible. It seems that we could learn from past mistakes and become more conscious of public health, but only time will tell.

Discussion Questions

Directions: Review the chapter in order to completely and correctly respond to the questions and prompts.

Keep this in the back (or maybe even front) of your mind: Science is system of methods that allows us to understand not only our planet, and ourselves, but of the entirety of the universe. Answering the following questions correctly will lead to a more robust understanding of how to apply science to ENVS problems.

1. What sorts of environmental health hazards are in your living environment? Why is that?
2. What is a chemical hazard that might be found in your home? Would you be willing to part with it?
3. Name a natural physical hazard that is in your area. Why is it an environmental physical hazard?
4. Name a disease, thought to be eradicated, that has reemerged. Why has it reemerged?
5. How do pesticides used thousands of miles away end up in the Arctic?
6. What contaminants might be in the soil around your home? Why?
7. What sort of food do you eat that might be subject to biomagnification? Why? If you do not eat meat or fish, imagine what someone you know might eat.
8. Have you seen evidence of waterborne contamination is your area? What was it like?
9. What is something that does not seem toxic at all but might be toxic in very large doses? Why?
10. Exactly what is meant by "the dose makes the poison"?
11. What is one health risk that you might face each day? Why?

Bibliography

Geetha, G., and Dhanasekaran, S., 2018, *Environmental hazards and its impact on human health: International Review of Business and Economics*, vol. 1, p. 64–67.
Lindahl, J.F., and Grace, D., 2015, *The consequences of human actions on risks for infectious diseases: A review: Infectious Ecological Epidemiology*, vol. 5, p. 1–11.
Pope, A.M., Snyder, M.A., and Mood, L.H., 1995, *Nursing, Health, & the Environment*: Washington, DC, National Academy Press, 288 p.
Sivanpillai, R., ed., 2016, *Biological and Environmental Hazards, Risks, and Disasters*: Waltham, MA, Elsevier, 466 p.

Credits

Chapter 8

Soil and Agriculture

The Foundation for Civilization

Objectives

- Develop an understanding the role soil plays in daily life.
- Come to an understanding of how weathering and erosion contribute to soil formation.
- Learn about soil, soil development, soil horizons, soil orders, and the underlying complexity of natural soils.
- Learn the science behind soil.
- Understand how agriculture has shaped human civilization.
- Come to an understanding of how agricultural development impacts natural systems as well as human society.

KEY WORDS AND TERMS

Agriculture (n.): The practice of raising plants for crops and animals for livestock for human use and consumption.

Horticulture (n.): The art and science of growing plants.

Outcrops (n.): A rock formation exposed on the Earth's surface.

Soil (n.): Regolith on Earth's surface from weathered and eroded material, combined with air, water, and organic matter.

Consider This . . .

How many things have you used today that are completely dependent on soil? If you only said "I ate something," then you have used something dependent on the soil. Since eating is vitally dependent on soil, and we need to eat to survive, you have in a sublime way used something today on which our entire existence depends!

Introduction

While it may be easy to take for granted, soil is possibly the most important feature of the Earth, behind water, with regard to human survival. In fact, soil is generally important for most organisms that live on the land, especially photosynthetic plants, which form the base of the food pyramid. Soil itself is a complex natural system that supports an important, complex system required by our equally complex civilization, global agricultural production.

Origin of Soils

Soil is a complex, natural, intermixed system comprising of air, biological components, chemical species, disintegrated bedrock (regolith), and water that occupies the extreme outer surface of Earth's lithosphere. Soil is an interface with the atmosphere, biosphere, hydrosphere, and lithosphere. Soil is, obviously, essentially to the existence of life on Earth.

Figure 8.1 Humus: Organic material

Soils are products of the breakdown of Earth materials, as well as living tissue. Over time disintegrated rocks and minerals may develop *in situ*, mixing with decayed organic materials to form soil. Much of the organic material that becomes part of soil is known as **humus** (Figure 8.1). Humus is a carbon rich, somewhat cohesive mass of material consisting of decayed and partially decayed organic material such as the remains of animals, fungi, plants, or other living things.

Weathering, Mass Wasting, and Erosion

Three processes are responsible for soil development:

- **Weathering**—The biological, chemical, and physical breakdown of rocks and minerals into smaller, and at times soluble, components. Natural forces at or near Earth's surface constantly act upon rocks and minerals, breaking them down.
- **Mass wasting**—The movement of minerals, rocks, and soil downslope due to gravity.
- **Erosion**—The physical movement of minerals, rocks, and soil across the Earth's surface due to gravity, ice, water, or wind.

There are different types of weathering. Multiple weathering processes could act on any exposed rocks, minerals, or soil surfaces at any given time. Some processes are synergistically related to others. For instance, a tree's growing roots are considered biological weathering, although the roots, opening cracks in the rocks, are also a form of physical weathering. The tree may also secrete chemicals that contribute to the chemical weathering of the exposed rock. Once Earth materials are weathered, they may be carried downslope through erosion or mass wasting until the materials are deposited in an area that is not greatly subjected to erosion or mass wasting. At that point, the disintegrated material continues to weather in place (also known as *in situ*) while it accumulates in that area as soil.

- **Biological weathering**—Physical and biochemical weathering processes resulting from the organisms that live in the soil and their activities. Examples of biological weathering include the biochemical secretions of algae, bacteria, lichen, and assorted microorganisms that live on rocks and in the soil. These chemicals hasten the breakdown of rocks and minerals into soil. Animals such as earthworms and voles burrow through the soil, helping to redistribute, break down, and even, in the case of earthworms, reprocess the soil. Some animals, such as ants and rats, even burrow into solid rocks as agents of direct weathering. Roots from trees and other plants are sent into soil, and even rocks, breaking those rocks down over time. Finally, many species of fungi make extensive homes in the soil, at times spanning entire acres, sending out filaments that form a symbiotic bond with certain types of plant root systems. These symbiotic relationships form nutrient exchanges in the soil, breaking down and reprocessing soil and soil chemical species. Biological agents of soil weathering cause chemical and physical weathering through biological activities and processes.
- **Chemical weathering**—This type of weathering breaks down rocks and minerals on a molecular level. Water is the most important agent of chemical weathering and transports ions and molecules, as well as acts as a solvent for dissolved chemical species. There are several main types of chemical weathering:
    - **Dissolution**—This occurs when soil chemicals dissolve as solutes within a solvent, that solvent usually being water. These acid/base reactions occur when low pH waters dissolve higher pH minerals. This is the process for the weathering of carbonates.
    - **Hydrolysis**—When water directly reacts with any substance. This happens when the hydrogen ions in water replace other ions in solution. This drives the weathering of feldspar to clay, an important component of soil.
    - **Oxidation**—This happens when any outer shell valence electrons are stripped from any given atom. These electrons may or may not be replaced depending on the nature of the reaction. This drives the production of iron (Fe) oxides.
- **Physical (or mechanical) weathering**—When rocks, minerals, and other Earth materials are physically broken down into smaller pieces without any changes to the internal chemistry of these materials. There are several notable types of physical weathering process:
    - **Frost wedging**—Water ice expands, increasing in volume. This expansion forces open cracks in rocks, hastening weathering.
    - **Thermal expansion/contraction**—Most natural materials, including rocks and minerals, expand upon heating and contract in colder temperatures. This puts stress on rocks and minerals, ultimately causing them to weather and break apart.
    - **Unloading**—As materials are removed from a rock surface, rocks expand, causing stress, which in turn results in fracturing over time (Figure 8.2). These fractured materials are then subject to erosion, especially in elevated topography.

Figure 8.2 Unloading on Stone Mountain, Georgia

Accumulation and Deposition

Once sediments are dislodged from bedrock, or **outcrops** of bedrock, they may be moved from their position. Depending on the climate and topography, they can either remain near the source area for the sediments or become transported away from the source area. Through mass wasting in topographically rugged areas, such as mountains, sediments may accumulate in valleys in between high elevation. These remain near the source area. Through erosion, sediments may be transported far from the source area to completely different environments of deposition. Over time, often hundreds to thousands of years, this **regolith**, accumulations of loose rocky materials, will accrue along with humus and other components to form true soil. Furthermore, the fate of soil may be within that of a **depositional** or **sedimentary basin**, an extensive regional-scale depression in the Earth, resulting in diagenesis of soils and sediments and in the production of sedimentary rocks.

GROUND TRUTH

Soil, Civilization, and Agriculture

Agriculture is the practice of soil cultivation and land use for growing crops and raising livestock for human consumption. Human beings have probably been cultivating plants and raising animals since before the beginning of the current Holocene epoch, but large-scale agriculture for human civilization began at the beginning of the Holocene with the agricultural revolution (Figure 8.3).

Cultivating plants and raising livestock for large groups of people probably began due to favorable climatic conditions after the ice ages of the Pleistocene epoch ended. Favorable conditions allowed for large-scale farming for the burgeoning cities and nations of the early Pleistocene. The raising of staple grain crops over large extents of land began at varying times across the surface of the world (Figure 8.3). Wheat cultivation began in the Middle East sometime around 11,000 years ago at the very dawn of the Holocene epoch. Later on, rice cultivation began around 9,000 years ago in China, the cultivation of maize (corn) began in Central America around 4,000–5,000 years ago, and so on. The origins of agriculture were wholly dependent on favorable environments and the soil that came with it.

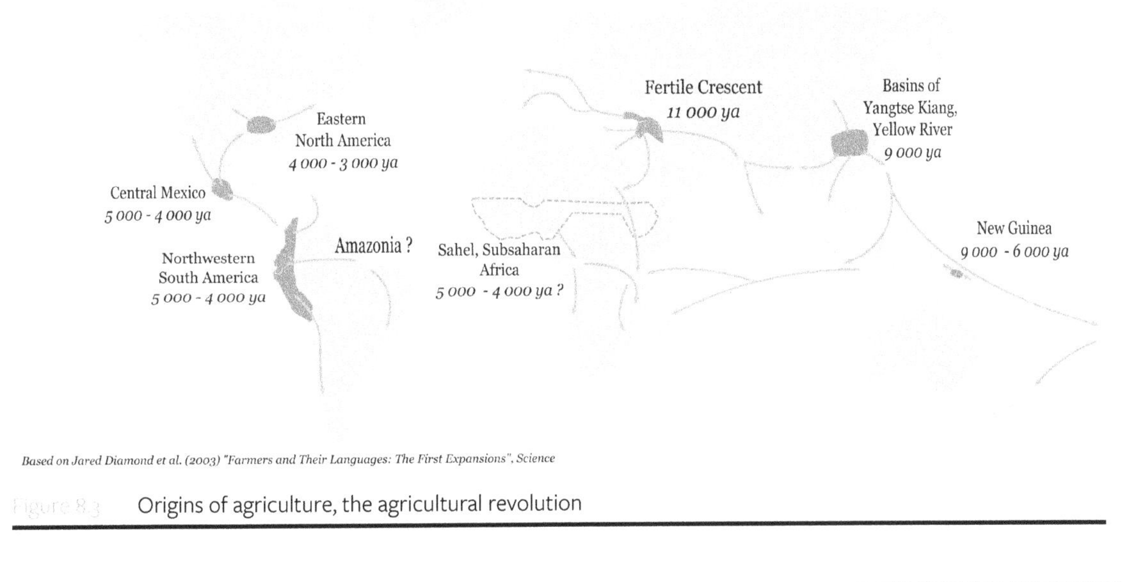

Figure 8.3 Origins of agriculture, the agricultural revolution

Soil Science

Soil science is an interdisciplinary scientific field directed at the study of soil, soil systems, and soil management for agriculture. Soil science can be divided into two main conceptual branches: **edaphology**, which is the study of soil related to its use, and **pedology**, which is the study of natural soils. Soil is complex, and scientists who study soil generally focus on a particular aspect of soil such as **agronomy**, the study of crop, soil, and water management for agriculture; **soil biology**, which is the study of soil as an ecosystem; **soil chemistry**, which is the study of soil's chemical characteristics; or **soil physics**, which is the study of soil's physical processes, usually related to pressure, temperature, and water content. Many different kinds of professionals work with soil, including archeologists, engineers, farmers, foresters, and people who work in regional planning.

The Five Soil Formation Factors

Soil, as a system, develops over time. Soil has distinct structures and components. Soil is not just random "dirt" that covers the Earth. For one to understand soil, one must also understand the major forcing elements behind the development of soil in natural, as well as at times anthropogenic, settings: the **five soil formation factors**:

- **Parent material**—The weathering of rock and mineral soil components from basic Earth materials. This is the most important soil formation factor. It is the basis for the chemical signatures within soil, including salts and metallic ions. To some extent parent material is also the basis for the inherent nutrient composition of soil. Depending on the bedrock from which soil is weathered from, it is also the factor in particle and particle size composition, such as clay and sand.
- **Climate**—This governs the rate of weathering through precipitation, wind, material properties over time, and gravity. Runoff, soil moisture, ice, and temperature-dependent weathering of Earth materials are all also climate dependent.
- **Biology**—Organisms and biological processes that take place within soil. Organisms such as earthworms and moles mix soil while moving and feeding within it. Organisms such as bacteria, earthworms, and fungi biochemically interact and influence soil composition.
- **Time**—Soil formation takes place constantly with different rates of formation occurring for different components. There are also differences between mature soils that developed over relatively long spans of time, and less mature, more recently developed soils.
- **Topography**—The overall shape of the Earth's surface, which in part governs the delivery of soil moisture, as well as the residence time of water. Slope also controls the accumulation of soil.

Soil Horizons

Soils are generally not randomly placed, unconsolidated masses of regolith and organic material. They are structured accumulations of Earth and organic materials that are placed on the surface of the Earth due to determinable physical, chemical, and biological factors.

Soils form **horizons**, layers of soil that occur for very specific reasons on and in the Earth. When examined vertically, these horizons comprise a **soil profile**, which is a cross section of all soil horizons present in local soils. Due to environmental and climatic factors, each of following six soil horizons does not necessarily need to be present in any given soil profile. Also, there are numerous subsets of each soil horizon, as detailed organizations like the **Natural Resources Conservation Service (NRCS)**, as well as other soil conservation services worldwide. Given the nature of soils and the methods of soil classification, there are over 70,000 different types of soil worldwide!

Here are the six basic soil horizons generally recognized by soil scientists, from the top down, uppermost to lowest (Figure 8.4). In more uncommon situations, there are a few other horizons that may be present in a soil profile, but again, these are rare. Note that the O, A, E, and B horizons are often referred to as the **solum**, or "true soil." These are the layers of soil that have undergone similar developmental processes. The solum is often the only section of soil that matters to people who grow plants since it is where plants will send their roots.

- **O**, also known as **humus**, or **litter**. The soil layer, at the top if present, that consists of organic material deposited by living things. This can include decaying plant matter, animal remains, feces, and any other deposits originating from organisms. Thick layers of litter often accumulate in tropical regions of high humidity, heat, and species diversity.
- **A**, also known as **topsoil**. This layer consists of weathered mineral components or regolith, with proportionally lesser amounts of organic materials. Topsoil is the basis for growth in agriculture and **horticulture**. Drainage occurs in topsoil and with that **leaching**. Leaching is drainage in which water, through its basic property of cohesion, transports minute particles (including clays), ions of salts, minute organic matter, and chemical species down to lower layers of the soil. Depending on the soil components, topsoil also retains a certain amount of water, which is used by plant life.
- **E**, also known as the **zone of eluviation**. This is not always present in soil but is often present in some well-drained topical or sub-tropical soils, as well as older or forest soils. The layer of eluviation is present when complete leaching of clays, chemical species including heavy metals, ions of salt, minerals, and organic matter occur in highly drained soils, often in environments that experience heavy precipitation. What is left behind, the eluviated layer, consists of mostly resistant, and inorganic, silica.
- **B**, also known as the **zone of accumulation**, or **subsoil**. At depths, usually a meter to a few meters, material leached from upper soil layers amasses in the zone of accumulation. This layer contains concentrations of clays, chemical species including heavy metals, ions of salt, minerals, and organic matter from the upper soil horizons.
- **C**, also called **parent material**. These are highly weathered rocks and minerals that may be found at depth or exposed. This provides the mineral source material for soil development.
- **R**, also called **bedrock**. This is the continuous mass of rock that comprises the outer surface of Earth's continental crust. It is the ultimate source for all soil parent material.

Soil Characteristics

There are other methods of soil classification in addition to that of the soil horizon system, or horizonation. Combined with soil horizon classification, these characteristics help to provide a complete understanding of a given soil.

Soil Color

Unlike characterizing minerals, color can tell you a lot about soils. Soils that are pale gray or white are often high in silica and heavily leached. These are generally infertile soils, with poor growing characteristics. On the other hand, soils that are black or dark brown generally indicate high levels of organic matter, which are generally fertile, making for good growing characteristics. Soils that are reddish in color often contain iron oxide, which can indicate acidic soils that are challenging to the growth of certain plants.

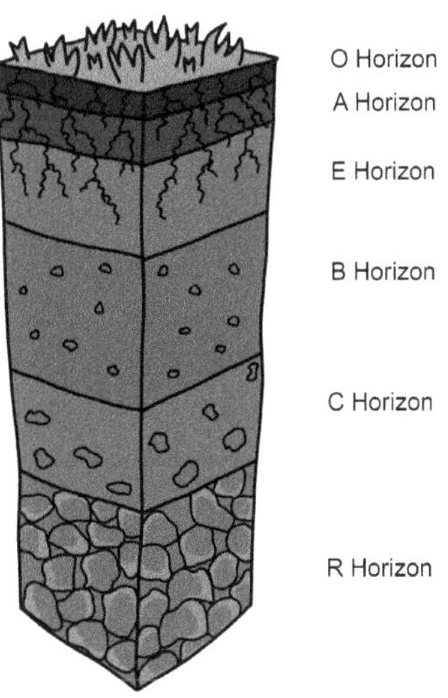

Figure 8.4 The six major soil horizons

Soil pH

Soil pH is an important characteristic concerning a soil's ability to support plant life. Certain plants grow well in soils with a pH lower than 7, acidic soils. Acidic soils are often found in relatively humid regions, such as the eastern part of the United States, because rainwater often contains carbonic acid. Plants, including cropland plants, that are tolerant of acidic soils include blueberries, collards, pecan trees, and certain pine trees such as loblolly pines.

Soils that have a higher pH, basic soils, often support many of the cropland plants that people eat regularly. These include asparagus, broccoli, lettuce, and others. These soils are often found in regions that generally experience lower rainfall levels, like the North American west.

Despite differences in pH plant support, the range for growth is fairly small, between a pH of 5 and 8, centered around neutrality, which is a pH of 7. Soil pH levels that are very low or very high generally do not support plants well, and in fact can be lethal to most plant life.

Soil Structure

Soil structure is controlled by the cohesiveness or "clumpiness" of soil, which is in turn often dependent on the clay content of a soil. Soil compactability is a function of soil structure. When dehydrated, highly compact soils can form bricklike structures, especially with high clay content. Highly compact soil may inhibit a plant's ability for water uptake through its root system. Repeated tilling or plowing of soil may result in compacted structures that resist soil drainage, which can effectively drown plants. Hard layers of compact soil from excessive plowing can result in plow pan or hardpan, which again acts as a barrier to water drainage. Often best management practices are required.

Soil Texture

Soil texture is an extremely important characteristic. It determines the transport and availability of both air and water within soil. Soil texture is simply the proportion of three different-sized particles in soil, mainly topsoil: sand, silt, and clay. A soil with a somewhat even mixture of the three particles, known as **loam**, tends to both allow for good drainage of soils in addition to retaining enough moisture to allow for plant growth.

The three soil texture sizes are sand, defined as rock or mineral particles that are between 2.0 and 0.05 mm in diameter; silt, defined as rock or mineral particles between 0.05 and 0.002 mm in diameter; and clay, often some form of clay minerals, defined as rock or mineral particles less than 0.002 mm in diameter. Incidentally, silt often originates from the flowing of lakes and streams, transporting organic matter onto floodplains, making it an important textural component. Sand allows for adequate transport, and clay allows for the retention of water, and valuable soil nutrients in the form of ions. Too much of one texture over the others makes for poor soil quality with regard to plant growth.

The soil texture triangle is a model that is used for analyzing the textural proportions of soil. It is a **ternary diagram**, a chart used to plot data in a system that consists of three components. In this case the three components are sand, silt, and clay particles. One uses sieves to separate the three soil texture sizes. Then each component is weighed. Each weight is divided by the total weight of all particle's then multiplied by 100 to yield the percentage of each texture. The textures are then plotted on the soil texture triangle to determine the overall soil texture type. This information can be used to determine the feasibility of land use or what sorts of amendments must be added to the soil to enable plant growth.

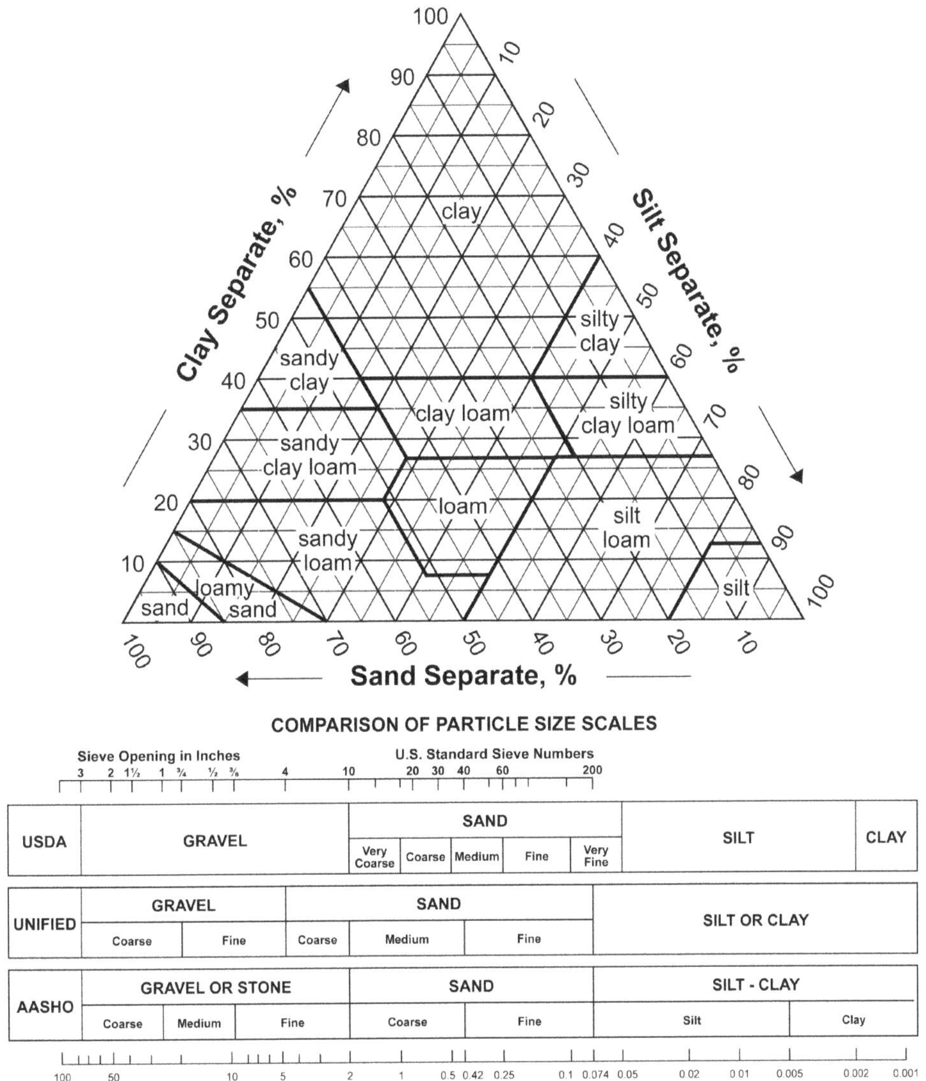

Figure 8.5 USDA soil texture triangle plus comparison of particle size scales

Cation Exchange Capacity

Cation exchange is a process that allows plants to obtain nutrients. Soils are often negatively charged, especially so for soils that contain large amounts of clay. Clay particles have a high surface area to volume ratio, and those surfaces tend to be negatively charged. Many of the nutrients that plants require for growth, such as calcium (Ca), potassium (K), and sodium (Na), are positively charged ions. These ions are often attached to the surface of negatively charged soil particles, particularly clay particles. Water content in the soil results in cation exchange between soil particles and the plant tissues, as water in transported from soil to plant life.

The ability of a soil to hold cations onto the surface of soil particles, which prevents them from leaching, thus increasing their availability to plants, is known as the **cation exchange capacity**. Knowing a soil's cation exchange capacity is useful for understanding soil fertility. This is greatest in fine soils, again soils that contain a lot of smaller diameter particles, generally soil textures that feature a lot of clays.

Soil Regions

There are so many different types of soil because of the vast regional differences in which soil is found. This regionality to soil is dependent on the differences of the five soil formation factors geographically. For instance, rain forests are areas of high species diversity and plant productivity, but the soils tend to be poor due to large rainfall amounts leading to excessive leaching in topsoil. In those kinds of ecosystems, plant productivity is tied to the nutrients in thick beds of humus. On the other hand, temperate grasslands have much lower levels of rainfall, leading to thicker, more fertile topsoil.

Another example of regionality is soil pH. Soils east of the Mississippi River in the United States are within regions of relatively high levels of precipitation. Precipitation in the United States' Eastern Seaboard is naturally acidic due to carbonic acid. This results in soil properties that support certain types of plants such as blueberry bushes, loblolly pines, and magnolia trees. Soils west of the Mississippi River are subject to less precipitation than the East Coast; in fact, in some cases they are found in areas of next to no rainfall such as parts of the desert Southwest. In addition to that, rainfall itself is less acidic west of the Mississippi River. This results in soils with higher pH levels. Plants that require basic soils thrive in these settings, such as asparagus, elm trees, and juniper trees.

The 12 Soil Orders

Regional differences in soils can be mapped. There are 12 major types of soils, with thousands of subtypes (Figure 8.6). These **12 soil orders** are part of the USDA system of **soil taxonomy**, a hierarchical method for classifying soils across not only the United States but the entire world. This system is somewhat like the older method for classifying living things. The major levels, from largest to smallest, are Order, Suborder, Great Group, Subgroup, Family, and Series. While there are 12 soil orders, there are over 4,000 soil series under those orders.

1. **Alfisols**—Moderately leached soils with a subsurface zone of clay accumulation. They tend to have basic pH levels and contain variable levels of aluminum (Al) and iron (Fe).
2. **Andisols**—Soils formed from volcanic ash and other eruptive volcanic products. They are generally younger than other soils.
3. **Aridisols**—These are soils that form in arid regions experiencing fewer than 90 consecutive days of moisture during the growing season, with very little leaching. These are often found in deserts. These soils often contain calcium carbonate ($CaCO_3$) yet very little organic. Soil development is slow.
4. **Entisols**—Recently formed soils with little to no development, and indistinct soil horizonation. These often form from river or beach sediments.
5. **Gelisols**—High latitude soils that have permafrost within 2 m of the surface.
6. **Histosols**—Soils that contain a large amount of organic materials high in carbon (C). Poorly drained. Often called peat.
7. **Inceptisols**—Recently produced soils originating from parent material. They feature soils horizons that are often weakly developed.
8. **Mollisols**—Soils with thick, well-developed A horizons formed in lower rainfall environment such as temperate grasslands. Tend to be higher in pH.

9. **Oxisols**—Intensely weathered soils that form in tropical and subtropical environments. Highly drained and contain high amounts of aluminum (Al) and iron (Fe) oxides, as well as clays. High levels of leaching lead to poor soil nutrients. Plants depend on thick layers of organic matter from decaying organisms for nutrients.
10. **Spodosols**—Forest soils with a subsurface accumulation of humus and metals. Tend to be acidic.
11. **Ultisols**—Strongly leached soils with a subsurface zone of clay accumulation. Found in tropical and subtropical regions. They tend to be acidic.
12. **Vertisols**—Clay-rich soils with high shrink/swell capacity. The shrinking and swelling of clay within the soil tends to form deep cracks in which surface materials settle, hence "inversion." Often forming from basaltic parent material, these soils tend to be basic.

Global Soil Regions

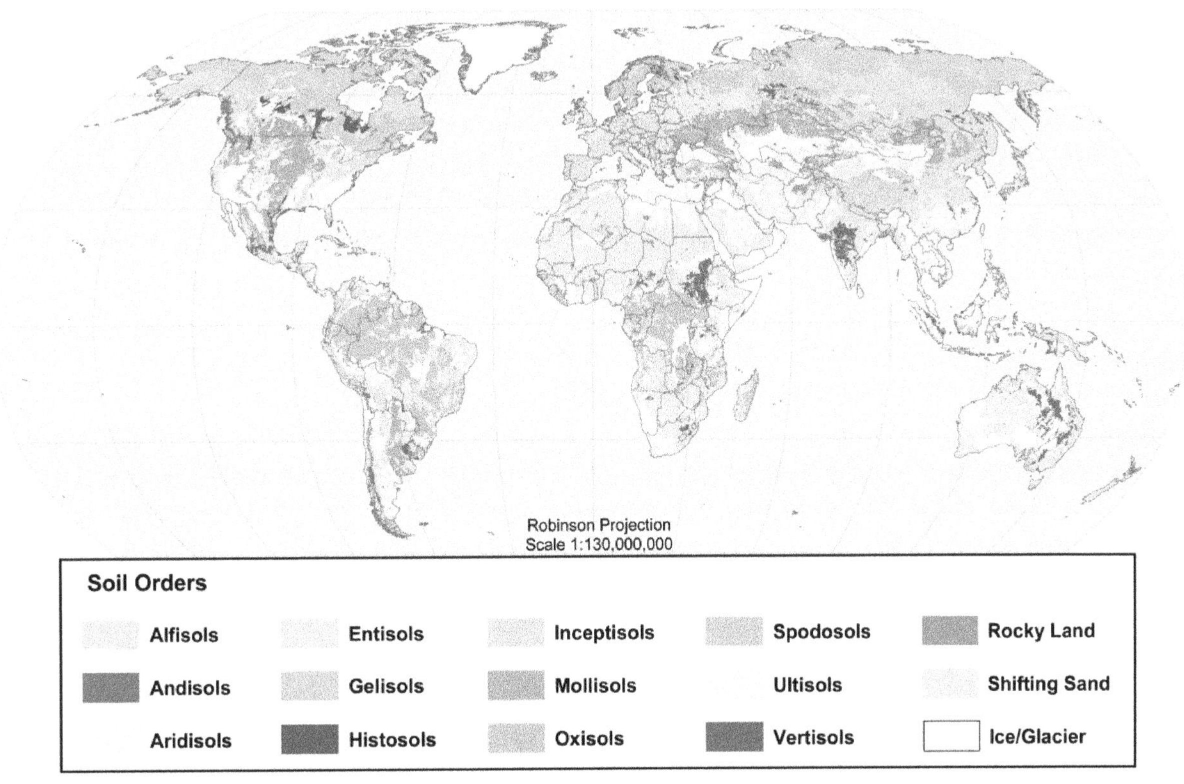

US Department of Agriculture
Natural Resources
Conservation Service

Soil Survey Division
World Soil Resources
soils.usda.gov/use/worldsoils

November 2005

Figure 8.6 NRCS global soil regions, including the 12 soil orders

Erosion and Surface Processes

The basis for soil is weathered parent material from rocks and minerals. Soil is generally a depositional material since it weathers from rocks and minerals in one place, only to be deposited in another. **Deposition** is the arrival of transported weathered material into a specific location. **Erosion** is the transport of weathered materials across the surface of the Earth, usually due to **fluvial** processes, stream processes, but also attributable to wind in arid environments. Fluvial erosion often transport weathered materials, rich in organic content, onto floodplains from floodplains. This helps enrich soils for farming. On the other hand, large-scale erosion contributes to positive feedback that damages ecosystems. Erosion is often the result of clear-cutting timber, excessive tilling, and overgrazing.

Erosion is driven in many different ways such as abrasion by the wind in arid regions; groundwater infiltration and springs; glacial processes; mass wasting events on steep slopes; overall rainfall and the resultant runoff; events that cause soil and bedrock erosion on the shores of lakes and rivers or in coastal areas; or terrestrial flooding. Erosion results is rate dependent and can accelerate if a driver of erosion is particularly excessive over time. Extremely powerful events can cause erosion, as can long-duration events such as lengthy storms. Erosion is generally more extensive on steep slopes. Erosion is also dependent on local climates.

Methods of Erosion

When plants are removed for development, such as construction or mining, erosion accelerates because plant root systems stabilize soil. Without stability, upper soil horizons are loose and subject to erosion. Loose soil can be removed from area due to winds, in arid areas, or by runoff in areas of relatively high precipitation. In areas of relatively high annual rainfall, soil erodes easily on steep slopes and the cut banks of highways. In the United States, soil conservation measures and best management practices led to an overall decline in erosion between 1982 and 2001, but erosion continues to a problem in the United States and elsewhere.

Types of Erosion

There are four basic types of erosion that are caused by rainfall:

- **Splash erosion**—This occurs when open soil is bombarded by raindrops. Splash erosion can lead to a cumulative effect in which a large amount of topsoil is moved, especially during high-intensity rainfall events. This loosened water-logged soil may then be carried downslope by surface runoff.
- **Sheet erosion**—This is erosion that can occur in areas of low rolling hills or over large-scale slopes. Loose soil is washed downslope over a large area onto other landforms or into streams. Sheet erosion can occur even on very gradual slopes.
- **Rill erosion**—In certain types of soil, runoff can form these small channels. As water continues to be transported by these rills, they can cut deeper into the soil. Around 0.3 m, deep rills become larger-scale erosive structures.
- **Gully erosion**—These form when rills become deeper than around 0.3 m. Gullies often form when runoff is excessively intense, cutting into subsoil. Gullies can be anywhere from 2 to 15 m deep. Gullies that are larger than 15 deep completely alter the geography of an ecosystem, forming badland or canyon structures.

Desertification

In more arid regions of the world, **desertification** is the result of large-scale erosion. Desertification is often defined as a loss of more than 10% plant productivity. Erosion driven by wind is a cause of desertification, as is climate change or weather pattern change, forest removal, loss of water resources, overgrazing, salinization, or soil compaction.

Desertification affects around 1/3 of the Earth's surface in over 100 countries and costs billions of dollars per year in natural resource loss.

The Dust Bowl in the United States during the early 20th century was a large-scale desertification event. During the late 19th and early 20th centuries settlers were farming extensively in the prairies of Colorado, Kansas, Oklahoma, New Mexico, and Texas. These farms loosened the topsoil in the region. When a drought occurred in this region, as well as a series of especially strong storms between 1934 and 1940, much of the topsoil was blown away in addition to a wide-scale loss of vegetation. Many farmers had to leave their land and ask for help from the government during the Dust Bowl. It occurred during the Great Depression in the United States, which was especially devastating for people in this area.

Soil Conservation

In order to safeguard against the possibility of overwhelming losses of soil resources, **soil conservation** measures are implemented. Soil conservation is a series of best management practices used to prevent the loss of soil resources. These best management practices are implemented worldwide in many regions. The **Natural Resources Conservation Service (NRCS)**, formerly known as the **Soil Conservation Service (SCS)**, of the USDA serves as a model of soil conservation.

The SCS was founded in 1935, in large part due to the Dust Bowl, to help farmers develop conservation plans for their farms. The mission of the SCS was, and continues to be as the NRCS, to work closely with landowners and farmers to assess their lands for possible issues, form strategic plans to ensure the viability of the land for use, whether for crops or rangeland, then implement those plans to assure stability and continued use.

The NRCS operates conservation districts that operate with federal authorization and funding but are run at the state level. Using this aid, local governments, with the help of landowners, set priorities for land use and reclamation. Since the renaming of the SCS to the NRCS in 1994, it has expanded its responsibilities to include pollution control and water management.

Methods of Soil Conservation in Agriculture

There are several, practical methods for conserving agricultural soil. These deal with methods of planting and tilling, as well as adjusting planting for methods on a seasonal basis:

- **Contour farming**—This is essentially plowing on hills at levels of equal elevation, plowing perpendicular to a hillslope prevents rill and gully erosion.
- **Crop rotation**—Alternating between different crops from season to season. Cover crops are used to protect soils in between the planting of main crops.
- **Intercropping**—Planting alternating crops side by side or in spatial arrangement. This helps to increase ground cover.
- **Reduced tillage**—Tilling the soil is minimized and a more "natural" growth method is used. The method known as no-till is even more minimal.
- **Shelterbelts**—Trees are planted on the edge of fields to reduce exposure to wind. These are also known as windbreaks.
- **Terracing**—Level platforms often with upturned edges are created on hillsides. These terraces act to hold water. These are often used in areas of rice cultivation such as in Arkansas in the United States and in Southeast Asia.

Challenges to Soil Conservation

There are quite a few challenges to soil conservation and soil loss with land use. One of the major solutions to soil loss due to erosion is to use plants and their root systems to hold in soil to slow down erosion. Erosion is often accelerated on slopes and stream banks, so it is a good idea to plant in such places.

Since **irrigation** artificially introduces water to land surfaces, it can be an agent of erosion. Irrigation is often used on arid, unproductive land in order to farm it. Extensive irrigation often leads to **waterlogging**, which is when the soil is overly saturated. This can suffocate plant root systems. Extensive irrigation can also cause the build-up of salt, **salinization**, in more arid areas when the water evaporates and ionic bonding occurs between the ions of salts.

Waterlogging can be reduced using **drip irrigation** in which water is targeted for individual plants. If other forms of irrigation are used, controlling the supply of water in order not to oversaturate the soil can be effective as well. For more arid regions, the use of water that contains reduced salts can be helpful in preventing salinization. Also, with regard to arid areas, plants that require relatively less water can be more effectively maintained for soil conservation.

Fertilizers are compounds that provide a boost to soil fertility. Inorganic fertilizers are often used on the enormous industrialized farms found across the globe. These inorganic fertilizers are composed of mined materials, or synthetic chemicals. They pose a problem to the environment because the chemicals with inorganic fertilizers contain nitrogen (N) and phosphorus (P) compounds that promote growth in all forms of plant life. This is good for the farm, but it might not be good elsewhere.

When N and P compounds are transported into rivers and streams via runoff, or groundwater, they can cause a runaway growth of algae and other plant life that can lead to **eutrophication**. Eutrophication is characterized by excess nutrients in bodies of water that cause plant life to dramatically bloom. These plants are consumed by heterotrophic microorganisms that deplete water bodies of oxygen, known as **anoxia**. Anoxic events in which oxygen levels drop below two parts-per-million (PPM) can cause widespread devastation to aquatic ecosystems, including killing off fish.

A method to reduce the possibility of eutrophication is to use organic fertilizers. These are fertilizers that are much like humus, or **compost**, the remains or waste products of organisms. These naturally deliver nutrients to plants in a much more balanced manner, causing the potential for over-fertilization to become greatly reduced. Of course, this type of farming takes extra effort on the part of the farmer and is often more expensive.

Both overgrazing and timber harvesting are challenging to soil conservation. Overgrazing occurs when herds of animals greatly reduce plant growth when they eat plants when grazing. This, coupled with soil compaction as the animals move around rangeland, provides serious soil conservation challenges. In fact, this is a leading cause of soil degradation worldwide. On the other hand, timber harvesting poses another type of challenge to soil conservation. Often timber is **clear-cut**, a practice in which all of the trees on timberland are cut down. This leads to exposed loose soil, which, especially on a slope, is prone to erosion. Modern timber harvesting methods include targeting trees for harvest, which prevents all of the trees from being cut down on timberland.

Agriculture and Agronomy

Agriculture is the practice of cultivating crops and raising livestock for human use. It is a vital practice for human civilization. Human beings, like all other animals, not only require food from plants, or animals, for survival but also utilize other plant and animal products for many other practical applications.

What Is Agronomy?

Agronomy is the multidisciplinary science of cropland plant growth utilizing meteorology, plant genetics, plant physiology, and soil science. The basis for the food pyramid in nature is photosynthetic plant growth. Human agriculture, despite its complexity, is not much different. Plants support agriculture, not only for human consumption, but also for the animals that we depend on and who depend on us.

Agricultural Production

Worldwide, food production is enormous, exceeding $4 trillion in total output in 2019. This ensures food production that currently exceeds population growth. Food production, crops, and livestock for the planet is industrialized through the use of biotechnology, crossbreeding, fertilizers, fossil fuels, fungicides, herbicides, irrigation, and pesticides. Farms are enormous in size, many of which exceed 100,000 acres, or 156.25 square miles. Many large farms, agricultural producers, across the world use high technology and automation to produce the most food possible annually. At the same time, starvation occurs. Around 10% of the global population does not have their food and nutrition needs met on a daily basis.

Food Security and Nutrition

Food security is the assurance that there is suitable and dependable food supplies available to all people at all times. This, of course, is not the case. This probably has little to do with supply and more to do with a number of other factors, such as regional conflicts, economic factors, national politics, and other factors. It is shameful but true. Many of these sorts of problems are likely due to one existential reason: "man's inhumanity to man."

Those who do not receive an adequate amount of food may experience different types of starvation. They may experience **undernourishment**, getting less than 90% of their daily caloric needs. In developing countries this is mainly due to economics. **Malnutrition** is a shortage of required nutrients, vitamins, and minerals on a daily basis. Often populations in hard-hit areas, regions experiencing conflict, may experience both malnutrition and undernourishment on a daily basis.

In the United States and other developed countries there is another type of problem, **overnutrition**. This is when a person receives more calories than their body requires. Much food available in the United States is higher in calories but disproportionally low in vital nutrients. As such, over 25% of adults in the United States can be classified as obese. Worldwide more than 300 million people are obese. Again, these problems with nutrition have less to do with food supply and healthy soil and more to do with other factors such as culture and economics. These factors also influence agricultural production.

Monocultures

Many of the crops grown using industrialized methods, or "factory farming," are **monocultures**. Monocultures are large expanses of land in which single crops are grown. Think corn in the United States, wheat in Central Asia, and rice in China. Growing crops in monocultures is extremely efficient and cost-effective. It also greatly increases agricultural outputs. At the same time, it greatly reduces an area's biodiversity. These monocultures are also susceptible to diseases and pests. If one spot within a monocultural farm becomes infected or infested, the organisms easily spread to the rest of the land.

Figure 8.7 Monoculture: Winter wheat

Monocultures also tend to narrow the human diet, which causes malnutrition yet are conversely responsible for excess calories per capita, malnutrition. This especially is especially true for cheap, easily available food that those who live below the poverty line consume. It also leads to the **food desert** effect. Food deserts in the United States are geographic areas that are a mile or more away from a grocery store or some other outlet that may provide wholesome foods that provide the nutrients that all human beings require. Human beings require a highly varied diet beyond the 15 or so crops, as well as the eight livestock species, that are supported by global monocultures. Anything less, in which groups of people suffer food insecurity, is tantamount to both environmental and social injustice.

Pests and Pesticides

A **pest** is any organism that damages crops. Pests can be anything from birds, bacteria, fungi, insects, or any other animal or animal-like organism that reduces the economic value of crops. **Weeds**, on the other hand, are unwanted plants that damage crops. Certain types of fungi are capable of damaging cash crops as well.

There are many chemicals available to combat pests, weeds, and fungal diseases. These take the form of **herbicides** to target weeds, **pesticides** that are used to combat pests (usually specific kinds of pests), **fungicides** that target specific fungi, and **insecticides** that target specific insects. Millions of tons of these chemicals are applied to agricultural land in the United States every year. Their use is expanding in developing countries.

There are several marked problems that arise when chemicals like herbicides and pesticides are used for large-scale agriculture. One of these is both air and water pollution. Some of the chemicals sprayed onto fields are composed of minute droplets, or dust-like particles. These chemicals may become entrained in winds and carried far away. Runoff from precipitation may also transport chemicals into waterways, potentially contaminating them.

Some pesticides may also become ineffective in doing what they were created to do. Some individual pests within a population may become genetically immune to a particular pesticide. These individuals may spread this immunity to their offspring, rendering the pesticide useless. This is especially effective with pests such as bacteria and certain insects that have short life spans. Thus, new generations breed rapidly. In response, chemical companies create more powerful chemicals to combat these pests in what has been called an **evolutionary arms race**. These toxins are then introduced into the environment. Rather than using chemicals, other biological agents may be used to combat pests, a process called **biocontrol**, but pests may evolve ways to combat biocontrol agents as well.

Careful techniques can be used to control pests, weeds, and other unwanted organisms using a suite of methods known as **integrated pest management (IPM)**. IPM takes some effort and management, but it can be effective. Rather than just utilize a small array of industrial chemicals, IPM can potentially utilize biocontrol, crop rotation, genetically modified crops, habitat alteration, manual pest removal, population monitoring, and even certain chemicals on a limited basis. IPM is scientific management and has been used successfully in many parts of the world. The EPA in the United States supports IPM and provides education and information on it.

The Role of Pollinators

Flowering plants, which constitute many agricultural crops, have a feature that has evolved to attract organisms that help them reproduce, the flower. Organisms that spread male sex cells to female sex cells in plants are called **pollinators**. These organisms include bats, hummingbirds, and insects, especially bees. Plant reproductive cells can be carried on winds, but pollinator organisms do much more of that work than the wind. Over 800 crop plants rely on insect pollinators, again particularly bees. In recent years honeybee populations plummeted, but these numbers have returned thanks to conservation efforts.

Genetically Modified Crops

In recent decades the use of **genetically modified (GM)** crops in agriculture has risen sharply. Genetically modified organisms are organisms that have been genetically altered through **genetic engineering**, the manipulation of genetic material. This is often done using **recombinant DNA**, the implantation of genetic material from one organism into a different organism. This results in transgenetic organisms, organisms that contain DNA from other species.

Genetic engineering of crops has known benefits, including disease resistance, higher yields, larger plants, and others. GM has also supported great economic returns for agricultural companies and growers. At the same time, GM is a fairly new technology with long-term results that are not completely understood. Critics note the possibility for negative impacts to ecosystems, as well as the possibility of damage done to native species. Pests could also become resistant to GM crops. At this time few dangers have manifested concerning GM, including the fears that GM could damage human health. This remains a controversial subject with a need for further research.

Crop Diversity

Since monocultures are vulnerable, it is important to preserve native plant diversity. This safeguards against crop failure. A lot of native diversity with crops has already dropped with the use of monocultures. For instance, wheat varieties in China have dropped from around 10,000 in 1949 to around 1,000 in the 1970s. One of the main reasons for monocultures is driven by economic markets. Monoculture crops are standardized, easy to produce, and easy to consume. Consumers tend to appreciate standardized, recognizable food as opposed to unfamiliar products with a more "natural" appearance.

Seed banks are entities and institutions that preserve seeds for future use, in addition to archiving seeds so that varieties are not lost. Seeds in seed banks are carefully preserved and even planted from time to time to ensure viability. The Svalbard Global Seed Vault on the Norwegian Island of Spitsbergen in the Svalbard Archipelago near the Arctic Circle is a prominent and successful seed bank (Figure 8.8). Currently there are over 1,100,000 seeds in the Svalbard Global Seed Vault. Regardless of the success of some seed banks, they are often underfunded and thus economically and operationally challenged.

Figure 8.8 Entrance to the Svalbard Global Seed Vault

Livestock

Livestock are animals that are raised for human consumption, whether it's for meat, labor, or some other use, such as wool from sheep. Although vegetarianism and veganism are viable life choices, humans evolved as omnivores, and many human beings still consume meat. Animals that are raised for meat and other purposes include cattle, chickens, pigs, quail, sheep, and turkeys. Animal products that people consume also include milk and eggs.

Global meat consumption and production has increased in recent years, likely a response to increased wealth, especially in the developed world. With this demand for meat, livestock producers have implemented new methods for raising animals. One of the major innovations is the use of **feedlots**. Feedlots are large warehouses designed to give livestock concentrated nutrients for meat production. This method is widely used for chicken and pork production; cattle farming remains mainly an open rangeland operation.

Feedlots provide certain benefits. Feedlots enable higher production rates, thus providing both consumer and economic benefits. Also, since animals are raised on smaller areas of land, impacts to open land are reduced. On the other hand, the use of feedlots come with challenges. It is nearly impossible for feedlots to not cause air and water pollution to varying degrees. The close quarters in which animals are raised in feedlots increases the chance and occurrence of disease outbreaks. In order to combat these diseases, animals are given few antibiotics and other preventative medications, which leads to poorer quality meat and nutrition issues. Finally, the often-brutal confinement of animals in feedlots raises some serious ethical questions.

Raising livestock is an intensive enterprise and impacts soil and water use. Raising crops to provide feed for animals, especially cattle, taxes cropland and soil. For every 1 kg of cattle meat, around 200 kg of feed is required. In addition, over 240 m2 is required to sustain one cow or bull, and beef requires around 750 kg of water to produce 1 kg of beef. Other types of livestock cause impacts of this sort, but beef requires the most resources. One possible solution to this dilemma is in raising free-range livestock that essentially live within an ecosystem.

Conclusion

Soil is a vital, required resource, not only for human health but for civilization itself. Natural soils have sustained terrestrial life for close to 450 My, so it makes sense that the foundation of human civilization was dependent on initial soil cultivation, which began thousands of years ago. Soil that supports crop- and rangeland continues to support human civilization, but with the current impacts on our planet's resources the challenge is come up with, and implement, sustainable methods for agriculture in the coming years.

Discussion Questions

Directions: Review the chapter in order to completely and correctly respond to the questions and prompts.

Keep this in the back (or maybe even front) of your mind: Science is system of methods that allows us to understand not only our planet, and ourselves, but of the entirety of the universe. Answering the following questions correctly will lead to a more robust understanding of how to apply science to ENVS problems.

1. What chemical weathering process results in the production of clay? What sorts of rocks or minerals are weathered to produce clays?
2. Why is regolith not technically soil?
3. What is the primary, initial soil formation factor? Why?
4. Why do clays, ions of salts, metals, and other soil constituents accumulate in the B horizon?
5. Classify the following topsoil from these textural components: 50% sand, 35% silt, and 15% clay.
6. Does rainfall contribute to carbonate dissolution in regions where limestone provides the parent material for soil? Why? What might be the products of carbonate dissolution?

7. Which of the 12 soil orders produces soils that are unstable and not suitable for large-scale plant growth? Why? There may be more than one.
8. What sort of human activity might hasten soil erosion? Why?
9. Name a common monoculture crop. Would it be possible to grow that crop in a more sustainable manner rather than as a monoculture? How?
10. Why is Spitzbergen Island a good location for the Svalbard Global Seed Vault?
11. What would be a good alternative to feedlots for raising livestock?

Bibliography

Fitzpatrick, S.D., Schroeder, P.A., and Endale D.M., 2015, *Creating Deep Soil Core Monoliths; Beyond the Solum: Southeastern Geology*, vol. 51, no. 2, p. 85–96.

Keys to Soil Taxonomy, 2014, 12th ed. United States Department of Agriculture, Natural Resources Conservation Service.

Withgott, J., and Laposata, M., 2019, *Environment: The Science Behind the Stories*, New York: Pearson, 7th ed., 768 p.

Credits

Fig. 8.1: Copyright © by Edafologia2.0 (CC BY-SA 4.0) at https://commons.wikimedia.org/wiki/File:Materia-organica-del-suelo.jpg.

Fig. 8.3: Copyright © by Ian Alexander (CC BY-SA 4.0) at https://commons.wikimedia.org/wiki/File:Centres_of_origin_and_spread_of_agriculture_labelled.svg.

Fig. 8.4: Copyright © by Lewi1224 (CC BY-SA 4.0) at https://commons.wikimedia.org/wiki/File:Soil_Horizons_Diagram.jpg.

Fig. 8.5: USDA Soil Survey Division Staff, "USDA Soil Texture Triangle Plus Comparison of Particle Size Scales," https://commons.wikimedia.org/wiki/File:SoilTextureTriangle.svg, 1993.

Fig. 8.6: USDA, "NRCS Global Soil Regions including The 12 Soil Orders," https://commons.wikimedia.org/wiki/File:Global_Soil_Orders_Map.jpg, 2009.

Fig. 8.7: Copyright © by Andy / Andrew Fogg (CC by 2.0) at https://commons.wikimedia.org/wiki/File:Monoculture_2_(140865977).jpg.

Fig. 8.8: Copyright © by Dag Terje Filip Endresen (CC by 3.0) at https://commons.wikimedia.org/wiki/File:Svalbard_Global_Seed_Vault_-_panoramio.jpg.

Chapter 9

Forestry and Land Management

Using Natural Resources Wisely

Objectives

- Come to an understanding of why forests are vital to life on Earth.
- Understand that forests are important natural resources providing a multitude of ecosystems services.
- Know that timber is an important resource that must be managed sustainably.
- Know that wildfires are part of nature as well as a significant hazard.
- Know that it is often necessary to set aside wilderness lands for a variety of reasons.

KEY WORDS AND TERMS

Forest (n.): A large area that is covered in trees and other plants, making up over 30% of Earth's land surface.

Timber (n.): Wood that is harvested by human beings and has a variety of uses, including construction, furniture, and paper pulp.

Park (n.): A large, natural public area that can used for recreation, research, or resources.

Consider This . . .

Even if you don't live in a forested area, you probably have come to depend on forests for something in your life. Whether it's paper to write on, furniture, or the framework of your home, forests provide these and many more benefits to not only human beings, but to life on Earth.

Introduction

Forests, large areas of land covered in trees, have existed in some form for close to 400 My (Figure 9.1). They have provided habitats for uncountable living creatures, ecosystems services since their development, and even energy resources such as wood for burning and coal throughout human history. They provide these and many more things to this day. Forests are also a place for recreation and contemplation of the beauty of nature.

Figure 9.1 Cairo, New York, site of the oldest known forest in the world with fossils dating back to 385 My.

Forestry Basics

Forestry is defined as the science and practice of the creation, conservation, and management of forests and associated natural resources. This also implies the scientific research of forests and other land resources, both wild and managed, for the benefit of humanity. Forestry is a multidisciplinary practice, including the economics, and policies behind forest and resource use, in addition to science. A **forester** is a professional who works within the field of forestry to some capacity.

Resource Management

Vital natural resources are often difficult to manage or are scarce. These resources are also often limited and variably renewable. **Resource management**, which is part of the discipline of forestry, is the practice of not depleting potentially renewable resources while harvesting them. Professionals who manage resources are influenced by economic, political, and social factors, with economics often being the main concern. The sound management of resources is also informed by science, which considers broader environmental concerns.

In a broad sense, resource management is necessary for all **natural resources**. Natural resources are constituents of Earth's systems that support life itself, in addition to supplying the needs of humans as well as human civilization. These vital services and ecosystems services include rangeland for livestock and animals in general; rocks and minerals for mined resources used for commerce, construction, economic value, industry, and technology; soils for agriculture, forests, and natural communities; wildlife and fisheries for biodiversity, game and non-game species, and marine species; and water, which supports life in general, including agriculture, water for drinking, and wildlife.

Forested Ecosystems

Forests are biological communities dominated by trees or other woody plants. Technically, forests are land surfaces that are 5 hectares, around 12.36 acres, in area with trees that are higher than 5 meters, around 16.4 feet, in height that maintain a continuous **canopy** of more than 10%. Canopy is the continuous coverage of leaves, or foliage, that is maintained in a forest.

Forests are also complex ecosystems housing great biodiversity. They even comprise entire biomes, although often discontinuously so. Forests cover around 30% of Earth's terrestrial surface, accounting for around 50% total plant productivity. Forests act as carbon sinks. Around 45% of the carbon associated with land is stored in forests.

Figure 9.2 Puzzlewood is an old-growth forest in the Forest of Dean, Gloucestershire, England

Important Resources

Forests provide many ecosystems services vital to Earth's biosphere in general. They provide habitats for many communities of organisms. Forests maintain air, soil, and water quality and generally play a large role in biogeochemical cycles. Forests are also economically important, providing wood for construction, fuel, general use, and paper production.

Primary and Secondary Growth

Most forests on Earth have been impacted by human use to some extent. Very few forests remain untouched by human activity, although there are still a few that remain what can only be called primeval. Many of these forests are in very remote areas of the world, although there are a few of these **primary forests** adjacent or relatively near human development. Primary, or old-growth, forests are those which have been uncut by humans. Forests that have been cut by human beings, then grown back to at least partial maturity, are **secondary forests**.

Forest Succession

Most forests have several stages of predictable growth, or succession, often following some sort of disturbance. **Primary succession** occurs when all or most of the vegetation has been stripped from a forest due to some major disturbance such as volcanism or intensive human development. **Pioneer species** such as lichen and grasses are the first to appear.

Secondary succession stems from the next stage of forest development or from a less devastating disturbance such as a flood or wildfire (Figure 9.3). This occurs with the development of shrubs and evergreens such as pines. This leads to a climax community of large deciduous trees such as oak and hickory.

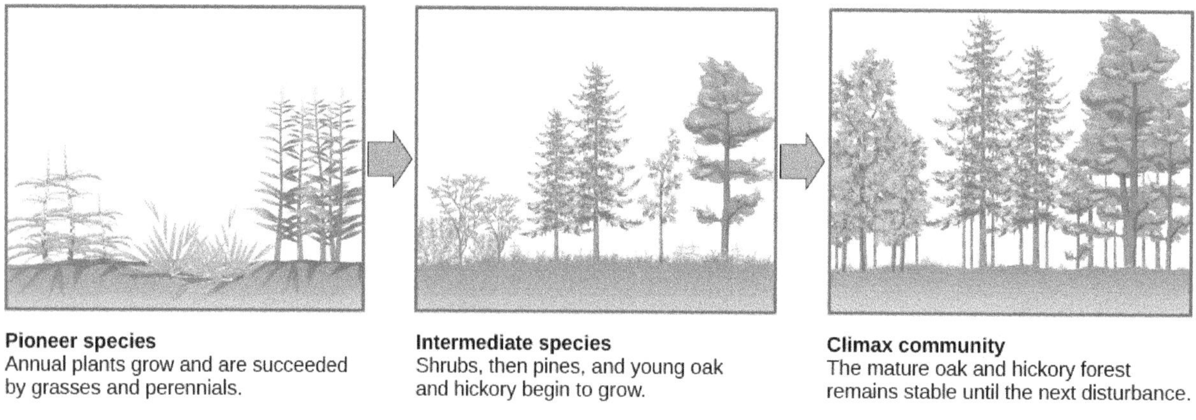

Figure 9.3 Secondary succession

Forest Management

A number of methods are used to manage forests and their ecosystems and resources. Many of these methods are applicable to the management of renewable resources on a broad scale. Overall, sustainability is key, yet at the same time economic needs must be met through resource harvesting.

Maximum Sustainable Yield

A **logistic growth curve** is a graphical model based on the logistic function (Figure 9.3). The model can be used for both populations and the abundance of a given resource such as fish, game, or **timber**. The curve plots the growth or availability of a resource, as the environment's carrying capacity for the resource is approached as the curve levels off. A logistic growth curve is s-shaped, showing a sharp increase in availability or population size until it ceases to grow at carrying capacity. Past 50% of the carrying capacity, growth slows down. This point on the curve marks **maximum sustainable yield**.

Maximum sustainable yield is the number of individuals of a population, trees for instance, that can be harvested without compromising the recovery of the population. The maximum sustainable yield varies from resource to resource, but it's generally around 50% of the carrying capacity. The logistic growth curve shows the most rapid growth occurring until maximum sustainable yield is reached. It is implied that rapid overharvesting of natural resources beyond 50% of the curve affects other species, with the potential to impact an entire ecosystem.

Ecosystem-Based Management

One of the issues with forest management, and natural resource management in general, is one of harvesting a product or resource in a way that does not greatly impact that resource's environment. Impacts to the overall environment often result in negative impacts on the availability of a natural resource. For instance, when all trees are stripped of an area that is prone to erosion, soil degradation can make it difficult to successfully replant trees to benefit from that area economically.

Good stewardship of natural lands comes with careful management that takes into consideration potential impacts on important ecosystems. These areas need to be understood at the ecosystems level, or even at scales larger than ecosystems, such as landscapes. Also, certain forested lands are so vital that protecting them from harvesting is necessary.

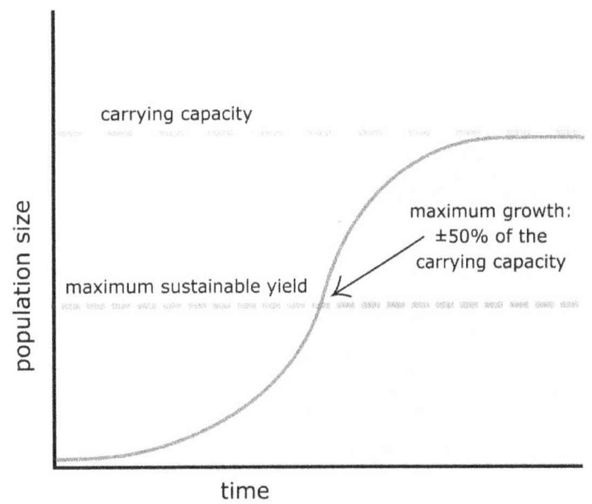

Figure 9.4 Basic logistic growth curve with maximum sustainable yield shown at 50% of a resource's carrying capacity

This type of **ecosystem-based management** is challenging for several reasons. Even when implementing conservation measures, economic benefit must be considered. Also, ecosystems, as well as larger systems, are complex with highly variable components, such as differential weather patterns, topography, and species abundance. Models of these ecosystems are difficult to create and assess because they are difficult to simplify. This is inherently a **scaling problem**. Environmental scaling problems exist due to the fact that modeling systems at smaller scales from data sets do not always transfer accurately to larger systems because natural systems, as well as impacts to natural systems, often do not scale linearly. Environmental systems (e.g., biogeochemical cycles, climate and weather patterns, magma emplacement, water movement on a large scale) exhibit highly variable behaviors at larger scales, at times much different from that which is quantifiable on a smaller scale and from local scale data sets.

Conflict and Management

Throughout history, wood products have been important to the development of human civilization. Prior to large-scale industrialization and synthetic products, wood was highly valued, along with mined products such as rocks and metals. Wood products continue to be highly valued. Recently, with concerns about environmental protection and sustainability, conflicts have arisen between loggers in the timber industry and those with environmental concerns seeking to halt the harvest of trees for various reasons, such as impacts to wildlife habitats. These conflicts have often been resolved in court and with compromises.

The *Northwest Forest Plan* of 1994 is a series of guidelines and federal policies governing land use in the Northwest United States that made compromises between the timber industry and preservationists regarding old-growth forests. While science based, the plan also made provisions for the economic harvest of trees by the timber industry. There were five main points to the **Northwest Forest Plan**:

1. Economic and human needs are of importance, and these needs are to be met.
2. The long-term health of forests, waterways, and wildlife is to be protected.
3. The focus is on sound scientific and ecologic management in accordance with responsible legal strategies and implementation.
4. Predictable and sustainable levels of timber and nontimber resources are to be met.
5. Federal agencies will work together.

Often, forested ecosystems are so complicated and vital, with high stakes for both ecological preservation as well as economic use, that more advanced methods of management are required. Systematically assessing and testing different methods through both management and science is known as **adaptive management**. Conditions with ecosystems are monitored and quantified, then adjustments are made accordingly. Throughout this process economic management is considered as part of the overall strategy for implementation. Adaptive management can be complex and time-consuming. It also potentially costs a lot of money. At the same time, depending on the ecosystems in question, and the budgets, these management strategies may prove to be highly beneficial for all concerned.

Timber Industry

Forested ecosystems provide a challenge for management on many levels. While they are valuable and vital natural resources on their own, harvesting is also important. Human beings rely on resources. People also need jobs and logging in that industry, the **timber industry**. The timber industry is concerned with the practice of forestry, logging and timber harvest, timber trade, and the economics of the lumber industry, primary forest production, wood products, and wood pulp for paper products. Around 2% of the labor force in the United States works in the timber industry.

Value of Forests

Forests and forested ecosystems are ultimately valuable to all living things on Earth, even at the level of the biosphere itself! Forests are some the richest ecosystems with regard to biodiversity, with structural complexity harboring said biodiversity. Forests also provide a variety of ecosystem services, including carbon storage, climate moderation, erosion prevention, oxygen release, reduction of flooding, slowing runoff, soil stability, and water purification.

Forests have always been invaluable to human civilization and human economies. The benefits of forests have and continue to be to provide fuel, shelter in a general sense and for building materials, transportation in the form of materials for boats, and paper products. They have helped society overall in achieving a high standard of living, as well as in the advancement of technology, either directly or indirectly.

Issues With Deforestation

Due to high demand, forests are often completely harvested from a given parcel of land, a practice known as **clear-cutting**. Clear-cutting has traditionally been the method used for harvesting timber on a large scale. Land in developing countries is clear-cut to boost the economies of those nations. It also quickly enables land to be obtained for both farming and settlement. This practice, when not managed, can lead to **deforestation**, which is the loss of forests. Deforestation is sometimes irreversible.

Deforestation results in the degradation of soil and the acceleration of soil erosion. It leads to the alteration of ecosystems. Once ecosystems are altered, deforestation can cause local species to decline. This can result in extirpation or even extinction. The loss of trees adds carbon to the air and can influence local temperatures. In extreme cases deforestation can lead to the decimation of civilizations. There is evidence that the original inhabitants of Rapa Nui (also known as Easter Island) engaged in a war that resulted in deforestation between the late 18th and 19th centuries. Nearly all of the inhabitants were ultimately killed, with the rest emigrating off of Easter Island. The descendants of the few survivors eventually returned to Rapa Nui, and now it is a viable, inhabited island.

Once timber is cleared from land it is no longer economically viable until trees are replanted, growing to maturity. This is currently the common practice, but in the past it was not always done in this manner. During the westward expansion of the United States in the late 18th and early 19th centuries, forests were clear-cut then essentially abandoned as timber companies moved westward. When something like this is done now, loggers can actually lose their jobs. In developing countries this can often lead to the exploitation of local peoples, with the economic benefits going to large multinational corporations.

Protective Measures

As timber was cleared and factories began operations throughout the 19th century it became clear to many that the environment they were living in was becoming devastated. This, in part, prompted the U.S. federal government to begin setting aside lands to be protected. This resulted in the formation of the **U.S. Forest Service** in 1905. The U.S. Forest Service still serves to protect around 8% of U.S. land area to be used for the greater good. This includes growing trees, protecting wildlife, and protecting watersheds, as well as ensuring timber supplies for logging, something that is allowed to a limited extent on U.S. Forest Service lands. In fact, several federal agencies manage land for public use in the United States in addition to the Forest Service such as the Bureau of Land Management, the Fish and Wildlife Service, and the National Park Service.

Timberland Management

Timber in the United States is harvested from both public and private lands. Certain areas of U.S. Forest lands are set aside for logging. U.S. Forest Service employees build logging roads and manage timber sales. Private companies log and sell the timber on the market.

Most timber in the United States is harvested on private lands, but millions of cubic feet of trees are harvested every year on public lands. Timber harvesting in the United States, and most of the developed world, is managed in accordance with maximum sustainable yield, which leads to stability in forests on public lands. Management of timber forests remains a complex practice reflecting ever-changing economic conditions.

Figure 9.5 Longleaf pine plantation in South Carolina, 1931

Over 50% of timber forests in the United States are on private lands, and much of that timber is on plantations. This is also the case for most of the world. **Plantation forests** are forests in which all trees are harvested at the same time, replaced with new seedlings after harvest. In this regard, entire forests are comprised of trees that are the same age (Figure 9.5). Emphasis is placed on fast-growing trees for maximum yield and maximum profit. Plantation forests are crops, not fully functioning forests within ecosystems. They are monocultures. Some plantations harvest trees through some form of selective management, maintaining a mix of species and ages of trees.

Timber Harvesting

There are a number of methods for harvesting timber, some of which have barely changed since their inception hundreds of years ago. Other methods have been developed recently to not only meet the demands of a large, modern population, but to implement sustainability as well as heightened levels of natural resource conservation.

Clear-Cutting

The predominant form of timber harvesting is clear-cutting, used for hundreds of years worldwide. Clear-cutting is simply cutting all of the trees down in a specific area or forest, and, in most cases, replanting seedlings to replace the harvested trees.

Clear-cutting is prevalent because it is the most cost-effective method of timber harvest. All trees in the area of interest are cut down and hauled away. While this method may mimic a natural disaster to some extent, it is devastating to local ecosystems and communities, even if the trees are in plantations. The disturbance to tree root systems and the heavy machinery cause soil erosion. The only recourse to clear-cutting is replanting seedlings to grow new trees, which is a common practice.

Other Methods

Other methods of timber harvesting are utilized other than clear-cutting:

- The **seed-tree method** is used to leave a few seed-bearing trees standing in a well-distributed manner, in order to reseed the forest. These trees are harvested at a later date.
- When **shelterwood cutting** is implemented, the forest is thinned, with some trees left behind to provide shade for seedlings.
- The **selection system** is used to harvest specific trees leaving others behind:
    - **Single-tree selection** is used to harvest individuals widely spaced apart.
    - **Group-tree selection** is used to harvest groups of trees but not an entire area or forest.

One must keep in mind that all methods of harvesting disturb habitats to some extent. They potentially alter forest structure, hasten runoff, increase erosion, increase flooding, and make landslides more likely on steeper slopes. To what extent these disturbances occur depends on the amount of harvesting and the level of management and best practices.

Salvage Logging

Salvage logging is the practice of removing dead trees after some sort of disturbance to the forest. It might seem like a responsible practice, but it often leads to some unwanted results.

Although basically dead timber, these trees still reduce erosion rates and protect the soil. Salvage logging also leaves behind flammable debris like branches and leaves, unless those are carefully removed. Standing dead trees, known as **snags**, provide habitats for many kinds of animals. In this way, removing snags can disrupt a forested ecosystem.

Figure 9.6 Cedar snags in the Saint Joe National Forest, Idaho, United States

Multiple-Use Policy

Public forests were originally intended as land set aside for recreation, ecosystem management, and wildlife preservation. Instead, they have become places in which logging is commonplace. This is despite public awareness of the importance of natural lands, as well as the public's propensity for outdoor recreational activity.

Forests in the United States are essentially still managed under **multiple-use policies,** which allow for practices like habitat preservation and mineral exploration to a certain extent, as well as ecosystem and wildlife protection. At the same time, despite critical outcry, public forests remain as primary places for timber production.

GROUND TRUTH

The National Forest Management Act of 1976

The National Forest Management Act (**NFMA**) is a federal law passed by the United States Congress in 1976. It is an amendment of the Forest and Rangeland Renewable Resources Planning Act of 1974 (RPA). The RPA was established in order to mandate federal reporting on the status of federal lands in the United States on a regular basis, either every five or ten years.

The NFMA established standards on how the U.S. Forest Service manages national forests, set requirements for the development of management in national forests and grasslands, and created directives for regular reporting on the status of renewable resources in all lands managed by the U.S. Forest Service.

Some of these guidelines included the following:

- Considering both economic and environmental factors in planning
- Requirements for species diversity
- Mandates for monitoring and ongoing research
- Allowing only sustainable harvests
- The protection of soils and wetlands
- Assessing all impacts before logging to protect resources

Wildfire Management

Wildfires

Wildfires, or "wildland fires," are defined as any uncontrolled burning of vegetation in the countryside or wilderness. Wildfires have characteristics that set them apart from other fires: extensive size, increased speed, or rate of spread, potential for a sudden change in direction, and the ability to jump gaps such as roads, rivers, and fire breaks. Wildfires are also characterized in terms of the cause of ignition, their physical properties such as speed of propagation, the combustible material present, and the effect of weather on the fire.

The Science Behind Wildfires

Fires occur due to combustion, or burning, which is a rapid, **exothermic** oxidation-reduction chemical reaction. **Oxidation-reduction** (redox) reactions are chemical reactions in which electrons are transferred between two chemical species. The chemical species in wildfires are generally some form of the chemicals within plant material, usually trees, which are high in carbon. This carbon readily reacts to heat and the other chemical species involved, oxygen. Exothermic reactions in turn release heat, analogous to energy, which often creates a positive feedback effect, allowing the fire to continue. A product of fire is a component known as **flame**: superheated gas that emits heat as well as light.

Essentially fire, including wildfires, require three components: fuel, heat, and oxygen. This is known as the **fire triangle** (Figure 9.7). Without all three of these components, a fire simply will not start. Removal of one or more of these components, variably depending on the nature of the fire, results in the fire being extinguished. Note that water, the most common method used to extinguish fires, reduces the heat of fires.

Figure 9.7 The fire triangle

Fires begin with **fuels**, combustible material such as wood or underbrush, that are subject to certain amounts of relatively intense heat as well as dehydration in order to undergo combustion. The converted products of combustion are often as combustible as the fuels themselves since these materials are converted to **char**. Char is the remains of woody plant material that is high in carbon content and thus is itself a fuel.

In a wilderness setting, either grassland or forest, the speed of a fire front is often dependent on wind velocity. In this manner, given their extent, wildfires are the cause for what is known as "fire weather." Wildfires in the wilderness require specific forest, topographic, and weather conditions. The greater the amount of water present in the environment, the less likely for a fire to start or begin, regardless of the origin.

Fire requires heat, usually in the form of some sort of spark like lightning or a carelessly tended campfire, applied to fuel to begin and a continuous application of the correct amount of heat to continue. Without fuel all but the most powerful fires will extinguish. At the same time, wildfires convert fuels, which are plant based, to char, which in itself is combustible. Fuels can be any dried flammable substance in addition to grasses or trees: old logs, dried underbrush, pine needles, and so on. Fire also requires oxygen since it is a redox reaction. A wildfire front, the moving line of combustion in a forest fire, acts to sustain a wildfire with enormous amounts of heat (often around 800°C) arriving before the fire itself in the forms of convection and thermal radiation.

Causes of Wildfires

There are quite a few potential causes for wildland fires. Wildfires originate through both human activities and natural processes. Some of the ways in which wildfires are caused include the following:

- Human induced
  - Accidental fires such as the careless tending of campfires and tossed matches
  - Arson, which is often based on financial gain through insurance fraud, but wildfires are also known to be acts of terrorism
  - Fires set for economic activities such as clearing for farming or equipment malfunction

- Naturally caused
    - Lightning strikes, which are a natural part of the life cycle of forests
    - Volcanic eruptions within specific geologic terranes
    - Rare meteor strikes due to the intense heat they cause
    - Coal-seam fires in which units of coal within sedimentary rock catch fire

Currently 10%–15% of wildfires are caused by natural conditions, and 85% to 90% of wildfires are due to some sort of human activity. Lately, many wildfires have become more intense and extensive, threatening entire populations of humans. These recent, intense wildfires are thought to be, in part, due to ongoing widescale climate change.

Role of Wildfires in Forested Ecosystems

Wildfires are a necessary, natural process in most forest ecosystems. Fire reduces over-accumulations of plant growth that can inhibit the growth of other species. The accumulations of undergrowth can also become fuels when these plants die and become dry, so the removal of such undergrowth is necessary. Wildfires can also stimulate the growth and the reproduction of plants, providing habitats for wildlife. For instance, the Western lodgepole pine requires the heat from fire to dislodge its seeds from pinecones.

Controlled Fires

The prescribed use of **controlled fires**, also known as controlled or prescribed burning, is a necessary measure in many managed forests (Figure 9.8). Controlled fires are purposefully set to eliminate overaccumulations of underbrush. The main reason for this is that when undergrowth proliferates, dies, and becomes dry it becomes a very combustible fuel. Over-accumulations of this fuel increase the chance of out-of-control wildfires. Other reasons to set controlled fires are to prepare sites for seeding and planting, disposal of logging debris, improved wildlife habitat, management of competing vegetation, disease control, and nutrient cycling. The U.S. Forest Service for the southern region has even published a manual on burning: "A Guide for Prescribed Fire in Southern Forests."

In 1910 Congress appropriated funds to aid in wildland fire suppression by the National Forest Service. Controlled fires are beneficial to a forest ecosystem, whereas unchecked wildfires evoke multiple costs: environmental losses, loss of life, and economic losses. The U.S. Forest Service spent $2.9 billion in fire suppression in 2022.

Wildfire Prediction and Prevention

The cost of unchecked wildfires would be considerably higher, especially in terms of environmental costs, were it not for prediction techniques. Several notable foresters, scientists, and engineers set standards for science-based wildfire research in the 20th century.

Wildfire Research

Pioneering forester Harry T. Gisborne developed a simple model known as the "fire danger meter" in 1931. In 1949, spurred on by the devastating Mann Gulch Fire near Helena, MT, Gisborne took it upon himself to collect a veritable mountain of data on forest fires and fire conditions. This effort cost him his life. The data he collected helped to begin a decades-long intensive research effort in wildfire prediction.

Figure 9.8 Controlled, or prescribed, burning in a forest

Ultimately, researcher Richard Rothermel and his team, 1971–1972, developed the Rothermel model, a tool so robust that it continues to be used today. Another researcher, Frank Albini, further refined Rothermel's model into the use of graphs known as "nomograms." These and other scientific modeling techniques help foresters fight fires to this day.

Remote Sensing

Today wildfire management is a problem on a global scale. Worldwide satellite remote sensing of fires continues to approach higher levels of accuracy and resolution. Despite this level of visibility there are inherent uncertainties in land-cover type, surface temperature, and wildfire areas. Trace gases and particulate species monitoring are of prime importance on the global scale.

This is why advanced satellite-based fire monitoring sensors such as **MODIS** were developed. MODIS (Moderate Resolution Imaging Spectroradiometer) is an imaging instrument aboard the Terra (EOS AM) and Aqua (EOS PM) satellites. MODIS scans the entire planet every one to two days in 36 discrete spectral bands. MODIS data is highly useful for inputs into a variety of real-time geographic, climatological, hydrological, and geological conditions, including the occurrence of wildfires.

Public Lands and Parks

Public lands, **parks**, and natural reserves are maintained to preserve pristine natural wildlands and to provide recreation for people. There are several reasons for establishing and maintaining public lands, parks, and natural reserves, including preserving outstanding natural features such as Niagara Falls; offering recreation to fishers, hikers, hunters, and tourists, something that also has economic value; the protection of utilitarian benefits such as drinking water; biodiversity protection; and many more.

National Parks and Reserves

National parks in the United States were established to protect public lands from resource extraction and development, while at the same time promoting recreation and research. The first national park in the United States, as well as the world, was Yellowstone National Park, which was established in 1872. It was followed by the Antiquities Act of 1906, in which the president of the United States can declare selected public lands as national monuments.

Wilderness Areas

Wilderness areas are lands that are off limits to any kind of development. Parts of these areas may be open to the public for reaction and other parts are open only to research. The goal is minimal impact upon the land. These lands are needed to ensure that people don't occupy and develop all natural areas. Wilderness areas have been established within the United States as well as other nations. They are overseen by the government agencies as well as private entities.

Certain wilderness areas are set aside as **biosphere reserves**. These are areas of land critical for the conservation of certain, often endangered, plants and animals.

Biosphere reserves are usually mapped out with three zones:

- A **transitional area** in the outermost zone in which managed activity is allowed, such as sustainable farming and limited human settlement
- A **buffer zone** within the transitional area allowing for minimal development, mostly supporting research and tourism
- A centrally located **core area** in which no development or disturbance is allowed

Conclusion

Over 30% of Earth's surface is covered by forests, or trees, in some manner. Since trees, or large tree-like plants, began to form forests 100's of millions of years ago, they have been integral to Earth's biosphere, atmospheric oxygen, and carbon budget. Forests preserve biodiversity and are invaluable for the resources and ecosystems services that they provide. Forested lands have also been the foundation for ecological awareness as well as for the development of parks, and other protective measures for our priceless wilderness lands, worldwide.

Discussion Questions

Directions: Review the chapter in order to completely and correctly respond to the questions and prompts.

Keep this in the back (or maybe even front) of your mind: Science is system of methods that allows us to understand not only our planet, and ourselves, but of the entirety of the universe. Answering the following questions correctly will lead to a more robust understanding of how to apply science to ENVS problems.

1. What is canopy, and why would it be important in forests?
2. What sorts of disturbances might interrupt forest succession?
3. Name another use of natural resources that might utilize maximum sustainable yield. Why is this?
4. Have you ever found value in, or enjoyed, a forest? Describe it.
5. How is plantation forestry damaging to a forest ecosystem and wildlife habitats?
6. What, to you, is the most sustainable method for harvesting timber? Why, and what sorts of wood might be harvested in that manner?
7. What is fire? How does it start?
8. We know some ways that wildfires are started. Think about how wildfires may become more intense. Explain.
9. How might instrumentation on satellites detect wildfires? Think about the range that is possible with sensing equipment.
10. Have you ever been to a national park or a large public park? What was that experience like?
11. What sorts of plants or animals might be protected in a biosphere reserve? Name a few.

Bibliography

Nema, S., 2020, *Forestry Science: Fundamentals and Terms*: New Delhi, NIPA, 308 p.
Withgott, J., and Laposata, M., 2019, *Environment: The Science Behind the Stories*: New York, Pearson, 7th ed., 768 p.

Credits

Fig. 9.1: Copyright © by Tyler A. McNeil (CC BY-SA 4.0) at https://commons.wikimedia.org/wiki/File:Town_of_Cairo_sign,_Cairo,_New_York.jpg.
Fig. 9.2: Copyright © by Reiner Tegtmeyer (CC BY-SA 2.0) at https://commons.wikimedia.org/wiki/File:Puzzlewood.jpg.
Fig. 9.3: Copyright © by CNX OpenStax (CC by 4.0) at https://commons.wikimedia.org/wiki/File:Figure_45_06_16.jpg.
Fig. 9.4: Copyright © by Woudloper (CC BY-SA 4.0) at https://commons.wikimedia.org/wiki/File:Logistic_growth_and_MSY_EN.svg.
Fig. 9.5: North Carolina Department of Conservation and Development, "Longleaf pine plantation in South Carolina, 1931," https://commons.wikimedia.org/wiki/File:Longleaf_Plantation_in_Eastern_S.C._Another_View_-_DPLA_-_48e7708c9a8c2d303892008cf05b323b.jpg, 1938.
Fig. 9.6: Copyright © by Ian Poellet (CC BY-SA 4.0) at https://commons.wikimedia.org/wiki/File:Cedar_Snags_3_-_St_Joe_NF_Idaho.jpg.
Fig. 9.7: National Park Service, "The Fire Triangle," https://commons.wikimedia.org/wiki/File:Image_of_a_fire_triangle;_the_three_sides_consist_of_oxygen,_heat,_and_fuel._(b1ddc4e5-7783-42e3-84fc-cadee85e5fce).png.
Fig. 9.8: Copyright © by USFWSmidwest (CC by 2.0) at https://commons.wikimedia.org/wiki/File:Prescribed_Burn_(6165766631).jpg.

Chapter 10

Freshwater Resources

The Crucial Compound

Objectives

- Come to an understanding of the magnitude in which we are dependent on water for our very existence.
- Understand the definition of freshwater and why it's important.
- Be able to calculate the extent of water's influence to some extent.
- Apply an understanding of the interconnectedness of Earth's water to situations.
- Understand why water conservation is so important.
- Know the benefits and drawbacks of the methods in which humans control water.
- Understand the ways water can become contaminated and why clean water is so important.

KEY WORDS AND TERMS

Downstream (n.): The direction in which water flows in a stream.

Freshwater (n.): Water that is potable and relatively free from contaminants and salt.

Groundwater (n.): Water that is held and transmitted beneath the surface of the Earth.

Hydrogeology (n.): The scientific study of groundwater.

Hydrology (n.): The scientific study of water in general.

Limnology (n.): The study of lakes.

Remediation (n.): The act of reversing or undoing environmental damage.

Upstream (n.): The direction from which water flows in a stream.

Consider This . . .

We are carbon and water. Water is where life began on Earth. Water is primordial, and we depend on it for our very existence. At the same time, the occurrence of water on and in Earth can be understood in order to ensure its continued support of life. We owe this precious substance that much.

Introduction

The fact that water exists in and on the Earth within all of the three basic phases of matter (solid, liquid, and gas) is a large part of why our planet is so unique. Water not only sits on Earth's surface, but it also flows across its surface, transporting particles and shaping the land. Ice also sits on Earth's surface, not only altering the rocks and soil but influencing the volume and temperature of the ocean. The ocean itself is fundamental to the existence of planet Earth. Water moves through Earth's atmosphere, influencing temperature and weather. Water is underground in aquifers, even in the chemical composition of many of the rocks and minerals within the Earth. Life itself would not exist on Earth without water. All of this occurs because of water's unique properties. Water is not only found on Earth; it is ubiquitous throughout the universe!

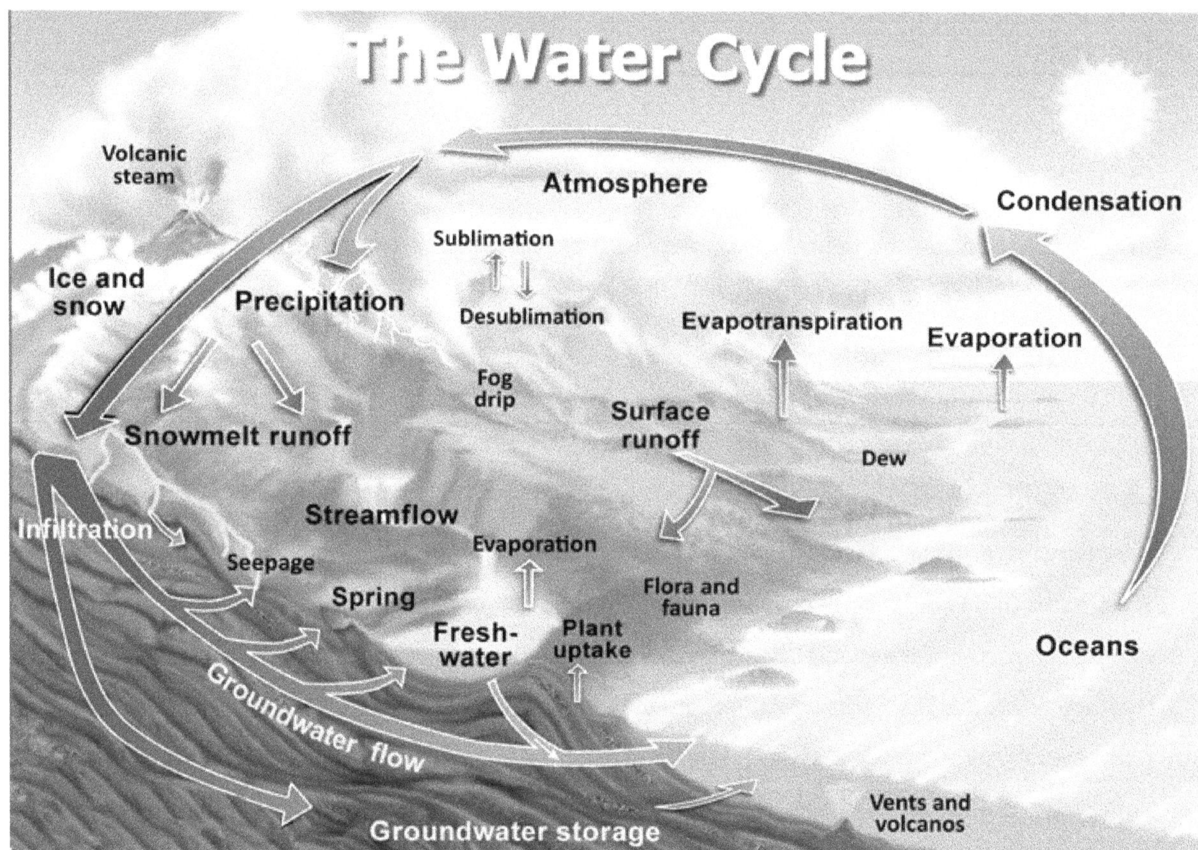

Figure 10.1 The hydrologic cycle; although there are many names for the flux of water on and in the Earth, these can be summarized to precipitation, infiltration, runoff, and evapotranspiration

The Hydrologic Cycle

The scientific study of water is called **hydrology,** and the branch of geology concerned with **groundwater** is called **hydrogeology**. Hydrology is a very broad-based discipline borrowing from meteorology, climatology, oceanography, geography, geology, glaciology, **limnology**, ecology, biology, agronomy, forestry, and other physical/chemical/biological sciences. Water resources development and water resources management are also integral to the functioning of developed and natural resources. Hydrological engineering is the branch of engineering that specifically deals with the machines and structures that enable modern human civilization to harness water in order to continue functioning.

The hydrologic cycle is a summary of the passage of Earth's water supply between different phases or sources, in flux:

- **Precipitation:** Condensed water vapor in the atmosphere that falls onto the ground as either a liquid (rain) or as a solid (freezing rain, hail, sleet, or snow)
- **Infiltration:** Precipitation that soaks into Earth materials like fractures in rocks, sediment, and soil
- **Runoff:** Precipitation that moves across the surface, or near surface, of the Earth
- **Evaporation:** Water that changes phase from liquid or solid to vapor on the surface of the Earth
- **Transpiration:** Water that is emitted, as a vapor, from plants on the surface of the Earth. For purposes of numerical calculations, evaporation and transpiration are often combined in the term **evapotranspiration**.

Sources of Earth's Water

Earth's water exists within a state of flux between various sources and sinks, as well as between the three phases of matter in which it occurs: solid, liquid, and gas. Numerous components of the Earth system contain Earth's water to varying extents. For instance, at any given time the ocean contains approximately 96.5% of all of Earth's water (Figure 10.2). The distribution of Earth's water, and its components, varies considerably across geographic space.

Since Earth's water is in a more-or-less constant state of flux, it is best described as a cycle: the hydrologic cycle, or water cycle. As in all cycles, there is no beginning or end to the hydrologic cycle, although the flux of all surface water is dependent on energy from the Sun. Groundwater, under Earth's surface, is driven mainly by Earth's gravitational field.

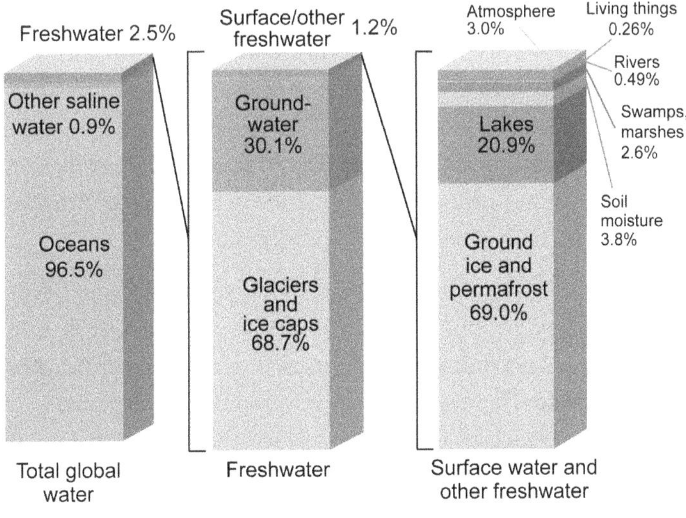

Figure 10.2 Earth's distribution of water

Earth's Surface Freshwater

Surface water includes all of Earth's streams and lakes, as well as soil water, wetlands, water vapor in the atmosphere, water that is within the bodies of organisms, and of course water that is in the ocean. Earth's **surface freshwater** does not include the ocean or other salty bodies of water. This water, generally free of the salts and other dissolved constituents, is also called **potable water**, which is another way of saying "clean" water that is generally drinkable water. **Non-potable water** is generally considered to be saltwater, which is not safe for human consumption, but also includes water that is contaminated from some source. Contaminated water does occur naturally, but a great amount of contaminated surface water comes from human activity in the form of water pollution.

Rivers and Streams

Streams are bodies of freshwater that flow across the surface of the Earth. There are many names for streams such as brooks, creeks, and **rivers**. These are all somewhat interchangeable, although rivers are usually considered the largest type of stream by volume. Rivers and streams flow according to the **hydrologic gradient**, or slope, within channels that form naturally in sediments, soil, or rock. The hydrologic gradient is the flow of water across the surface of the Earth with respect to the change in height over the length of flow. It can be summarized in this formula:

$$i = \frac{h_1 - h_2}{L}$$

where i is the hydrologic gradient and h is the **hydrologic head**. The hydrologic head is the position of water on or under the Earth, so in the formula $h1$ is the higher position and $h2$ is the lower position. L is simply the length at which the water travels. The hydrologic gradient is unitless, meaning it's the difference of position in height over length, in which length is canceled because it is the same term in the numerator and denominator of the fraction.

Streams and the lands surrounding streams are part of a **riparian** system; the word "riparian" refers to everything in the stream's system. There are several components to riparian systems, such as a **floodplain**. A floodplain is a relatively flat area of land adjacent to the banks of a stream. Floodplains in large river systems are often used for agriculture since periodic flooding delivers nutrient-rich silt to them.

The **cross section** of a river or stream is a measurement of water volume, stream bottom, and water velocity from one bank to another. Measuring a stream's cross section can be important with respect to water conservation. The lowest part of a stream cross section is the **thalweg**. The thalweg is often connected to flow in groundwater systems that feed into streams. Throughout history, a river or stream's thalweg has often been used to demarcate the natural boundary between nations.

Drainage Basins and Networks

Areas of land that contribute water, as runoff or some other flux, to streams are called **drainage basins**, or **watersheds** (Figure 10.3). A **drainage divide** is an elevated length that divides drainage basins. The drainage divide is the tipping point between adjacent drainage basins with respect to whether precipitation drains, usually as runoff, into one basin or another.

There are different scales for drainage basins. At its most basic, a drainage basin is a small stream from its **headwaters** to its **mouth**. The headwater(s) of a stream is its source. A stream's headwaters can be a point at which water collects from precipitation in an upland area, a spring, a lake, or even another stream. The mouth of a stream is the place at which it empties into another body of water, which can be another stream, a lake, a wetland, or the ocean.

Rivers form networks of drainage basins through **tributaries**. A tributary is when the mouth of a smaller stream drains into a larger one.

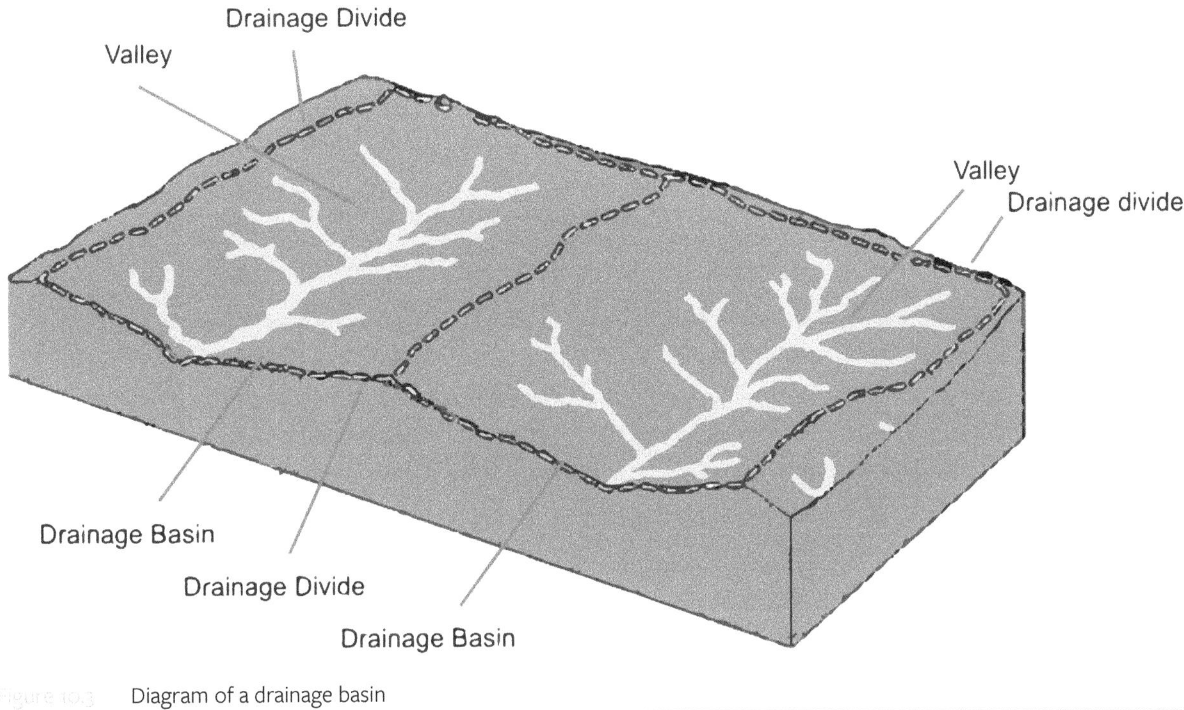

Figure 10.3 Diagram of a drainage basin

Streamflow

Streamflow is characterized primarily by velocity. Stream water usually flows in a predictable, uniform manner in that each component of the water moves smoothly past the layers with little mixing. This is called **laminar flow**. On the other hand, past a certain threshold factor, related to speed, water exhibits **turbulent flow**. Water under turbulent flow constantly changes direction, layers mixing in a seemingly random manner.

The overall velocity of a stream is mostly controlled by two factors: the **gradient**, or slope, of the stream, and **channel characteristics**. The slope of a stream is the change in elevation over distance. The steeper the gradient, the faster the stream, and past a certain speed water flow automatically turns turbulent. You can see this is mountain streams on steep slopes, or for a more extreme example, in waterfalls. Channel characteristics include a number of factors, such as channel or hydraulic roughness, shape of the channel, and size of the channel. Channel roughness is the frictional resistance of the streambed, which can include geological factors such as fractures in rocks and sediment, as well as vegetation lining the channel. The shape of the channel can alter water flow through changes in the channel's angle. Channel size, or width, can alter streamflow as well as increase or decrease the number of laminations or layers. The different channel characteristics all act together to affect streamflow.

Stream velocity is often called **discharge**. Discharge is the volume of water in a stream passing a certain point of reference over time. It is the velocity of a volume of water. Knowing the discharge is important for flood control and other matters of watershed management.

FYI

Manning's Equation and Reynold's Number

Manning's equation, or Manning's formula, governs open channel flow. It is named after Irish engineer Robert Manning who, in 1890, reformulated a previous set of equations derived by French engineer Philippe Gaspard Gauckler. Manning's equation applies to uniform, laminar flow. In general, it is applicable to all streams, including rivers:

$$V = \frac{k}{n} R_h^{\frac{2}{3}} S^{\frac{1}{2}}$$

where V is the stream velocity. The right side of the equation includes k, which is a conversion factor, 1.49 used for English, or Imperial, units, and 1 is used for SI, or metric. This conversion factor k is over n, which is the Gauckler–Manning coefficient, an empirically derived number based on a number of factors such as channel roughness and shape. The Gauckler–Manning coefficient is a small number, usually a decimal fraction rarely over 1. R_h is the hydraulic radius, which is the stream cross section, A, over the cross-sectional length of channel or the wetted perimeter, P. S is the slope of the stream.

Manning's equation does not apply to turbulent flow, such as rapids. At the point of turbulence, water becomes governed by a different, and much more complex, set of equations. The point at which turbulent flow takes over streamflow is governed by the Reynold's number, named after British engineer Osborne Reynolds, who popularized it from a formula originally derived by the influential Irish physicist George Stokes.

The Reynold's number is somewhat specific in describing the transition from laminar to turbulent flow based on its magnitude. Reynold's numbers below 2,000 indicate laminar flow, 2,000-4,000 is transitional with some elements of both laminar and turbulent flow, and Reynold's numbers above 4,000 represent turbulent flow:

$$R_e = \frac{\rho u L}{\mu}$$

where R_e is the Reynold's number, ρ is the density of the fluid (the fluid being water, which in SI is 1 g/cm_3), u is the streamflow speed, L is the length of flow being measured, and μ is the dynamic viscosity, which is the resistance to movement of the fluid.

Stream Profile

A **stream profile** is the longitudinal, cross-sectional view of a stream as viewed from the headwaters, or source, to the stream's mouth. Streams of any size can be characterized in this manner, from the largest to the smallest (Figure 10.4).

A stream's profile changes as it courses downslope. The headwaters of a stream are usually within some upland area, and the mouth of a stream is usually within a lowland area. This transition from high to low gradient controls a stream's channel characteristics and its velocity. These changes are factors that affect a stream's profile. Stream channel roughness and slope decrease from **upstream** to **downstream**. Channel size, discharge, and flow velocity all increase from upstream to downstream.

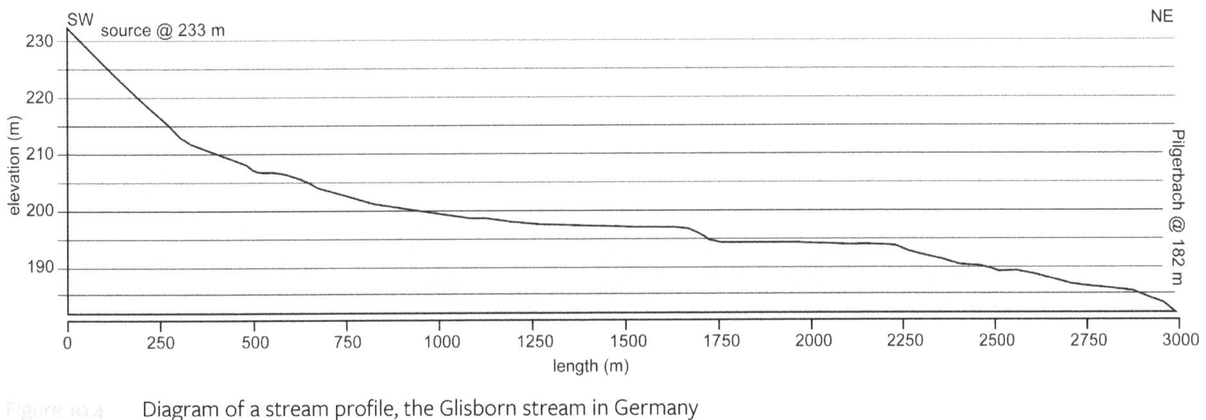

Figure 10.4 Diagram of a stream profile, the Glisborn stream in Germany

Sediment Transport

Streamflow constantly affects materials in the stream's channel as well as along the streambank. This action of flowing water dislodges particles and sediments transporting them downstream in some manner. In turn, runoff from the land, over the floodplain and into the stream, transports materials as well. Much of this transported material is in the form of chemicals, such as the dissolved ions of salts and metals.

Stream erosion loosens and lifts particles through **abrasion** and **dissolution**. Abrasion occurs when particles and sediments transported by a stream collide with rocks, sand, or other sediments, dislodging them or wearing them away. Dissolution refers to minute stream constituents on the microscopic or molecular scale, which have been solutes within the solvent, water. Smaller sediments such as clays and chemical species are transported far downstream, whereas larger sediments require faster streams for transport.

Waterfalls are rivers or streams that descend over a body of rock in an upland area. Waterfalls eventually erode rocks down to the lower stream, thus ceasing to exist. The famous Niagara Falls, in upstate New York, will eventually end as the falls completely erode the rock it courses over.

Figure 10.5 Oxbow lakes in Alaska

Oxbows are erosional features that form in rivers and streams coursing through very flat regions. Oxbows are extreme bends in a stream as they meander through soil and sediment of low gradient. Ultimately erosion will cut through the meanders, causing the stream to form a neck cut-off and causing the stream to flow straighter than before. The isolated, u-shaped body of water that is left behind is an **oxbow lake**.

Erosion works on streams and their watersheds slowly over time, but erosion and transport can occur quite suddenly as well. A large volume of stream sediments are dislodged from land surfaces due to storms. These storms can also alter the geometry of the stream bank and channel.

STREAM LOAD AND DEPOSITION

Transported material within a stream is called the stream's **load**. There are three types of load: **bed load**, which is the transport of larger sediments that rarely leave the streambed; the chemicals in solution are the **dissolved load**; and smaller particles like silt and clay are the **suspended load**. A stream's **capacity** is the maximum load that a stream can transport. Capacity is dependent on a number of factors such as stream size and velocity. In this regard, a large, fast-moving stream will have a high capacity. Stream **competence** is related to capacity. Competence is the maximum particle size that a stream can transport and is directly dependent on velocity.

When the velocity of a stream drops, sediments fall to the streambed or along the sides of the stream as stream competence drops. Sediment deposition is based on the particular competence of a stream at the time of deposition; thus, stream sediments tend to be well sorted, meaning sediment particles are all of similar size. Stream sediments that are deposited within or near stream channels are known as **alluvium**.

Sediment deposition leads to a number of unique depositional forms. Channel deposits include **bars**, which are deposited within the channel and are composed of sediments or gravels. **Deltas** are large areas of sediment deposition that occur when velocity drops to near zero in a very flat area, usually at the mouth of a river at the ocean. Deltas consist of many slow-moving branches of a stream meandering around the mounds of deposited sediments. Deltas have been important throughout history as places of agricultural abundance due to the high levels of silt and organic matter that are deposited. In this way, deltas are also carbon sinks. Braided streams are much like delta, but they tend occur in upland areas where fast-moving streams flow over rocky basement, diverting the stream into multiple channels over some distance.

Streams deposit sediments along floodplains as well. In areas of low stream velocity these sediments can accumulate to form deep channels. **Natural levees** form in streams parallel to the channel through successive flooding over time. **Back swamps** form adjacent to streams cut off from streams by natural levees or some other landform. Periodic flooding, as well as high water tables, contribute water to back swamps. **Yazoo tributaries** are streams that are limited in length that form parallel to the main stream due to periodic floods.

Alluvial fans are depositional landforms that are generally found in arid areas. They develop where a high-gradient stream discharges through a narrow valley, usually controlled by some sort of geological feature made out of rock such as a canyon. These **unconsolidated**, meaning non-homogenous, deposits slope outward down into a valley in a broad arc.

Stream Network Patterns

On a large scale, as seen on maps, stream networks form interconnected patterns of drainage. The most common of these stream networks are the following:

- **Dendritic patterns**—These patterns appear like neural networks on maps and develop on top of highly uniform bedrock.
- **Radial patterns**—These develop on isolated, singular landforms like volcanoes.
- **Rectangular patterns**—Stream networks develop this pattern on highly jointed, or uniformly weathered, bedrock.
- **Trellis patterns**—These develop on bedrock with different rates of weathering.

Lakes

A lake is a large body of open standing water, usually freshwater, although some lakes are salty. Ponds are smaller bodies of water, although the use of the words "lake" and "pond" are somewhat interchangeable. Some lakes can be classified as **inland seas**, exceptionally large lakes that are sometimes salty. These inland seas are expansive, containing biota adapted to conditions associated with large, open bodies of water. Unlike streams, lakes and ponds are not composed of water that flows in generally one direction. Water flow varies considerably in lakes. Lakes and ponds are often connected to streams, but there are exceptions.

Lake Zones

Lakes have distinct zones from the shores of a lake to its depths. Each of these zones feature their own unique habitats for flora and fauna. Note, this zonation is similar to that in the ocean:

- **Littoral zone**—This region consists of the circumference around a lake, including streams that may empty into the lake and adjacent wetlands.
- **Benthic zone**—This is the bottom of a lake and includes the habitats of many different organisms, including a multitude of invertebrates such as crayfish and mussels.
- **Limnetic zone**—This is the open portion of the lake's water in which sunlight penetrates, allowing for photosynthesis.
- **Profundal zone**—This is the bottom of deep lakes in which sunlight does not penetrate; there are fewer organisms here, and organisms that dwell in this zone are highly specialized for life in darkness.

Lake Chemistry and Nutrients

Unlike rivers and streams, lakes are often sinks for nutrients and chemical species. This is because lake water has longer residence time in lakes than in constantly moving streams. These often nutrient-rich waters support great biodiversity. Due to this residence time, lakes can also be sites of nutrient imbalances that can affect entire ecosystems.

Lakes and ponds at high altitudes are often oligotrophic: They have low-nutrient and high-oxygen conditions. This condition is not necessarily out of balance with high-altitude ecosystems; in fact, it's somewhat common. There are not a lot of nutrient delivery systems to bodies of standing water at high altitude, but there is an abundance of wind, which delivers oxygen to waters at high altitudes. These ecosystems are unique and often quite healthy.

Bodies of standing water, like lakes and ponds, that are at lower elevations may become eutrophic. Eutrophic bodies of water have high nutrient content but low oxygen conditions. This is often due to excessive nutrients (mostly N and P) being delivered to lakes and ponds in runoff and streams that feed into a given lake. These nutrients often come from fertilizer in agricultural runoff. When photosynthetic organisms, like algae and cyanobacteria, in lakes and ponds process these nutrients, they tend to thrive and bloom. When this happens, heterotrophic microorganisms devour the autotrophic microorganisms in mass feeding events. This heterotrophic biota uses up most of the **dissolved oxygen (DO)** in the body of water, causing oxygen levels to drop to unsafe or even lethal levels.

Wetlands

Wetlands are systems that combine elements of open water and land. They are permanently, or semi-permanently, flooded land surfaces. They often exist because the surface of the Earth is at or below the water table. Wetlands are extremely valuable ecosystems that feature an abundance of biodiversity. They provide a multitude of ecosystem services. Wetlands slow down the negative, and erosive, effects of rapid runoff rates. They act to control flooding. Since they are bodies of standing water on top of rock and soil, wetlands recharge aquifers. Wetlands also filter pollutants both geometrically and through the abundance of microorganisms that live in them.

Throughout human history, wetlands have been drained to make way for agriculture or human settlements. It is only recently that scientific evidence has brought to light that this practice may not be the most productive one. In recent years the water filtration properties of wetlands are actually the reason that agriculture and industry has promoted the engineering of **artificial** or **constructed wetlands**, designed filter pollutants, treated wastewater, and controlled runoff.

There are several different types of wetlands:

- **Bogs**—Ponds overlaid with thick, floating mats of vegetation. The formation of a bog is supported by voluminous amounts of plant life that grow during spring and summer in higher latitudes, followed by long, cold, dormant winters. Each growing season adds plant life to cover a bog. As such, these are successive systems, much like forest succession. **Peat**, a woody, low-oxygen material that is the starting point for the development of coal, can often be found in **peat bogs**, sheltered in thick floating mats of vegetation.
- **Marshes**—Flooded land surfaces in open, non-wooded areas. The plant life in marshes is dominated by grasses and other herbaceous plants. The soil in marshes is rich and supports an abundance of biodiversity. Organisms in **salt marshes** are adapted to saline waters. Salt marshes are found on the coast and are important buffers for hurricanes and other maritime storms, in addition to being areas that support enormous marine biodiversity.
- **Swamps**—Flooded land surfaces within forests. Swamps are dominated by trees. Like marshes, swamps are generally home to a vast amount of biodiversity and provide many ecosystems services, like all other wetlands.

Groundwater

Groundwater is any precipitation that is not taken up by soil water that does not evaporate, flow into waterways, or get taken up by organisms. Soil water is held in the spaces between soil particles and is in a high state of flux. **Aquifers** are porous sponge-like formations of rock, sand, or gravel that hold groundwater. Some aquifers are local and small scale, occurring near streams that are supported by **crystalline bedrock**, which consists of igneous or metamorphic rock. This is the case with groundwater in north Georgia. Other aquifers are enormous in extent, such as the Upper Floridan Aquifer that is beneath south Georgia, South Carolina, south Alabama, and northern Florida. Groundwater can be very old, especially if trapped in deep aquifers, up to thousands of years old.

Groundwater is the largest reservoir of fresh water that is readily available to humans. Large farms and agricultural consortiums depend on groundwater from large, regional aquifers for growing. Groundwater is also an erosional agent that causes the dissolution of carbonate rocks. This leads to the formation of caves and sinkholes. Another function of groundwater is that it equalizes and supports rivers and streams. All water on Earth is ultimately connected, and often connected through groundwater.

Groundwater Distribution

Groundwater is distributed, at depth, within various zones. These waters are slowly fed by precipitation that falls to the ground. **Recharge** is when water sinks into the Earth from precipitation or runoff. The process of water being transported downward through gravel, regolith, sediment, and soil is called **infiltration**. As water sinks into sediments, soil, or some other media it is filtered because of abundance of particles and narrow spaces. This is known as **percolation**. Groundwater movement is generally exceedingly slow. Even water infiltrating soil near the surface of the Earth is measured in centimeters, or inches, per hour.

Vadose Zone

The volume of land containing water from the Earth's surface downward until water saturates all pore space is called the **vadose zone**, or **unsaturated zone**. This zone contains water, but it not saturated with water. Water within the vadose zone cannot be pumped into wells because the **pore space**, the empty space between particles, is occupied by both air and water. This means that there is tension between the two fluids, imparting an overall negative pressure to most of the vadose zone. This is why pressure in the unsaturated zone is less than that of atmospheric pressure. Any saturation that occurs in the vadose zone is due to flooding and precipitation on the Earth's surface and is relatively transient.

SOIL MOISTURE AND THE CAPILLARY FRINGE

Moisture that is held within the pore spaces of the vadose zone is known as **soil moisture**. Soil moisture is important for plant growth. Soils containing 20% or more clay not only aids in the retention of moisture for plant growth but also aids in the delivery of essential nutrients through chemical species dissolved in soil moisture.

The **capillary fringe** is a belt of soil moisture that extends upward from saturation. Increasingly greater volumes of water are found with depth in the capillary fringe, held by the surface tension of water in tiny passages between grains of soil or sediment. In soil or sediment containing smaller particles with tiny pore spaces, the capillary fringe can be several feet thick. In sediment composed of larger particles like gravel with large pore spaces, the thickness of the capillary fringe is negligible. Directly below the capillary fringe is where groundwater saturation begins.

Phreatic Zone

The zone below the vadose zone where water completely saturates pore space is the **phreatic zone**, or **saturated zone**. Water within the phreatic zone can be pumped upward using wells because pore space is saturated and the overall pressure is greater than that of atmospheric pressure. Water within the saturated zone can arguably be called "true" groundwater. The upper limit of the phreatic zone, where pressure is equivalent to that of atmospheric pressure, is the **water table**.

Connection to Surface Water

Freshwater systems are all connected to each other in some manner. From infiltration and runoff to residence within the phreatic zone, there is continuity with Earth's water systems, even with regard to the ocean itself.

Gaining, Losing, and Ephemeral Streams

The interaction between groundwater and surface water constitutes a vital, dynamic connection in the hydrologic cycle. This can be seen when groundwater interacts with streams. There are several types of interactions between groundwater and streams:

- **Gaining streams**—These types of streams gain water from the influx of groundwater through the streambed. The surface of a gaining stream is often the water table itself, adjacent to the stream. This connection between groundwater and surface water also supports lakes and wetlands.
- **Losing streams**—These types of streams lose water through outflows from the stream bed. Losing streams are often found in arid areas with little recharge and very low water tables.
- **Ephemeral streams**—These type of streams are temporary, usually forming from high levels of precipitation. Once the rain stops water in ephemeral streams, it infiltrates downward into the vadose zone, in addition to evaporating.

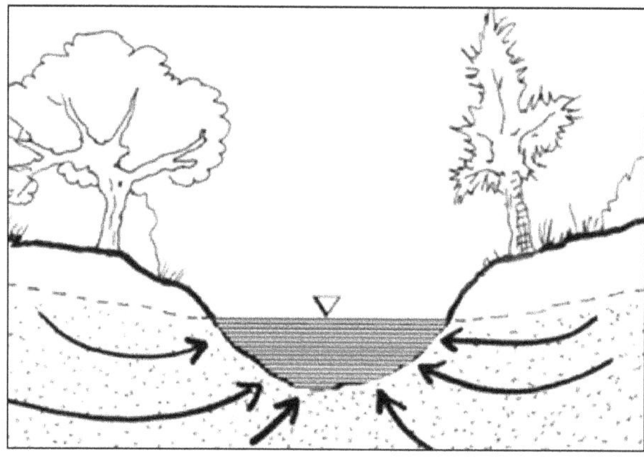

Figure 10.6 Diagram of a gaining stream

The Hyporheic Zone

Beneath and surrounding most streambeds is a zone of shallow groundwater and surface water exchange called the hyporheic zone. This zone is an ecotone that supports a variety of biological and biochemical processes, including the following:

- A habitat from many microscopic organisms and the support of fish and aquatic plant species
- Controlling the rate of chemical species exchange between groundwater and stream water
- Normalizing stream water temperature
- The reduction of pollutants dissolved in stream water

Groundwater Movement and Storage

Groundwater does not "flow" like surface water. In fact, its movement is usually exceedingly slow. Groundwater, due to its lack of speed, depth, and characteristics, is often stored for many years beneath the surface of the Earth.

Porosity and Permeability

Porosity is the percentage of pore spaces within regolith, rock fractures, sediments, or soil. This is a considerable delimiter of how much water can be stored within the phreatic zone. Porosity is dependent on the type of Earth materials in which water can be stored; because of this, large variations in porosity can occur between different types of regolith, rocks, sediment, or soil.

Permeability, on the other hand, is the amount or rate at which water is transmitted within the saturated zone. Permeability, in part, controls the rate in which water can be removed from the phreatic zone.

Groundwater Layers

There are several types of layers of Earth materials that can either support or inhibit the movement of groundwater within either the phreatic or vadose zones:

- **Aquifer**—This is a large permeable unit of groundwater that allows for the free transmission of water. It is essentially the entire phreatic zone on a large scale. Some aquifers are extensive. The Ogallala Aquifer, one of the world's largest known aquifers, extends under the states of South Dakota, Nebraska, Wyoming, Colorado, Kansas, Oklahoma, New Mexico, and Texas.
- **Aquitard**—This is a groundwater unit that hinders groundwater movement. These are composed of variably permeable materials like some clays.
- **Aquiclude**—A groundwater unit that almost completely hinders the movement of groundwater. They are often composed of packed clay or crystalline bedrock.
- **Perched water tables**—A groundwater unit that is above the water table within the vadose zone. These hold water because the units are composed some sort of impermeable material.

FYI

Darcy's Law

Darcy's law is the equation that governs the flow, or transmission, of water within the phreatic zone and other saturated media. Water beneath the water table does not flow; it is transmitted. The governing equation is a potential equation similar to the transmission of electrons due to voltage potential in a battery. Darcy's law was formulated by the French engineer Henri Darcy in the early 19th century (Figure 10.7). Darcy's law is important in the oil industry, as well as in the agricultural industry, which is dependent on water wells as a source of irrigation. There are variants to it, depending on differential conditions but this is a basic form:

Figure 10.7 Henri Darcy

$$Q = -KA\left(\frac{\Delta h}{L}\right)$$

where Q is the total of the volumetric groundwater transmission and K is the hydraulic conductivity of a media due to porosity and other factors. In rocks and soil K is dependent on the nature of the material. A is the cross-sectional area of measurement. The hydraulic head is Δh. The hydraulic head is the change in position of groundwater transmission with respect to height. L is the length of groundwater transmission. Δh over L forms the hydraulic gradient, or slope, of groundwater.

Scales of Movement and Storage

Groundwater systems exhibit enormous variations in movement and storage. Some systems are localized at the water table near a singular stream's edge, while other systems are enormous, beneath entire regions on the surface of the Earth. The storage of groundwater is highly variable too. Some groundwater systems are ancient and prohibitive in their depth. Australia's Great Artesian Basin is 1.7 million miles in extent and close to 2 miles deep in places. Extracting water from it for use in arid Australia is essentially a mining operation, which of course has raised environmental concerns. The age of some of the water in the Great Artesian Basin is upward of 2 million years.

Types of Aquifers

There are two main types of extensive aquifers, both of which are controlled by the geology of a particular region:

- **Confined aquifer**—This type of aquifer occurs when water-bearing, porous rocks are trapped between layers of less permeable substrate like clay. Confined aquifers are under a lot of confining pressure. Sometimes the pressure that a confined aquifer is under causes its pressure to be great enough to cause a surge onto the surface if a well is drilled into it. This is known as an **artesian aquifer**.
- **Unconfined aquifer**—When an aquifer has no upper layer confining it, it is unconfined. Water that infiltrates down into an unconfined aquifer is recharged by surface water supported by precipitation.

Springs

Natural **springs** are features in which the water table intersects the surface of the Earth. These can also be formed from a perched water table within the vadose and can occur within fractured rock directly after precipitation. Groundwater-fed streams often form the headwaters of rivers and streams.

Hot springs are springs that are heated beyond the general water temperature of the ground. These springs are usually located near a heat source such as a body of magma. **Geysers** are intermittent hot springs by large underground chambers containing groundwater. This water is heated until past the boiling point, then expanded and erupted onto the surface with great force. Hot springs and geysers are generally only found in locations of active volcanism.

Carbonate Systems

Groundwater is often somewhat acidic. This is because it contains weak carbonic acid from **meteoric water**, water derived from precipitation such as rain or snow. Precipitation contains carbonic acid from an interaction between carbon dioxide and water vapor in the atmosphere. Being low in pH carbonic acid that enters groundwater systems containing high pH calcite, such as limestone, reacts to form calcium bicarbonate, which a soluble material.

As bedrock formed of carbonates dissolves, various features are formed within the Earth:

- **Caves and caverns**—Most caves and caverns are formed from the dissolution of carbonates like limestone. Caves are cavities within the Earth, sometimes with an accessible opening to the surface. Caverns are a system of caves. Most caves and caverns are formed below the water table and end up becoming exposed to the surface due to tectonic uplift or some other geologic process.

- **Sinkholes**—Features on the Earth's surface in which groundwater dissolves the rock layers to form an open pit. Sinkholes are a hazard, having caused millions of dollars in property damage in the past and at least one death.
- **Karstic landscapes**—Topography that has been shaped by the effects of acidic water dissolving carbonate bedrock. Karstic topography, or karst, can cover entire regions. Some of the features in these landscapes include sinkholes, irregular terrain from chemical weathering, and streams that disappear into the Earth's subsurface.

Water Use and Impacts

Water is a vital, necessary resource, and as such its occurrence and use impacts human society, even human civilization, in a critical manner. Litigation between political entities and even wars have been fought over water. As such, the wise and equitable use of water, water conservation, is critical to our very existence!

Water Use Across the Earth

Water is not equally distributed over the Earth's surface, as well as within the Earth's subsurface. Different regions of the Earth possess immensely different amounts of groundwater, surface water, and precipitation. Some regions of the Earth possess an abundance of water from precipitation, while some areas are arid. Many high-density populations, in critical areas near the equator, face not only harsh climates but a critical scarcity of water.

Due to excessive groundwater withdrawal, shifting weather and climate patterns, and scarce infrastructure, many of the Earth's regions are precariously short of clean, potable water. Consumption of water in the past has been unsustainable, and the effects are currently being felt by many. There are two main modes of freshwater use: **consumptive** and **non-consumptive**. When water is withdrawn from groundwater or surface water and not made potable again, it is consumptive. The water goes somewhere of course, but it is often contaminated as industrial wastewater or sewage, not readily usable in the foreseeable future. Non-consumptive use is when water is reclaimed, or temporarily removed from groundwater or surface water. Hydroelectric dams are an example of non-consumptive use. When river water flows through turbines in a hydroelectric dam, it simply flows downstream, essentially untouched.

Water Wells

A **water well** is a hole, pit, or drilling operation designed to extract water from beneath the water table. Water wells need to reach or be drilled beneath the water table. The nature of the unsaturated zone precludes water being made available for wells.

Groundwater Withdrawal

Water wells, or just wells, act upon the water table when in operation as water is being extracted. The water table locally lowers around the circumference of a well that uses an electrical pump due to the absence of water around it. This is known as well **drawdown**, a lowering of the water table. The local lowering of the water table around a well shaft is called the **cone of depression**. Drilling operations often have to be closely monitored and carefully managed in order to not cause drawdown to exceed the capacity of a unit of groundwater.

Groundwater Contamination and Impacts

Interactions with surface water as well as the breaching of units containing chemical pollutants in the subsurface can cause groundwater drilled for use to become contaminated. Any accidental breach of an aquifer by industrial wastewater or sewage can easily contaminate well water. One of the problems with polluted or contaminated groundwater is that the water is not on the surface, thus it is hard to put through **remediation** and decontamination efforts. At the same time, extensive aquifers within porous sedimentary rocks can effectively clean contaminated water that transmits through them.

Excessive groundwater withdrawal in coastal areas may cause **saltwater contamination**. This is when saltwater that has seeped into the subsurface becomes drawn into wells, thus contaminating the water supply with, by definition, non-potable water.

Groundwater withdrawal can also cause **subsidence**. This occurs when groundwater that structurally supports the ground is withdrawn too fast and causes the land surface to sink. Groundwater is used to irrigate California's San Joaquin Valley, a region of major agricultural production. This has caused decades of land subsidence, as many as nine meters in one spot!

Water Structures

Human beings have engineered many features to extract, contain, and divert water over the millennia. The practice is likely as old as civilization itself. Some of the oldest known dams are over 3,000 years old.

Dams

Dams are any obstruction placed in a river or stream for the express purpose of using the stored water. Dams are used for flood prevention, drinking water, irrigation, and electricity generation. There are over 40,000 dams that have been erected across the world. At this point it close to infeasible to erect more dams since most of Earth's major waterways have already been dammed to some extent. Rivers that are available for damming are too remote, in places like northern Canada and Russia, to do so.

There are benefits to damming rivers and streams, but there are drawbacks as well. Dams provide water for irrigation and municipal use. They also are used to turn turbines as hydroelectric dams to provide electricity. Dams provide large bodies of water, called reservoirs, for shipping and recreation. On the other hand, the damming of rivers alters or even destroys entire biological communities. Dams often create situations for pollution, including sediment pollution. Also, dams that are built of concrete materials run the risk of catastrophic failure.

Reservoirs

The body of water formed from damming a river or stream is called a **reservoir**. Reservoirs can be thought of as artificial lakes, and many of them are as large as extensive natural lakes. Reservoirs are the bodies of water that provide irrigation and recreational opportunities. Reservoirs can also be very deep and over time acquire the same attributes as natural lakes such as a littoral and a benthic zone. Many reservoirs are used to stock fish.

Water Diversions

Other means to control natural waters include diversions. These diversions include **dikes** and **levees**. Dikes are embankments or walls used to prevent flooding. Levees are ridges that are built to prevent flooding along some length of a waterway. It is understandable why these structures are built. They are generally built to prevent floods, which are natural disasters that are dangerous to human beings. At the same time, flooding is a natural process, delivering nutrient-rich sediments to floodplains. People often choose to build at the expense of the land, water, and themselves.

Flooding caused by Hurricane Katrina in 2005 caused over $100 billion in damage and close to 2,000 fatalities in New Orleans and the surrounding areas (Figure 10.8). An extensive series of dikes and levees have been built in that area to control the Mississippi River by the U.S. Army Corps of Engineers over the years. This is because New Orleans is a busy port and one of the sites of shipment for oil from the Gulf of Mexico. Those dikes and levees failed to hold back the water swelling over the land.

Figure 10.8 View of flooded New Orleans, Louisiana, 2005, in the aftermath of Hurricane Katrina

Canals are another form of water diversion. Canals are artificial waterways cut into the Earth. Canals are often used for agriculture and drainage.

Floods

The most common and most destructive geologic hazards are **floods**. Floods occur when water courses over usually dry land surfaces. They occur naturally as precipitation levels exceed the ability of stream channels, lakes, and other bodies of water to contain rising waters. Floods also renew land surfaces through the delivery of nutrients from bodies of water. Floods can also be caused by human engineering efforts that fail, such the breach of dikes and levees and catastrophic dam failure.

There are several types of floods:

- **Regional floods**—These are periodic events that occur due to the probability of a flood occurring within a given year. They usually occur naturally and periodically, with the only concern being that people tend to build with these areas anyway. There are different categories of regional floods, such as the 100-year flood in which a flood has a 1% chance of occurring in any given year. These figures are used by insurance policies for landowners, which should be enough reason to pay heed.
- **Flash floods**—These are local floods due to unusually heavy rain. Since these floods are often sudden and unexpected, they are the floods that do the most damage to human life and property.
- **Ice-jam floods**—In relatively cold regions of the Earth, such as upper latitudes, ice can dam rivers and streams, causing upstream flooding. When the ice melts or breaks, this can also cause downstream flooding.
- **Dam failure floods**—This type of flooding is pretty straight-forward. A dam catastrophically fails, releasing water downstream. These floods can be exceedingly dangerous.

Freshwater Pollution

Pollution is the release of matter or energy into the environment, causing damage to the health and well-being of living things, including humans. If contaminants are not kept out of rivers and lakes, they can become concentrated, building up in toxicity. Water pollution and contamination is hard to clean up or remediate. This is especially true for groundwater pollution, which, by definition, is difficult to reach yet has a major impact on surface waters. Water pollution can take two forms: **point-source** and **non-point source**. Point-source pollution can be attributable to a single source such as an open sewage pipe. Non-point-source pollutions originates over an entire area, such as all of the contaminated runoff with a city.

Nutrient pollution from fertilizer causes eutrophication in bodies of water, especially lakes. This leads to excessive blooms of autotrophs, in turn leading to invasions of heterotrophic biota, which deplete water bodies of oxygen, causing **hypoxia**, a catastrophic drop in oxygen levels. Varieties of other chemicals from industrial processes can cause untold damage to natural waters, including hydrocarbons from petroleum products. In the 1960s the surface of the Cuyahoga River in Ohio caught fire due to the presence of so much oil. As recently as 2014 in Flint, Michigan, water supplies were so severely contaminated that water became undrinkable.

Waterborne diseases and pathogens cause more loss of life than any other form of water pollution. Poorly treated human waste and fecal matter from livestock operations are the main cause of this pollution. Some of the possible water contaminants include fecal coliform bacteria, cholera, typhoid, and hepatitis A. These problems are especially pronounced in the developing world where many sources of water are constantly contaminated and unclean. This is especially true in rural Africa and Asia. That is why the work of outreach organizations that help provide medicine, clean water, and sewage treatment is so important.

Toxic chemicals that contaminate water come from both natural and human sources. Human error contributes chemicals such as pesticides, petroleum products, synthetic chemicals, and toxic metals to pollution sources. Mining contributes constituents such as acid mine drainage and arsenic and lead to water pollution. Arsenic and other metals can also come from natural sources. Water supplies in Bangladesh have been contaminated by arsenic in the past, shed from rocks transported downstream by the Ganges River.

Conclusion

Without water, life as we know it would not exist, yet water plays an even larger role in Earth's natural systems. Water shapes the surface of Earth, the very soil sitting atop continental crust that we all live on. It is part of the tectonic forces that move Earth's lithosphere and has a role in the development of many volcanoes. Salt water fills Earth's ocean basins; these very oceans have been the foundation of human civilization. Deep beneath our feet water is locked into molecular bonds within the minerals that make-up much of Earth's solid rocks. Earth is truly a water-world!

Discussion Questions

Directions: Review the chapter in order to completely and correctly respond to the questions and prompts.

Keep this in the back (or maybe even front) of your mind: Science is system of methods that allows us to understand not only our planet, and ourselves, but of the entirety of the universe. Answering the following questions correctly will lead to a more robust understanding of how to apply science to ENVS problems.

1. Using Figure 10.2, calculate the percentage of all of Earth's water in rivers.
2. Using the hydrologic gradient formula on page four, calculate the hydraulic gradient from the headwaters of the Chattahoochee River (elevation 3,550 feet) to the town of Helen, Georgia (elevation 1,444 feet), a distance of 5 miles, in feet per mile.
3. Have you ever seen turbulent flow in water? Why did it look the way it did?
4. Have you ever seen a stream depositional feature? Describe how it may have formed.
5. Describe eutrophic and oligotrophic lakes. Where might they be found?
6. Describe the similarities and differences between bogs, marshes, and swamps.
7. How and why does the capillary fringe form? Think about some of the properties of water that you've learned.
8. Give an example of a material that is porous and an example of a material that is permeable. Can something be both?
9. Have you ever seen a spring? If so, why was it there?
10. Do some wells not need to be pumped? Why or why not?
11. Give an example of a point source for water pollution in your area. How can it be remediated to stop polluting your community?

Bibliography

Bear, J., 1972, *Dynamics of Fluids in Porous Media*: New York, Dover, 764 p.
Fetter, C.W., 2001, *Applied Hydrogeology*,: Upper Saddle River, New Jersey, Prentice-Hall, 4th ed., 598 p.
Withgott, J., and Laposata, M., 2019, *Environment: The Science Behind the Stories*: New York, Pearson, 7th ed., 768 p.

Credits

Fig. 10.1: John Evans and Howard Periman, USGS, "The Hydrologic Cycle," https://commons.wikimedia.org/wiki/File:Watercyclesummary.jpg, 2013.
Fig. 10.2: USGS, "Earth's Distribution of Water," https://commons.wikimedia.org/wiki/File:Earth%27s_water_distribution.svg.
Fig. 10.3: Copyright © by SilentSword22 (CC BY-SA 4.0) at https://commons.wikimedia.org/wiki/File:Figure_of_a_drainage_basin.svg.
Fig. 10.4: Copyright © by Tecty (CC BY-SA 4.0) at https://commons.wikimedia.org/wiki/File:Glisborn-2.jpg.

Fig. 10.5: Copyright © by Oliver Kurmis (CC BY-SA 2.0) at https://commons.wikimedia.org/wiki/File:Nowitna_river.jpg.
Fig. 10.7: Source: https://commons.wikimedia.org/wiki/File:Henry_Darcy.jpg.
Fig. 10.8: Commander Mark Moran, Lt. Phil Eastman, and Lt. Dave Demers, "View of Flooded New Orleans, Louisiana 2005, in the Aftermath of Hurricane Katrina," https://commons.wikimedia.org/wiki/File:Katrina-new-orleans-flooding3-2005.jpg, 2005.

Chapter 11

Oceanography and Marine Biology

The Cradle of Life

Objectives

- Come to an understanding of the vital influence that the Earth's oceans have on life.
- Know that life itself came from the ocean, in large part.
- Understand what the composition of ocean water is and why that makes it unique.
- Know about the structure of ocean water, as well as the topography and bathymetry of the ocean.
- Understand the vastness of life and living ecosystems in the ocean.
- Know that the oceans, while vast, are sensitive to the impacts that people place on them.
- Work through some possible solutions for marine conservation.

KEY WORDS AND TERMS

Headland: A landform that rises higher than the rest of shore, jutting from the shoreline into the ocean.

Island: An isolated piece of land in an ocean or sea, smaller than a continent and without certain continental features.

Maritime: Connected to the sea, or living in or found near the sea; also relating to commercial marine or naval matters.

Perpendicular: At a 90°, or right, angle from a straight feature.

Storm surge: A sudden rising of the sea due to atmospheric pressure changes or powerful storm winds.

Consider This . . .

Earth is not just a water world, it is an ocean world. As such, we are all connected to the ocean, regardless of where we might be on Earth's surface. The ocean makes planet Earth what it is, with a continuous influence, not only the atmosphere and the lithosphere, but the biosphere life itself.

Introduction

The **ocean**, singular for the great interconnected expanse of ocean water that covers around 71% of the Earth, or the **oceans**, plural, is ancient. It is some 4 Gy (billion years) old, having probably formed at the beginning of the Archean Eon, if not earlier. It should be no surprise that the first signs of life in the stratigraphic record can also be dated to that time. The oceans not only support life on Earth but are also where life originated. Incidentally, the ocean and the oceans are somewhat interchangeable terms, both of which will be used here.

While there is really only one ocean, there are five separations of this world's ocean divided into the "oceans" (Figure 11.1). The boundaries between these oceans have grown over time due to a variety of cultural, geographical, historical, and scientific factors. There are five oceans, generally recognized by most nations, as well as international scientific organizations:

- **The Arctic Ocean**—This ocean is north of the Atlantic and Pacific Oceans and North America and Eurasia. It is completely north of the 60° north latitude line. It is the world's smallest ocean.
- **The Atlantic Ocean**—The second-largest ocean on Earth, this ocean is located between Europe and Africa in the east and North and South America in the west. It is mostly between the 60° north south latitude lines, and it is bisected by the Mid-Atlantic Ridge.
- **The Indian Ocean**—This ocean is the third-largest on Earth. It is approximately 20% of the Earth's water surface. Its boundaries are southern Asia in the north; the Arabian Peninsula and Africa in the west; the Malay Peninsula, Sunda Islands, and Australia in the east; and the Southern Ocean in its southern reaches. Much of the Indian Ocean is in the tropics, which are between the equator at 0° and the Tropic of Capricorn at 60° south latitude.
- **The Pacific Ocean**—This is the world's largest ocean and covers a third of Earth's surface. It is bounded in the east by North and South America, and in the west by Asia, Australia, and New Zealand. It is mostly between the 60° north and south latitude lines. This ocean is ringed by the **Circum-Pacific Belt**, otherwise known as the ring of fire, in the east, west, and north. The Circum-Pacific Belt is an almost continuous belt of subduction zones and tectonic activity giving rise to around 66% of the world's volcanoes. Around 90% of the world's earthquake activity occurs in the Circum-Pacific Belt.
- **The Southern Ocean**—This is a newly classified ocean. It is the world's fourth largest ocean and completely surrounds the south polar continent Antarctica. It is bounded by the Atlantic, Indian, and Pacific Oceans in the north. Most of this ocean is south of the 60° south latitude and is generally very deep, on average 4,000–5,000 m.

Seas are smaller than oceans. They are generally adjacent to, and bounded by, land to some extent. A **bay** is bounded by an inward curve of the land. Bays are generally smaller than seas. A **gulf** is similar to a bay except it is bounded by land except for one opening.

Figure 11.1 Map of Earth's oceans

The Ocean

The scientific study of the ocean is called **oceanography**, and the scientific study of marine organisms, as well as marine ecosystems, is called **marine biology**. The oceans not only cover a significant portion of Earth's surface but also hold around 97% of Earth's water. The ocean is the primary component of the hydrosphere, influencing the atmosphere, biosphere, and lithosphere. Oceans are connected to freshwater on land through runoff and streams that empty into it.

From a geological standpoint, the ocean resides atop the Earth's low-lying oceanic crust. At the same time, raised continental surfaces can become flooded to form **epeiric** or **epicontinental seas**. In the past, for instance during the late Cretaceous period when the Earth had no polar ice caps, epicontinental seas covered many of the planet's continental land surfaces. Currently, due to water being taken up in polar ice caps, there are few standing epicontinental seas. Continental shelves, the flooded edges of tectonically inactive continents, probably can't be considered seas. There is a sunken mass of continental crust called Zealandia that is part of New Zealand, but it is covered by the Indian Ocean. The Baltic Sea and Hudson Bay are technically epicontinental seas.

Ocean Water

As the ultimate repository for all of the weathered and eroded rock and minerals from continental landmasses, the ocean contains an abundance of solutes and suspended materials, not to mention countless microorganisms. Ocean water contains 96.5% actual water. The rest consists of various sorts of dissolved chemical species, dissolved gases, dissolved metals, and the ions of dissolved salts. The ocean developed its salty nature early after it formed 4 Gya. Currently, hundreds of millions of tons of dissolved terrestrial chemical species and sediments are delivered to the ocean from rivers, streams, and runoff every year. As such, ocean water as a whole is buffered to be basic rather than acidic, at a pH of around 8.

Of vital importance to marine organisms is dissolved carbon dioxide (CO_2) and **dissolved oxygen (DO)**. The ocean, on average contains about 7.5 PPM (parts per million) DO and around 10 PPM dissolved CO_2. Oxygen is added to the system from bacteria, plants, and atmospheric diffusion. Carbon dioxide and oxygen are important to marine systems because they are vital to living organisms. Marine photosynthetic plants use CO_2 to produce their food, and marine heterotrophs, such as fish and squid, require oxygen for breathing.

Below 4 PPM DO marine animals experience health problems and will generally swim away from that portion of the ocean, if they can. At 2 PPM DO or below, marine creatures will generally die. The oceans contain an abundance of nutrients like nitrogen (N) and phosphorus (P) that are vital to plants, but these nutrients can also be problematic. Excess plant nutrients, especially N, lead to eutrophication, which lowers DO to dangerous levels, inducing hypoxia in the water column.

A significant portion of ocean water consists of ionic species from dissolved salts, on the order of between 10,000 and 35,000 PPM. Predominant among these ions are chloride and sodium. There is about 19,000 PPM chloride ion (chlorine, Cl^-) in the ocean on average, and about 11,000 PPM sodium (Na^+). Other ions of dissolved salts include magnesium (Mg^{2+}), potassium (K^+), and sulfate (SO_4^{2-}). In a related manner, the ocean is Earth's largest carbon sink, from atmospheric CO_2 and carbonate rock dissolution.

Controls on Vertical Structure and Temperature

The ocean is vertically structured according to conditions at depth. Ocean temperatures decline with depth, somewhat rapidly (with cold polar waters being the exception) to about 1,000 m. At a depth of 1,000 m, ocean water is subject to only a slight steady decline until it reaches close to freezing. The drop in temperature is between 10–20° C within the first 1,000 m of the ocean.

Ocean water becomes dense with depth, partly due to the decrease in temperature and partly due to an increase in salinity. Colder, heavier water sinks, and lighter, warmer, less-saline water remains on the surface. These changes in water density and composition have led to layering in the ocean:

- **Surface zone**—This is the zone at the ocean's surface in which sunlight penetrates. This allows for photosynthesis. This region is where waves occur, initiated by wind. This zone is warmed by sunlight and has a consistent density.
- **Pycnocline**—This is the zone below the surface in which ocean water density increases and temperatures decrease rapidly with depth until around 1,000 m below the surface, where sunlight does not penetrate. At this point temperature barely increases with depth, and ocean water density, as well as pressure, steadily increases.
- **Deep zone**—This is the deepest part of the ocean's water. Here, the waters are near freezing, and the density of ocean water literally makes it "thick." Water movement is sluggish in this zone or even nonexistent. This region of the ocean is unaffected by sunlight, surface temperatures, and surface weather.

Due to the high specific heat capacity of water, the oceans mediate temperatures on land, particularly on the coast. It takes much more energy to change the temperature of water than it takes for Earth materials, so ocean temperatures remain steady. In this way, the oceans absorb and emit heat, which can regulate temperatures on land, inducing mild weather. On the other hand, excess heating of the ocean during summer months can lead to the production of powerful tropical storms.

Currents

The ocean's surface zone is replete with ocean **currents**. Currents are riverlike flows of water in the ocean that are driven by density differences, gravity, heat, and wind. These currents act to transport heat signatures, nutrients, and the larvae of numerous marine species throughout the ocean. Unfortunately, currents can also transport contaminants and pollutants across the ocean.

Some currents have a vast influence on the environment. The **Gulf Stream** originates near the tropics of the Northern Hemisphere, then is sent to Northern Europe across the Atlantic Ocean. The water, carrying a warm heat signature, moderates the climate of Europe, making it temperate. If one looks at a map of Europe, especially Northern Europe, one would notice that the continent is at a rather high latitude, especially with regard to the relatively mild climate for that region. The moderate climate is due to the delivery of heat by the Gulf Stream.

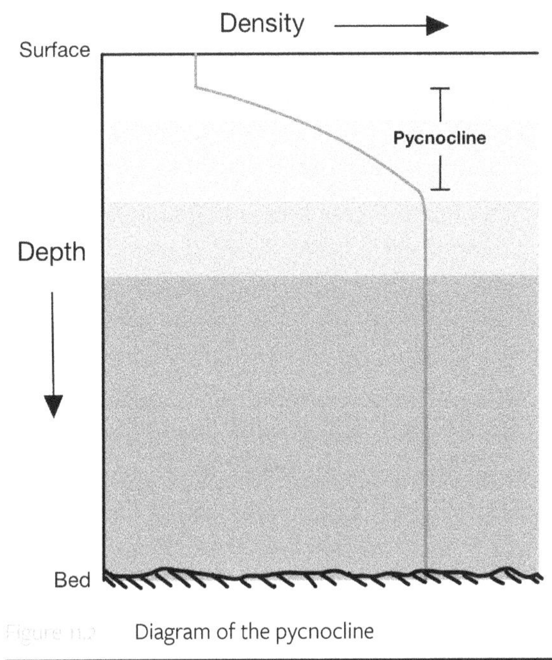

Figure 11.2 Diagram of the pycnocline

VERTICAL CURRENTS

Currents need not move horizontally; vertical currents do exist. Water at depth that replaces wind-blown surface water is known as an **upwelling**. These vertical flows of nutrient-rich water ensure high primary production upon delivery to the surface. **Downwellings** are vertical currents that get sent downward when surface currents converge. These downwellings delivery oxygen-rich water from the surface to the ocean bottom.

Ocean Zones and Regions

The depth-dependent regions of the oceans and seas are similar to that of large lakes. In the ocean, though, the scale tends to exceed that of even the largest lake.

- **Littoral zone**—This is the zone of the ocean from high tide, on the shoreline, to the edge of the continental shelf. In shorelines with little to no continental shelf, this zone ends at the drop to an oceanic trench, or to the ocean floor.
- **Photic zone**—This is the well-lit top layer of the ocean in which sunlight penetrates. Solar radiation in this surface zone generally supports high primary productivity.
- **Pelagic zone**—This region of the ocean contains all of the habitats and ecosystems between the ocean's surface and ocean's floor.
- **Benthic zone**—These are all of the habitats and ecosystems occurring on the ocean floor.

- **Abyssal plain**—This is the deepest part of the ocean and the flattest surface on Earth. This zone includes all of the habitats and ecosystems below 3,000 m depth, a zone in which no sunlight penetrates.

As should be apparent, the various zones and regions of the ocean overlap considerably. For instance, the benthic zone, being the floor of the ocean, is well within both the littoral zone and the abyssal plain.

Topography and Bathymetry

The bottom of the ocean is its **topography**, its geographic features. The topography of the ocean is as complex as that on land, with some differences, as land is bounded by the atmosphere whereas the ocean is bounded by water. The seafloor is rugged and contains many topographic features, such as the abyssal plain, mountain ranges including mid-ocean ridges, sediments and debris, submarine canyons, trenches, and underwater volcanoes.

The **bathymetry** of the oceans are the measurements of its depths. There are several features in the ocean that are depth dependent by definition. **Continental shelves** are the gently sloping edges of continents covered in seawater. Some tectonically active coastal areas have very short continental shelves or none at all. The continental shelves along the tectonically inactive Atlantic Ocean are very broad. The **shelf-slope break** is the sudden increase in slope at the edge of the continental shelf. The **continental slope** is a gradient between the shelf-slope break and the ocean's floor. The area from the continental shelf to the continental slope is known as the **continental margin** (Figure 11.3). The aforementioned **abyssal plain** is the deepest part of the ocean floor, below 3,000 m, and the flattest place on Earth. Within the abyssal plain, at the subduction zones between lithospheric plates, there are the **ocean trenches**, the deepest places on the surface of the Earth. The Challenger Deep, part of the Marianas Trench near the east coast of the island of Guam in the Pacific Ocean, is the deepest place on Earth at around 11 km (6.84 miles) straight down.

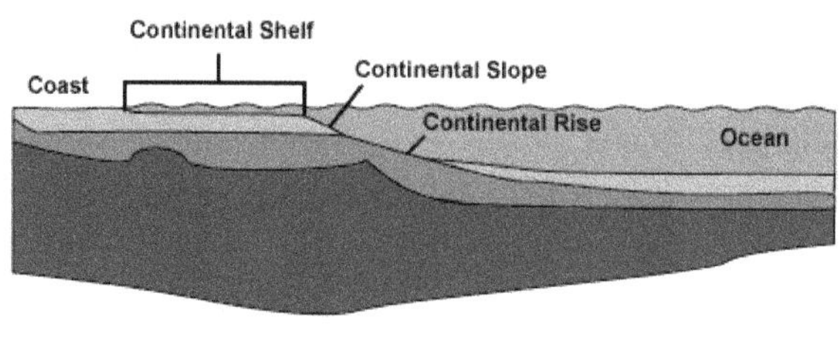

Figure 11.3 The continental margin

The Coastal Zone

The **coast** of the ocean and seas consists of everything from the furthest extent of **maritime** features on land to the shore. The **coastline** marks the boundary between the land on the coast and the sea. Within the coast is the **shore**, which is the area between the lowest tides and the upper boundary of the coast affected by ocean storms. The **shoreline** is the line that marks the contact between the land and the sea.

The Shore

The shore of the ocean is where land interacts with the ocean. As such, there are several divisions of the shore, all of which feature the unique habitats of marine and coastal organisms:

- **Backshore**—This area is from the highest tidal line landward to the upper limit of the coastline.
- **Foreshore**—This is the area exposed at the lowest tide to the highest tidal reaches.
- **Nearshore**—This area is between the lowest tidal shoreline and the point at which waves break during those lowest tides.
- **Offshore**—This is seaward of the nearshore boundary.

Beaches are accumulations of sediments on the shore of the ocean, as well as on the landward portion of lakes. The relatively flat platform of sand, or other sediments marked by a sudden but low-magnitude decrease in slope, is called a berm. The beach face is the wet, slightly sloping surface from the berm to the shoreline. Due to the nature of deposition, beaches are mostly composed of sand, but this is not always the case. A beach can be composed of any number of combinations of boulders, clay, gravel, silt, and sand depending on the type of coastline as well as terrestrial contributions.

Ocean Waves

Ocean **waves** are movements of water on and near the surface of the ocean due to winds. Since there is no topography in the open ocean, wind-driven waves can become quite powerful. Waves provide most of the energy that alters and modifies the shoreline. Waves are translators since they derive energy and motion from winds.

There are two main components of ocean waves. The crest of a wave is the top of the wave. The **trough** of a wave is the low region between two crests. There are also ways waves can be measured: **wave height**, **wavelength**, and **wave period**. Wave height is the total height of a wave from trough to crest. Wavelength is the distance between two wave crests. The amount of time that is takes for one wavelength to pass before at another, at a fixed point in space, is one wave period. This is the same thing as wave frequency. The height, length, and period of a wave are dependent of several factors such as the overall speed of the wind, the length of time that the wind has blown, and **fetch**, which is the distance that wind has blown over open water.

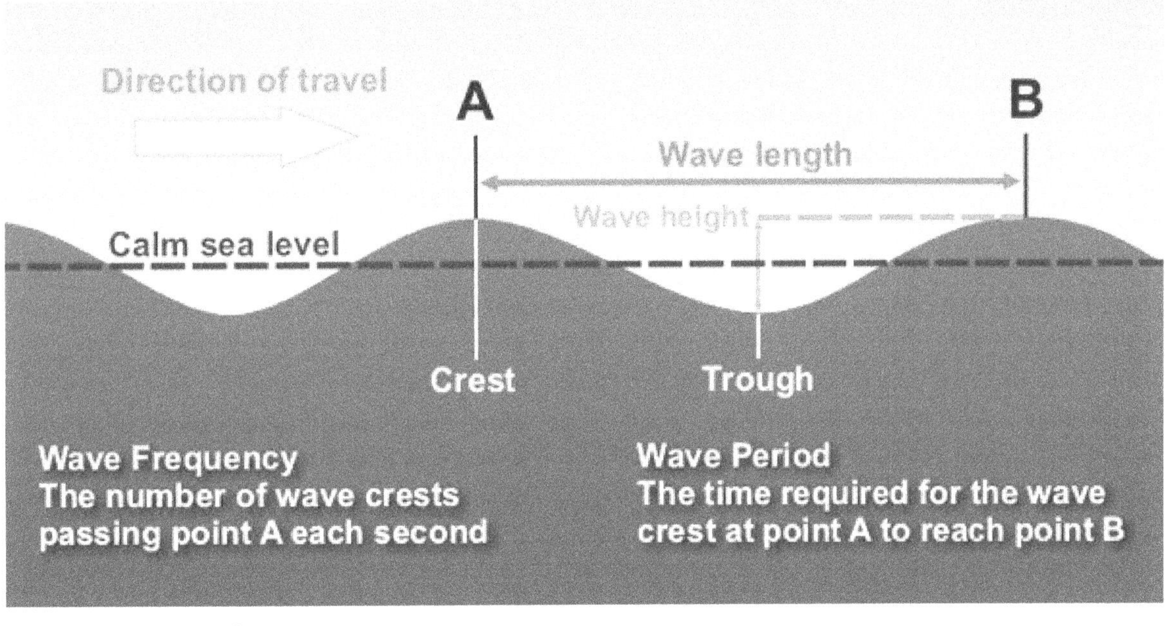

Figure 11.4 Diagram of a water wave

There are several types of waves significant to movement on the ocean's surface. **Oscillatory waves** are waves in which energy propagates forward through the water but the water itself does not move in position. This is the classic circular wave motion. **Translatory waves** occur when the floor of a water body intersects with one half the wave height. At this point, forward wave energy is transferred upward. These translator waves are responsible for the monstrous height of waves during tsunamis. Since in the open ocean tsunamis travel near the speed of sound that energy is translated upward near the shore, accounting for the height of tsunami waves, which are often over 20 m.

Even under normal conditions, ocean waves that crash on the shore are very energetic. When waves break on the shoreline it is known as **surf**. These waves become turbulent once they reach the shore; then they fold over, crashing into the land as surf. These breaking waves exert a lot of force. They are agents of erosion on the shore due to wave impact and pressure, as well as abrasion from rock fragments and other sediments carried by the ocean.

Movement Along the Shoreline

Waves usually do not approach the shore perpendicularly. Instead, they usually break on the shore at an angle. Waves traveling toward the shore through shallow water with a smoothly sloping bottom, like that on a continental shelf, tend to become parallel to the shore. The change in direction of ocean waves is due to wave refraction, which is the bending of waves. Because of this refraction, wave energy tends to get concentrated with features that jut out from the shore rather than the shore itself. These waves cause erosion, and over time that erosion leads to the straightening out of shorelines. These **perpendicular** to semi-perpendicular waves cause beach sediments to zigzag down the shore, something known as **beach drift**.

The oblique change in wave direction due to refraction also produces what are known as **longshore currents**. These occur in the surf zone and flow parallel to the coast. These longshore currents also move fine suspended sand and smaller sediments through the water on the coast, and they roll larger sediments, including gravel, along the ocean bottom.

Shoreline Features

There can be quite a number of features along the shoreline in addition to beaches. The occurrence of these features is controlled by currents, shoreline rocks, wave intensity, and whole Earth coastal activity. Whole Earth coastal activity includes rising, sinking, or stable coasts. **Rising coasts** occur because of tectonic uplift, large-scale ocean regression, or some sort of local uplift. **Sinking coasts** are due to tectonic lowering, large-scale ocean transgression, or local lowering of the land. **Stable coasts** are somewhat quiescent and not subject to any sort of local or large-scale change in the elevation of the coast.

Features along the shoreline occur due to deposition, such as beaches, and erosion. Shoreline erosion is caused by any number of factors, such as the amount and type of local tectonic activity, geometric configuration of the coastline and other near-shore landforms, prevailing wind patterns and overall weather patterns in the area, proximity to the mouths of rivers and the sediments that those rivers deliver, and the topography and composition of the coast.

Shoreline features include the following:

- **Baymouth bars**—A depositional feature, these are sand bars that form due to longshore currents. These almost, or completely, enclose the mouth of a bay.
- **Sea arches**—When powerful waves continually batter shoreline caves in softer rock, such as limestone, these features can be formed. Over time as the **headland** erodes it can become separated from the shoreline forming an arch.

- **Sea stacks**—These are pieces of the headland that become separated due to the eroding action of waves.
- **Spits**—These are shoreline depositional landforms that occur when a longshore current causes sediment to accumulate off a headland.
- **Tombolos**—These are spits that become connected to a nearby **island**.
- **Wave-cut platforms**—Flat areas of rock cut by the erosive action of ocean waves.

Barrier islands are islands composed of sand and sediment that form parallel to shorelines (Figure 11.5). They occur along many of the world's shorelines but are especially concentrated along the East Coast of North America, as well as the North America's Gulf Coast, particularly Texas. These islands seem to form on coastlines along continental shelves that do not experience particularly extreme tidal range, but in actuality their formation is not yet completely understood. Barrier islands occur anywhere from one to 30 km offshore. They can be a few kilometers to over 180 kilometers long. Lagoons and salt marshes, places of great biodiversity, often form between barrier islands and the shore. Barrier islands protect the coastlines along which they are formed.

Figure 11.5 Sapelo Island, a barrier island off the coast of the state of Georgia

Tides

Tides are regular increases and decreases in the height of the ocean. They are caused by the gravitational pull of the Moon and to a lesser extent the Sun. There are two main types of tides, **spring tides** and **neap tides**. Spring tides occur when the gravitational forces of the Moon and the Sun are in conjunction with one another. This produces the highest and lowest tides, a large tidal range. Neap tides occur during the first and third quarters of the Moon, when the gravitational forces of the Moon and the Sun are offset. This produces the shortest tidal ranges, the lowest high tides and the highest low tides. Tides can also be influenced by the shape of the coastline and the shape of the ocean, or sea, basin.

There are regular patterns to the occurrence of tides. When there is one high tide and one low tide daily, it is known as a **diurnal tidal pattern**. Two high and two low tides per day is known as a **semidiurnal tidal pattern**. **Mixed tidal patterns** are irregular, with high and low tides that fluctuate daily.

Tidal currents are composed of horizontally flowing water driven onto land by rising and falling tides. There are two types of tidal currents. **Flood currents** advance with the rising tide, and **ebb currents** move toward the sea as the tide falls. **Tidal streams** are features in which streams are cut into the shore through the action of rising tides on sand and sediments. These can branch off on especially flat shores to form **tidal deltas**. **Tidal flats** are areas of the coast affected by tides.

Marine Biology

The ocean in general teems with life, even in the polar regions. There is strong evidence that life itself began in the ocean some 4 billion years ago. Biotas within the oceans are complex, and life is everywhere. Organisms can occupy every level of the ocean due to the presence of water, whereas on land it is restricted to land and constrained by the Earth's atmosphere. Even birds and other flying creatures do not occupy the atmosphere the way that creatures of the ocean occupy every level of the sea.

Open Ocean Biota

Within the ocean, microscopic **phytoplankton**, single-celled microscopic plant-like organisms, within the pelagic and photic zones comprise the base of the trophic pyramid, providing the oceans with abundant primary productivity. In fact, the most productive organisms on Earth are those that live within tropical algal beds and reefs. The bulk of producer organisms consist of algae, cyanobacteria, and protists. **Zooplankton**, microscopic single-celled animal-like organisms, feed off of those organisms, which in turn feed fish, jellyfish, smaller metazoans, and whales. Predators of the ocean include fish-eating birds, large fish, sea turtles, sharks, and squid, plus other cephalopods. There are countless phytoplankton and zooplankton organisms within the ocean.

Creatures of the Abyss

Within the ocean's depths from around 3,000 meters down to the abyssal plain there exist organisms that are well adapted to those extreme conditions. Animals at those depths have adapted to extreme the lack of light and extreme pressures. Many scavenge carcasses and other detritus that drifts down from above. Some of those animals are predators, while others have symbiotic feeding relationships with bacteria. Some organisms at those depths carry bioluminescent light-producing bacteria. Hydrothermal vents at mid-ocean ridges are host to active and unique ecosystems. These places are the habitats for many different, uniquely adapted organisms such as chemosynthetic microorganisms, tubeworms, and specially adapted shrimp.

Open Ocean and Coastal Ecosystems

There are countless ecosystems unique to the ocean. These ecosystems can be completely within the ocean, from the ocean's surface down to the abyssal plain. There are also transitional ecosystems along the coast.

Coral Reefs

Coral reefs are located within shallow tropical and subtropical seas. These reefs are formed by small, primitive organisms within the phylum *cnidaria* called **corals**. The coral organism in and of itself is merely a tiny polyp. Corals create their own homes using calcium carbonate, from which they extend tiny tentacles to filter feed on microorganisms in sea water. Corals live in symbiosis with algae called *zooxanthallae*. They are colonial organisms, so they build their homes in clusters around each other, often millions of other individual corals, forming coral reefs. These reefs are attached to bedrock within the oceans and can become enormous. The Great Barrier Reef off the coast of Australia is 2,300 km long.

Coral reefs are vital ecosystems, home to thousands of different organisms. These organisms live within the nooks and crannies of the coral reef. Coral reefs also protect shorelines by absorbing waves, and thus protect against the damage caused by storms.

Currently, coral reefs are dying off due to several factors. Coral bleaching occurs when *zooxanthallae* are not present. Since corals are in symbiosis with *zooxanthallae,* the corals die off. This is called coral bleaching because white patches are left behind when the organisms die. Nutrient pollution can inhibit coral growth when the various types of out-of-control algae cover the corals. Acidification of oceans can prevent corals from building their homes, and finally some forms of fishing utilize cyanide to capture fish. This kills other organisms, including corals.

Kelp Forests

Kelp is a type of large, thick, brown algae growing from the base of continental shelves. It forms dense strands that grow in large bunches called **kelp forests** along certain, temperate coasts, in places like Japan and the northwest coast of the United States. Kelp provides food and habitats for many aquatic organisms. Kelp forests absorb wave energy and protect shorelines from erosion. They also provide products for human beings and have economic value. They can be harvested to be eaten. Kelp can also be used in cosmetics, paints, paper, and soaps.

Intertidal Ecosystems

Intertidal or **littoral ecosystems** are found on the shores of the oceans and seas. These ecosystems are found between the highest high tide and the lowest low tide. The organisms that live in these places live part of the time on dry land and part of the time submerged. Despite the apparent challenges to living in these places, intertidal ecosystems are areas of thriving biodiversity.

Within rocky shorelines are crevices and pools, which are the habitats for numerous organisms such as anemones, barnacles, crabs, mussels, sea urchins, starfish, and tunicates (sea squirts). These, and many other, organisms are adapted to daily changes in moisture, salinity, and temperature.

Salt Marshes

Salt marshes are found along coasts in temperature latitudes, often near estuaries and between barrier islands and the continent (Figure 11.6). Like other marshes, salt marshes are flat areas flooded by water, in this case seawater. They are subject to tides, and portions of a tidal marsh are above water during low tides. They are places of high primary productivity in which hearty, salt-tolerant plants like spartina grow. Salt marshes also provide crucial habitats for birds and commercial fish and shellfish. They filter contaminants and pollution and stabilize the shore against **storm surges**. Unfortunately, many salt marshes are endangered due to development and fishing operations.

Figure 11.6 A salt marsh off Cumberland Island in the state of Georgia

Mangrove Forests

Along many tropical and subtropical sandy coastlines are **mangrove** trees. Mangrove trees grow in mangrove forests along these coasts. Mangroves are uniquely adapted to life on the shore. They have a unique root system in which some roots curve upward for oxygen and some curve downward into the water, then into the floor of the ocean for support. Mangrove forests are vital ecosystems and serve as the habitat for many coastal organisms. They also serve as nesting areas for birds. Mangrove forests act to filter pollutants, protect against storm surges, retain soil and sediments, and slow runoff.

They also provide nurseries for commercial fish and shellfish and contain the raw materials for construction materials, medicine, shrimp farming, and tools. Therein lies the problem. For these reasons and due to shoreline development, mangroves have been exploited. Around half of the original mangrove forests are now gone.

Estuaries

Estuaries are partially enclosed bodies of water in which any number of rivers empty. They are a transitional zone between freshwater and saltwater systems, and as such are ecotones. Estuaries provide critical habitats for many organisms, including birds and shellfish. These bodies of water also provide spawning grounds for anadromous fish (fish that spawn in freshwater and mature in seawater) such as salmon. Estuaries provide critical commercial fishing opportunities for crab, fish, and oysters. Many estuaries, such as the Chesapeake Bay, the largest estuary in the United States, have been negatively impacted by overfishing, as well as development and pollution.

The Ocean and Civilization

Human beings have depended on the ocean for thousands of years. It can be argued that civilization exists, in part, due to our dependence on the world's oceans. The oceans have provided people with the ability to move themselves, goods, and services across vast distances long before motorized transportation. The ocean has had an incalculable impact on human civilization, and human beings have had major impacts on the ocean as well.

Energy and Mineral Resources

Beneath the oceans, below the ocean floor, are vast deposits of oil and natural gas, most of which are currently being drilled. Despite safeguards put into place over the years to prevent them, oil spills still occur. Oil spills are devastating to marine ecosystems and the fishing industry. One of the largest recent oil spills, the Deepwater Horizon oil spill of 2010, cost a total of $61.6 billion in damages and took 11 lives.

The oceans can also provide renewable energy resources. Some of these are currently being used, such as vast offshore wind farms. Oceanwater heating, tidal, and waves are also promising forms of renewable energy that are in development.

On the shoreline and beneath the sea are vast deposits of economically valuable mineral resources. These include calcium carbonate, gravel, sand, salt, sulfur, and silica. The oceans also hold deposits of copper, gold, silver, and zinc. Valuable manganese nodules litter the depths of the ocean's floor, but these prove difficult to reach, making it an infeasible mining operation, currently.

Marine Pollution

Coastal cities dumped trash and untreated sewage along the shorelines of the United States even into the middle of the 20th century. Unfortunately, this practice continues throughout the world. From all land surfaces, chemicals, excess nutrients, oil, and plastic make it into the ocean. From boats, especially cruise ships, raw sewage and trash is regularly dumped. Fishing gear is abandoned once it's used up in the fishing industry; unfortunately, that trash makes its way into the ocean.

Plastic is especially problematic as a marine pollutant. Most plastic is non-biodegradable, meaning it persists in the environment. These plastics are transported by ocean current and waves until they make their way to some solid surface, to the bottom of the ocean, or to an organism. Marine wildlife will sometimes ingest it, or sometimes get tangled up in plastic. Both often kill organisms.

GROUND TRUTH

Trash Island

Trash Island, also known as the **Great Pacific Garbage Patch,** is actually a collection of "trash islands" spanning the Pacific Ocean from North America to Japan (Figure 11.7). Though these three collections of garbage floating in the ocean are called islands, one cannot walk on them. They are collections of waste bounded by the North Pacific Subtropical Gyre and do not actually have solid surfaces. A **gyre** is a gigantic, swirling, circular ocean current. At the center of a gyre are calmer waters, which is part of the reason that trash islands exist.

There are currently three trash islands, the Eastern Garbage Patch, the Western Garbage Patch, and the one in the Subtropical Convergence Zone. They are mainly composed of large plastic debris, an assortment of odd garbage such as fishing gear and appliances, and **microplastics**. Microplastics are tiny bits of plastic that have physically been broken down from larger plastics or were manufactured to be small in the first place.

Currently, and unfortunately, the Great Pacific Garbage Patch and other marine garbage patches are growing in size. Scientists and other experts are working to understand these trash islands, and hopefully find a way to at least reduce them and their impacts in the future. In a worst-case scenario, these features are a testament to the irreversible impacts that humans can have on the world's oceans.

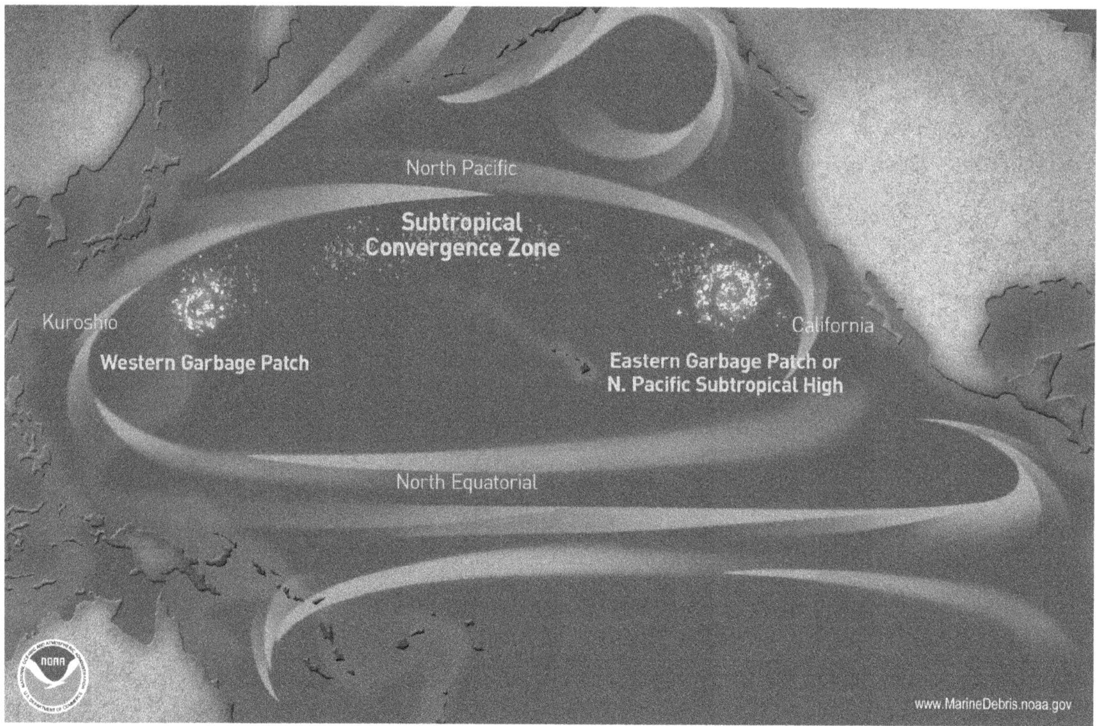

Figure 11.7 Location of the Great Pacific Garbage Patch

Chemicals end up in the oceans from many different sources. The aforementioned oil spills continue to be a global problem. Oil, in large amounts, coats and poisons wildlife. A lot of this oil comes from smaller sources such as small amounts spilled from boats every day. A large amount of oil that leaks into the oceans comes from natural seeps at the bottom of the ocean. These are continuous, and there is little that we can currently do about it. In fact, this has probably been going on for millions of years.

Heavy metals are ocean pollutants as well, especially mercury. Mercury contamination comes from coal combustion as well as other industrial sources. Like other heavy metal toxins, mercury bioaccumulates and biomagnifies through organisms in marine feeding systems. Mercury is dangerous to young children and pregnant or nursing mothers. Mercury toxicity affects the kidneys, lungs, and nervous system. Regrettably, one must be careful when eating fish that are top feeders such as sharks, swordfish, and tuna. One must generally limit one's intake of these fish due to possible high levels of mercury.

Certain marine algae secrete powerful toxins, which can be dangerous when algal blooms occur. Algae often blooms uncontrollably due to nutrient pollution from agricultural runoff. One type of photosynthetic microorganism, *dinoflagellates*, causes **red tide**. Red tide comes from *dinoflagellate* blooms, which discolor ocean water with a reddish pigment. These organisms are toxic and poison wildlife, as well as any humans who enter the water. These red tides also negatively impact fisheries and the tourism industry.

Fishing the Ocean

The oceans have provided a bounty of food for human populations since prehistory. As vast as they are, the oceans are currently vulnerable to not only pollution but to the fishing industry. Humans are putting pressure on marine resources. Around half of the world's fish populations are exploited to the brink of population collapse. For instance, during recent years the world's supply of cod, a popular fish, nearly crashed to the point of extinction. The problem was that juvenile individuals were being fished. Cod are long-lived, up to 20 years. Once young, pre-spawning-age cod were fished, no new cod were spawned. It took a concerted effort and wise stewardship of this resource to bring cod back from the brink of extinction and back as a widely available food source.

Factory Fishing

During the 20th century and into the 21st century, the fishing industry has become increasingly industrialized. **Factory fishing** is the industrialized use of huge vessels that use the latest technology to capture large volumes of fish for global markets. Many of these ships even contain facilities to process, freeze, and package fish while at sea.

There are several methods used for catching fish depending on the intended catch:

- **Driftnets**—Long nets that float in the ocean attached to a series of buoys, flotation devices. These nets are used for schools of fish such as herring, mackerel, and sardines, as well as for sharks.
- **Longline**—A series of hooks and other capture devices floating in the ocean attached to buoys. This is used for swordfish and tuna.
- **Trawling**—Large net that is dragged along ocean bottoms. This is used for groundfish like cod and certain pelagic fish.

Just one of the many problems associated with large-scale fishing, like factory fishing, is **by-catch,** which is the accidental capture of unintended and unwanted animals. The use of driftnets drowns marine mammals and reptiles such as dolphins, seals, and turtles. Longline fishing kills turtles and thousands of seabirds every year. Possibly the most damaging technique, bottom-trawling, destroys entire marine communities. Industrialized fishing, in the large sense, depletes enormous amounts of fish each year. This continues to be a critical problem because the world's populations depend on seafood, primarily large populations in Asia.

Figure 11.8 A large factory fishing vessel

Marine-Protected Areas and Reserves

Government and conservation efforts have worked to set aside as much of the ocean as possible for protection. **Marine-protected areas (MPAs)** have been established on the coastlines of many developed nations over the years. Some fishing and resource extraction is allowed in these areas but on a limited basis, subject to strict guidelines. **Marine reserves**, on the other hand, are areas where fishing and extraction are prohibited. This promotes the continuance of viable, robust marine ecosystems. It also improves the fishing industry because fish that spawn, protected, in reserves often disperse into the wider ocean. These reserves work to reduce habitat destruction as well as promote healthy marine ecosystems.

Conclusion

The Earth's oceans are almost as old as the Earth itself, a testament to the ubiquitous nature of water within the Universe. As such the oceans have been integral to the development of Earth's systems from the formation of rocks and minerals to the movement of tectonic plates to the vast maritime ecosystems beneath their surface. Having covered a great part of the Earth's surface for billions of years, the Ocean continue to harbor secrets, including quite possibly the secret of the origin of life.

Discussion Questions

Directions: Review the chapter in order to completely and correctly respond to the questions and prompts.

Keep this in the back (or maybe even front) of your mind: Science is system of methods that allows us to understand not only our planet, and ourselves, but of the entirety of the universe. Answering the following questions correctly will lead to a more robust understanding of how to apply science to ENVS problems.

1. What are Earth's five oceans, and what are some of their characteristics?
2. What is the impact on carbon dioxide in the ocean, and why is it important?
3. How does dissolved oxygen (DO) enter the oceans? Why is DO important for fish?
4. Think about the various ions of dissolved salts in the oceans. How might they have gotten there?
5. What is the pycnocline, and why did it form?
6. Why is the Gulf Stream important to places off the North Atlantic coast?
7. Describe the various zones and regions in the ocean's waters.
8. Why would sand and sediment collect on one shore of an island rather than another? Think about beach drift.
9. Describe a shoreline ecosystem that you may have been to, or describe one that you have read about.
10. Why is mercury contamination a big problem in the ocean?
11. Can you think of ways to fish more humanely and sustainably?

Bibliography

Levin, H.L., and King, Jr., D., 2017, *The Earth Through Time*: Hoboken, NJ, Wiley, 11th ed., 590 p.
Thomas, D.N., and Bowers, D.G., 2021, *Introducing Oceanography*: Edinburgh, Dunedin Academic Press, 2nd ed., 192 p.
Withgott, J., and Laposata, M., 2019, *Environment: The Science Behind the Stories*: New York, Pearson, 7th ed., 768 p.

Credits

Fig. 11.1: Copyright © by Pinpin (GNU FDL) at https://commons.wikimedia.org/wiki/File:World_map_ocean_locator-en.svg. A copy of the license can be found here: http://www.gnu.org/licenses/gpl.html
Fig. 11.2: Copyright © by Ebmwtreng (CC BY-SA 4.0) at https://commons.wikimedia.org/wiki/File:Pycnocline_illustrative_drawing_03_13_22.jpg.
Fig. 11.3: Interiot, "The Continental Margin," https://commons.wikimedia.org/wiki/File:Continental_shelf.png, 2006.
Fig. 11.4: NOAA, "Diagram of a Water Wave," https://commons.wikimedia.org/wiki/File:Water_wave_diagram.jpg.
Fig. 11.6: Copyright © by Trish Hartmann (CC by 2.0) at https://commons.wikimedia.org/wiki/File:Salt_marsh_Georgia_US.jpg.
Fig. 11.7: NOAA, "Location of the Great Pacific Garbage Patch," https://commons.wikimedia.org/wiki/File:Pacific-garbage-patch-map_2010_noaamdp.jpg, 2010.
Fig. 11.8: Copyright © by Dennis Jarvis (CC BY-SA 2.0) at https://commons.wikimedia.org/wiki/File:Nova_Scotia_DSC_4331_-_M.V._Northern_Osprey_(2489700687).jpg.

Chapter 12

Weather and Climate

The Atmosphere in Which We Live

Objectives

- Understand that Earth's atmosphere is ancient, but it has not always been like our current atmosphere.
- Know that Earth's atmosphere is complex and well structured.
- Come to an understanding of how much we depend on our atmosphere for our very existence.
- Know that pollution comes from many sources, both natural and anthropogenic.
- Know that there are things we can do to help combat air pollution.
- Understand that climate change has been occurring ever since Earth had an atmosphere.
- Know that the current level of climate change is accelerated to a certain extent.
- Come to an understanding of how to use critical thinking to understand the science behind climate change.

KEY WORDS AND TERMS

Anthropogenic (adj.): Any outcome that is generated by human activity.

Cloud (n.): A mass of condensed water vapor or ice particles in the upper atmosphere.

Condensation (n.): Water that collects on surfaces from water vapor after a drop in temperature.

Equator (n.): The geographic line that divides the Earth into the Northern and Southern Hemispheres at its greatest circumference. It is 0° latitude.

Precipitation (n.): Water, in the form of liquid water, or ice particles that falls to the ground.

Remediation (n.): The act of reversing or halting environmental change.

Sublimation (n.): A solid that changes phase directly to gas.

Thunderstorm (n.): A powerful storm that brings heavy rain, hail, strong winds, lightning, and thunder.

Wind (n.): The movement of air in the atmosphere.

Consider This . . .

We live in a thin, fluid sheath surrounding our planet Earth, the atmosphere. While, on our scale, the atmosphere is vast, if one was to look at it from the perspective of a layer of the Earth, it is extremely thin and even fragile looking (Figure 12.1). Perhaps this is a good approach to take as we begin our exploration of Earth's weather and climate systems.

Figure 12.1 The Earth's atmosphere as seen from space

Introduction

The overall term for the scientific study of Earth's atmosphere is **atmospheric science**. Much like other broad scientific fields, such as biology and geology, there are more specific fields of study within the atmospheric sciences. **Meteorology** and **climatology** are two significant sub-disciplines of the atmospheric sciences. Meteorology utilizes atmospheric chemistry and physics to enable specialists to understand current atmospheric events and make short-range predictions of the weather.

One question that is often asked is "What is the difference between **weather** and **climate**?" As Mark Twain said, "If you don't like the weather in New England, just wait a few minutes." Weather refers to atmospheric change, often on a local level over a relatively short period of time, whereas climate is concerned with long-term conditions within the atmosphere often over large areas, or even globally. Records of ancient climates or **paleoclimates** tell us a lot about Earth's past leading up to our present time, much like paleoenvironments for ancient ecosystems and paleontology tell us a lot about ancient life.

Earth's Atmosphere

The Earth's **atmosphere** is the thin layer of gases surrounding Earth's surface. Terminating at the **Kármán line**, the total atmosphere is generally regarded as around 100 km thick, which is a little over 1.5% of Earth's total radius. The Kármán line is the generally accepted end of Earth's atmosphere and the beginning of outer space. It is somewhat arbitrary since, at that altitude, the atmosphere is extremely sparse and next to nonexistent. The Earth's atmosphere is a gaseous fluid that recycles and transports chemicals, nutrients, and water. Secondary chemical reactions occur within the atmosphere, such as the production of carbonic acid (H_2CO_3) from the reaction of carbon dioxide (CO_2) with water vapor (H_2O). Due to both its chemical constituents and its physical nature, the atmosphere moderates Earth's climate.

Origin of the Atmosphere

Within several hundred million years after its formation, the Earth's atmosphere had formed. It was completely different from our current atmosphere, being more like the composition of gases released from volcanoes: carbon dioxide, hydrogen sulfide, methane, sulfur dioxide, carbon monoxide, and the like. As the Earth cooled and was pelted by comets and other bodies during the violence of the ancient solar system, water was delivered to the planet, which ultimately formed the hydrosphere. This, in turn, brought water vapor to the atmosphere through evaporation. Around 3.5 or so billion years ago, life began, as indicated by the fossil record. With this life came photosynthesis, which releases oxygen. Ultimately, about 2 Gya, oxygen began accumulating in Earth's atmosphere, something known as the **Great Oxygenation Event**. By around 541 Mya, at the start of the Cambrian period, oxygen levels were close to current levels.

Composition of the Atmosphere

Earth's current atmosphere is composed of a variety of gases (Figure 12.2). These gases are grouped together as **permanent gases**, gases that do not fluctuate in concentration, and **variable gases**, gases that fluctuate in concentration on a regular basis. These gases are often measured as **mole fractions**, the concentration of a particular chemicals species molecules within a given volume. The permanent gases are as follows:

- **Nitrogen (N)**: 78.08% atmospheric composition. Atmospheric nitrogen is found in the triple-bonded N_3 molecule. For all intents and purposes, this form of nitrogen is inert since it is exceedingly difficult to split it apart. On the other hand, denitrifying soil bacteria on Earth are capable of doing this naturally, making that nitrogen available for plants, which require nitrogen, and in turn life on Earth. Nitrogen is the basis for the nitrogen cycle.
- **Oxygen (O)**: 20.95% atmospheric composition. Oxygen is very reactive, so it is rarely found outside of a molecule. Atmospheric oxygen is usually in the form of the O_2 molecule. Oxygen is an important oxidizer and vital to respiration for all living things, including plants. Heterotrophic organisms require it for cellular respiration in order to break down food for energy.

- **Argon (Ar)**: 0.93% atmospheric composition. Atmospheric argon is in the form of argon-40, of the daughter products of potassium-40 decay within the Earth's crust. This makes argon slightly radioactive. Argon is a **noble gas**, a gaseous element that has a completed outer valence shell of electrons. Due to this, it does not readily bond with other atoms. Argon is not **bioavailable**, meaning it has no known function in biological processes, although it is used in some human industrial processes such as welding.

Other permanent gases in Earth's atmosphere are only found in traces amounts. Of these gases the only one of note is neon (Ne), a noble gas, which comprises 0.0018% of the atmosphere. Helium (He) and hydrogen (H) also found in trace amounts, as are two other noble gases, krypton (Kr) and xenon (Xe).

The variable gases are found in Earth's atmosphere at much smaller concentrations than the permanent gases. All variable gases are **greenhouse gases**, gases that absorb heat from solar radiation emitted from the Earth. These gases then redirect heat back to Earth's surface.

- **Carbon Dioxide (CO_2)**: 0.04% atmospheric composition. Carbon dioxide is vital for photosynthesis, which provides the raw materials for all trophic relationships on Earth. Thus, carbon dioxide is vital for life itself. Carbon dioxide is a major component of carbonate rocks and is a dissolved component of natural waters. It is an important part of the carbon cycle. The most prevalent greenhouse gas, the concentration of atmospheric carbon dioxide, has increased within recent years. It is the fourth most prevalent gas by concentration in the atmosphere behind argon, oxygen, and nitrogen.
- **Methane (CH_4)**: 0.002% atmospheric composition. Methane is one of the simplest hydrocarbons. It is found underground in stratigraphic deposits of fossil fuels as well as beneath the seafloor as **methane clathrates**, methane that is trapped in ice. Some microorganisms, archaea, use methane in their life cycles, and certain animals like cattle emit methane as part of their digestive processes. Wetlands are the world's largest natural source for methane. Methane is part of the carbon cycle.
- **Water Vapor (H_2O)**: 0%–3% atmospheric composition. Water vapor is vital to most biological processes, particularly for photosynthesis. It is present throughout the planet within the hydrosphere and as part of the hydrologic cycle. Water evaporates from surface waters, which can increase locally due to **precipitation**. These are all reasons there is a high degree of water vapor present in the atmosphere at any given time, which varies geographically across the Earth. Water vapor often reaches a saturation point in the atmosphere, condensing into liquid water, something other atmospheric gases do not do. Water vapor freezes, turning into ice high in the atmosphere, during precipitation events, atop tall mountains or in high latitudes.

Other variable gases in Earth's atmosphere include nitrous oxide (N_2O) and ozone (O_3) in trace amounts. Both nitrous oxide and ozone are naturally present in the Earth's lower atmosphere. Nitrous oxide is part of the nitrogen cycle. Ozone is important higher in the atmosphere as part of the **ozone layer** that absorbs incoming radiation. Both gases are part of **anthropogenic** energy and industrial processes, and both gases have been increasing in atmospheric concentration in recent years. **Chlorofluorocarbons** (**CFCs**), a class of variable gases, are present in the atmosphere in trace amounts as well. CFCs are completely anthropogenic and have had a number of commercial and industrial uses, such as in refrigeration and as propellants. Due to the fact that CFCs have a detrimental effect on the ozone layer, their use was generally banned in the late 1980s, and their atmospheric concentrations have been waning in recent years.

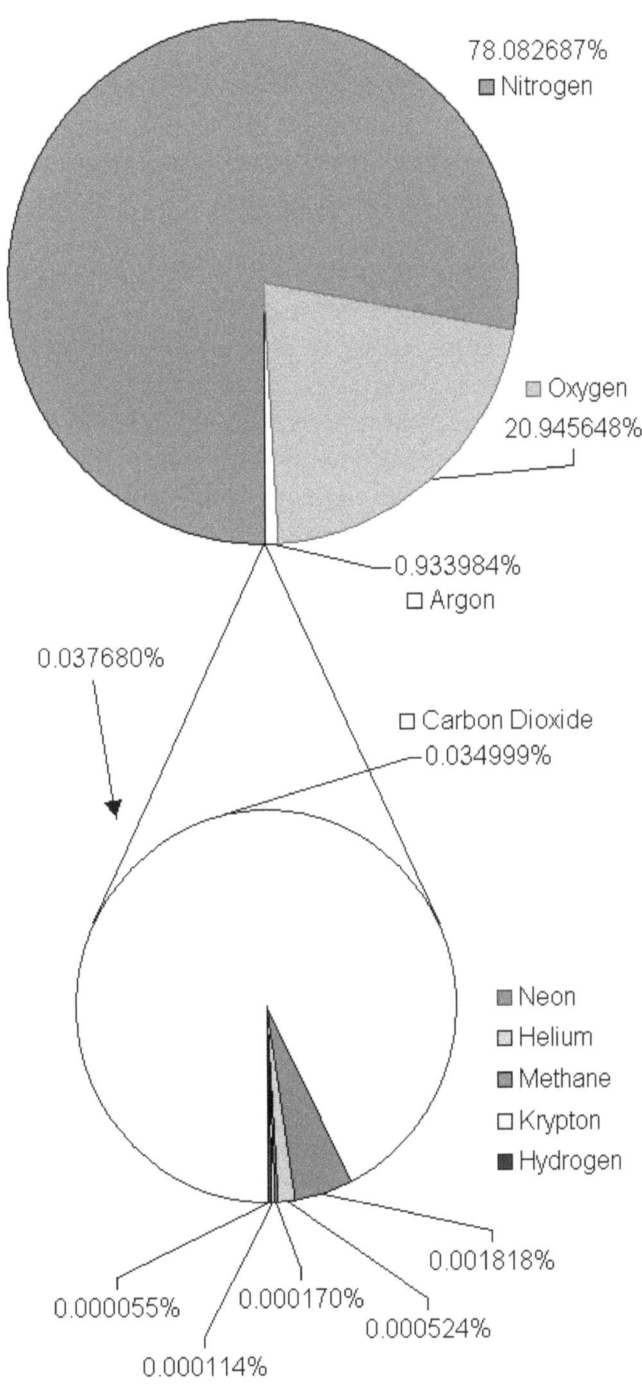

Figure 12.2 General gas composition of Earth's atmosphere

Figure 12.3 Earth's atmospheric layers

Atmospheric Layers

Earth's atmosphere is structured in layers due to compositional, pressure, and temperature differences. There are five layers to Earth's atmosphere, not including the transitional layers between the main layers (Figure 12.3):

- **Troposphere**: 0 to 6–20 km above Earth's surface. The troposphere is the layer of the atmosphere that contacts Earth's surface. The height of the troposphere varies and is much higher near the equator. What people generally think of as the "atmosphere" is the troposphere since all living things on Earth live within or under the troposphere. Weather, as we know it, occurs within the troposphere, in part due to **atmospheric convection,** meaning that air travels vertically as well as horizontally. Air is densest and air pressure is highest in the troposphere, and most of Earth's water vapor is found in this layer of the atmosphere. Temperatures steadily decline with height in the troposphere to below freezing at the top. The boundary layer between the troposphere and the next layer is called the **tropopause**.
- **Stratosphere**: 6 to 50 km above Earth's surface. Around 19% of Earth's atmospheric gases are in the stratosphere but only trace amounts of water vapor. Earth's **ozone layer** is within the stratosphere between 15 and 30 km up. The ozone layer, composed of the gas ozone (O_3), shields Earth from solar ultraviolet radiation. Air does not vertically mix in the stratosphere, and temperatures increase with height to above freezing. This is because the ozone layer acts as a boundary. The boundary layer between the troposphere and the next layer is called the **stratopause**.
- **Mesosphere**: 50 to 85 km above Earth's surface. Air becomes very thin in the mesosphere but not thin enough for it to protect Earth from most meteors. Meteors are slowed down and often burn up in the mesosphere. Temperatures decrease steadily with height in the mesosphere to below freezing. The boundary layer between the troposphere and the next layer is called the **mesopause**.
- **Thermosphere**: 85 to 600 km above Earth's surface. Air is extremely thin in the thermosphere and becomes even thinner with height. This is inversely proportional to temperature in this layer, which steadily climbs to around 2,000° C at the top. This is because this is the atmospheric layer in which solar radiation is absorbed by atoms and molecules, causing a sharp increase in temperature. At the same time, astronauts who may go outside of their spacecraft do not burn up because the concentration of gases is not enough to affect them. The **ionosphere** is within the thermosphere between 60 to 300 km above Earth's surface. The ionosphere is a layer of ions formed from incoming solar radiation. Radio signals are reflected off it, making it possible to send them around the world. The boundary layer between the troposphere and the next layer is called the **thermopause**. The thermosphere also contains the **Kármán line** at 100 km above Earth's surface. This boundary was calculated by the mathematician and aerospace engineer, Theodore von Kármán, to be the line in which "outer space" essentially begins because at this point conventional flight is not supported by the atmosphere.

- **Exosphere**: 600 to 10,000 km above Earth's surface. The exosphere ends where outer space truly begins, because at around 10,000 km above its surface the Earth's influence on atoms and molecules is less than that of the influence of the Sun. The air is so thin in the exosphere that atoms and molecules rarely collide. The temperature is around a steady 980° C or so in the exosphere due to insolation. Again, particles are so scarce there that astronauts are fairly safe from the temperatures, as is their equipment.

Properties of Earth's Atmosphere

The Earth's atmosphere, especially within the troposphere, has a number of unique properties These tropospheric properties lead to weather phenomena across the planet.

Atmospheric pressure is the air pressure on the surface of the Earth that is produced by a column of air. Air becomes denser, with atoms and molecules more tightly packed, closer to the Earth's surface, so air pressure is greatest closest to the Earth's surface. Atmospheric pressure decreases with altitude. Eighty to 90% of all atmospheric air molecules are concentrated within the troposphere. Pressure differences, or **pressure gradients**, across the surface of the Earth are linked to weather phenomena such as storms. Air pressure is measured as force per unit area.

Relative humidity is the ratio of water vapor in the air, the **actual vapor density**, over the amount that the air can possibly contain, the **saturation vapor density**, expressed as a percent, so the ratio is multiplied by 100. Saturated vapor density is also known as the **dewpoint**. The saturated vapor density is directly proportional to decreases in temperature, meaning that the saturated vapor density decreases as temperatures decrease. This results in **condensation** in the upper atmosphere and at the surface of the Earth when the air is humid and the temperatures are relatively cool. Condensation of this sort, near the Earth's surface, results in **fog** (Figure 12.4). Lower saturation vapor density, due to lowered temperatures, is why fog usually occurs in the morning.

The formula for relative humidity (rH) is as follows:

$$rH = \frac{actual\ vapor\ density}{saturation\ vapor\ density} x100$$

Temperatures across the surface of the Earth are highly variable, and temperatures with height depend on conditions within Earth's atmospheric layers. Temperatures are largely controlled by insolation, as well as location across the surface of the Earth.

Figure 12.4 Fog curls around the peaks of Mt. Lu (Lushan) in Jiangxi province, China

The Influence of the Sun

The geospatial association between the Earth and the Sun regulates the amount of solar energy that the Earth receives from the Sun in any given place and at any given time. The Sun's energy is the primary driver of Earth's weather and climate, including air movement and temperature. Solar energy also controls evapotranspiration and thus is the primary driver of the hydrologic cycle, which in turn is part of Earth's weather and climate patterns. The influx of solar radiation is greatest at the **equator**.

Seasonal Reasons

The Earth's seasons are due to the **axial tilt** of the Earth. The current axial tilt of the Earth is 23.4° from the rotational axis. This tilt changes over time in a period of some 41,000 years, with a variance of between 22.1 and 24.5 degrees. Currently the axial tilt is decreasing.

The axial tilt is the reason for Earth's seasons. As Earth orbits the Sun, the tilt of either the northern or southern hemisphere is subject to either more direct or less direct insolation. When one hemisphere is subject to more direct sunlight, that hemisphere will be warmer than normal. At maximum insolation, when sunlight strikes the Earth most directly, this leads to the seasons we call summer, in the northern hemisphere, or winter in the southern hemisphere. Each hemisphere switches as the Earth orbits the Sun. Thus, it is the least oblique or more direct insolation that warms the Earth the most in one given **latitudinal hemisphere**, while the opposite latitudinal hemisphere experiences colder temperatures due to more oblique or indirect insolation. Latitudinal hemispheres are the Northern and Southern Hemispheres divided by the equator.

Convective Circulation

As the Earth's surface is heated by the Sun, it heats the air above it. This warmer air expands and rises. Rising air is transported across the surface of the Earth by **winds**, as it continues to rise. As air rises it cools and contracts, becoming denser, ultimately sinking to Earth's surface to continue the cycle anew. This pattern of vertical convection influences both weather and climate patterns.

Convective circulation also delivers water vapor to the upper troposphere. As relatively warm air rises it carries water vapor, as well as heat, near Earth's surface into the upper atmosphere. Water vapor in the upper troposphere condenses and releases heat, cooling. This condensed water forms **clouds**. When the upper atmosphere cannot contain condensed water, it falls to the Earth as some form of precipitation. This is especially pronounced in the tropics since lower latitudes are relatively warmer and more humid than upper latitudes.

Air Masses and Fronts

Air masses with similar characteristics like air density, moisture, and temperature move across the surface of the Earth, driven by prevailing winds. The boundaries between these air masses are called fronts. There are several types of fronts. Between 30 and 60 north latitudes, essentially the continental United States, these fronts are pushed by **westerlies**, winds that blow from the west. The main ones are as follows:

- **Cold front**—A boundary between a moving mass of cooler, denser drier air and a warmer, moist air mass. The colder air mass replaces a warmer air mass. Cold fronts bring precipitation, often intense rain, and sometimes **thunderstorms**.
- **Warm front**—A boundary in which warm, moist air displaces colder, drier air by essentially sliding over the colder air mass. Warm fronts move slower than cold fronts and often bring light rain.
- **Occluded front**—Often a cold front that overtakes a warm front. The two air masses curve counterclockwise, sometimes bringing intense thunderstorms. Drier conditions usually follow an occluded front.
- **Stationary front**—A stalled front between two masses that are similar. Stationary fronts tend to persist and often bring rain.

Atmospheric Pressure

Air pressure across the Earth differs due to prevailing conditions. Centers of pressure that are different from the surrounding build up from time to time. When **high-pressure centers** occur, air tends to spiral downward in a clockwise manner (it's the opposite in the southern hemisphere), moving away from the center as air descends. These centers tend to bring drier conditions and fair weather.

Low-pressure centers cause air to be drawn upward, spiraling in a counterclockwise manner, in the northern hemisphere, in a configuration called a cyclone. As air moves toward the low-pressure center and upward, the system causes the air to become turbulent; these systems often bring clouds, precipitation, and even thunderstorms. Extremely intense low-pressure systems can induce convection, which can cause very powerful storms, **hail**, and **tornadoes**. Hail is precipitation in the form of ice that forms within the updrafts of powerful thunderstorms. This is why hail usually occurs during warmer months, because surface heating is one of the main drivers of thunderstorm formation. Tornadoes are columns of air that contact the ground, rotating violently. Tornadoes are often caused by large, intense low-pressure thunderstorms called **supercells**.

Thermal Inversion Layer

Rising tropospheric air usually rises upward, unimpeded, gradually becoming colder as altitude increases. This induces a cycle of vertical mixing with warm air rising after cool air falls. **Thermal inversion layers** are layers of cool air between two layers of warmer air (Figure 12.5). Inversion layers essentially trap warmer air underneath, restricting its circulation. These inversion layers often occur because of topography that isolates lower land surfaces as well as rapid heating from paved surfaces. This often occurs in the Los Angeles Basin.

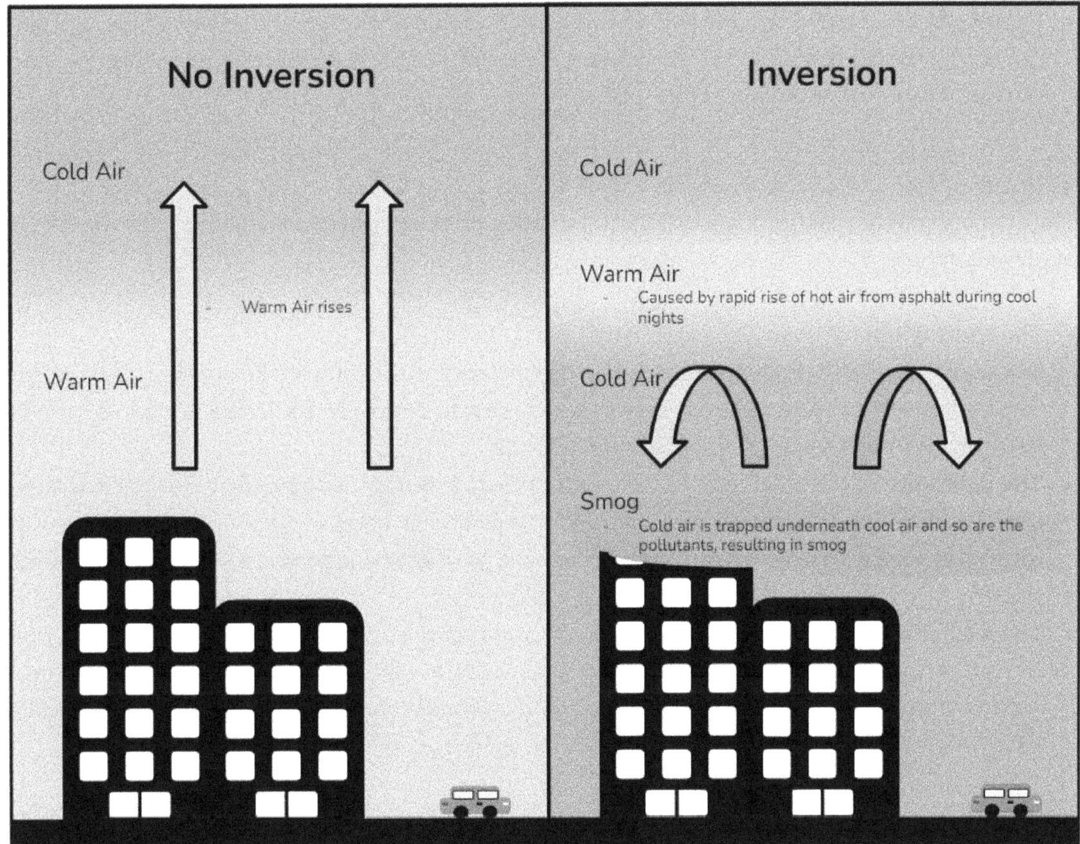

Figure 12.5 Thermal inversion compared to a lack of thermal inversion in an urban setting

Climatic Patterns

Large-scale, convective currents cause regional climatic patterns based on latitudinal position. These affect overall circulation, moisture distribution, and weather patterns.

Global Circulation Patterns

Global circulation patterns are large-scale movements of air called atmospheric **cells**. These occur between specific latitudes and are responsible for weather on a global scale. There are three atmospheric cells:

- **Hadley cells**: Between the 0° and 30° north and south latitudes. Hadley cells are tropical circulation patterns that occur and terminate at the equator. Hot, rising surface air is blown toward the equator. This rising air cools, and the moisture that it carries produces clouds at the equator. These clouds release water, resulting in heavy rainfall near the equator.
- **Ferrel cells**: Between the 30° and 60° north and south latitudes. Warm air is driven northward across the surface of the Earth from one of the 30° tropical latitudes, the tropic of Cancer in the northern hemisphere and the tropic of Capricorn in the southern hemisphere. This air rises, cools, and converges around the 60° latitude. This doesn't necessarily cause storms. In fact, energy in this system is moderate, so the wind and weather that it produces is highly variable.
- **Polar cells**: Between the 60° and 90° north and south latitudes. Air rises at the 60° latitude, descending at 90° toward the North or South Pole. These are the weakest of the cells and have little impact on the extremely cold weather at either of the poles.

Wind Patterns

The interaction between atmospheric cells and Earth's rotation produces global wind patterns between the key latitudes of 0°, 30°, and 60°:

- **The trade winds**: Between 30° and 0° north and south latitudes. These winds tend to blow to the southeast in the northern hemisphere and to the northwest in the southern hemisphere. They are called "the trade winds" because hundreds of years ago these winds were used by mariners to cross the Atlantic Ocean in the tropics toward North and South America.
- **The doldrums**: At the 0° latitude. At the equator the trade winds converge, causing a drop in wind energy, often resulting in no winds at all. They are called "the doldrums" because mariners in the past knew that entering those waters ran the risk of getting stuck in a place where no winds could fill the sails of their ships.
- **The westerlies**: Between 30° and 60° north and south latitudes. Called "the westerlies" because these are winds that blow from west. The overall tendency is to blow to the northeast in the northern hemisphere and to the southeast in the southern hemisphere. The westerlies are the strongest global wind patterns and are associated with the type of weather found in temperate zones.

Air movement across the Earth tends to become deflected from a straight path, bending to the right in the northern hemisphere and to the left in the southern hemisphere. This deflection is caused by wind patterns interacting with the Earth's rotation. It is called the Coriolis effect, which can also contribute to the formation of strong storms like hurricanes.

Air Pollution

Air pollution is cause by **pollutants** that become airborne. Pollutants are biological agents, chemicals, or substances that can harm people and other organisms, as well as cause environmental damage. When one considers air pollution **outdoor**, or **ambient**, **air pollution** is what one usually ponders. Outdoor air pollution occurs in open spaces. Outdoor air pollution happens globally; it has many forms and many causes. Many governments, particularly in the developed world, have enacted strict policies to combat air pollution in recent years, but it continues to be a problem, especially in urban areas. Countries in the developing world, particularly in equatorial areas, do not necessarily have such policies in place, so air pollution tends to be more of a problem in those nations.

Classifying Outdoor Air Pollution

Air pollution can be emitted by mobile or stationary sources. Mobile sources are usually associated with vehicles such as cars, trucks, and airplanes. Although there are many low-emission electric vehicles in use today, many motorized vehicles still run on fossil fuels, mainly petroleum products.

Knowing where air pollution comes from is important for **remediation**. **Point sources** are the specific locations from where pollutants are discharged, such as coal-fired power plants, factories, or even a singular smokestack (Figure 12.6). Nonpoint sources come from broad areas and multiple sources, such as motor vehicles within a city.

Figure 12.6 Point-source air pollution smoke rising from plant tower

It is important to understand the method of origination with regard to how pollution is generated. There are two main classes on pollutants in this regard. Pollutants that are directly harmful as they originate from their source are **primary pollutants**. Examples are soot or nonreactive gases. **Secondary pollutants** are formed when two or more chemical species interact within the environment, for instance when hydrogen sulfide and water react to form sulfuric acid.

Natural Air Pollutants

Natural air pollutants are emitted from natural processes and sources. They are often beyond the scope of human beings to eliminate. People can only be aware of, and prepare for, the release of natural pollutants. The following are the major forms of natural air pollution:

- **Dust storms**—Windblown fine particulate matter that becomes dispersed across the land. Dust storms generally occur in arid areas and are sometimes violent and dangerous. Erosion from desertification is associated with dust storms due to unsustainable agricultural practices. This was the cause of the Dust Bowl in the American West in the 1930s. Every year dust is blown across China from dust storms in the Gobi Desert to the north, and dust is blown from Africa to the Americas by the trade winds.
- **Methane**—Produced by the digestive processes of certain animals and from rotting conditions within wetlands. Methane is usually not a serious pollutant because it occurs in low concentrations. It can be an irritant when air is stagnant and often brings a foul odor.
- **Pollen**—Plant sex cells that are often distributed by winds. Although a necessary part of functional ecosystems, pollen can be a nuisance, coating outdoor structures and machinery. Pollen is also an allergen and a serious health problem for many people.
- **Volcanoes**—The emission of gases and particulate matter from volcanic vents, fissures, and lava itself. Volcanoes emit a suite of gases, including water vapor (H_2O), carbon dioxide (CO_2), sulfur dioxide (SO_2), hydrogen sulfide (H_2S), nitrogen (N), argon (Ar), helium (He), neon (Ne), methane (CH_4), carbon monoxide (CO), and hydrogen (H). High concentrations of these gases make air unbreathable. In 1987, volcanic carbon dioxide was the cause of death for around 1,700 people in West Africa. Carbon dioxide in high concentrations is heavier than air, hugging the ground and making it inescapable. Carbon monoxide is a toxic, invisible, odorless gas that is lethal in high concentrations. Volcanoes also emit particulate matter in the form of aerosols. Aerosols are solid or liquid particles that become airborne. Volcanic aerosols block sunlight. High amounts of aerosols sent into the upper atmosphere can effectively lower the average temperature of the planet. The "year without a summer" in 1816 was the result of aerosols emitted by the volcanic Mount Tambora in Indonesia. Silicate particles from volcanoes in the upper atmosphere can also abrade aircraft engines, possibly causing engine failure.
- **Wildfires**—Outdoor fires that occur in woodlands and grasslands. Since 1983, the National Interagency Fire Center, a United States federal agency, has documented around 70,000 wildfires per year, although this number seems to be growing. Wildfires release clouds of soot and ash, which can travel for miles depending on wind speed. They also release a complex mixture of gases depending on the fuel, but predominant among these are carbon dioxide, carbon monoxide, and methane. The frequency, intensity, and length of wildfires is currently increasing on an annual basis, with anthropogenic climate change as one of the factors.

Anthropogenic Air Pollution

There are many types of anthropogenic air pollutants being emitted on the planet today. Pollutants come from all sorts of enterprises. **Industrial** air pollution comes from petrochemical plants, steelmaking, and synthetic chemical factories. Large-scale agriculture is industrial in scope and highly dependent on fossil fuels. Ammonia, methane, and nitrous oxide are emitted from large feedlots where cattle, chickens, or pigs are raised. Air pollution from **commercial** sources include fossil fuels used to run businesses, chemicals used for building materials, and chemicals used for cleaning that emit vapor. **Residential** sources for air pollution are more diffuse and difficult to identify but can include cleaning materials and fossil fuels.

Most forms of anthropogenic air pollution can be reduced or eliminated by reducing at the source. Use of low toxicity and natural materials can help on a consumer level. On the commercial and industrial level, compliance with legislation from regulatory agencies is especially vital. For instance, through compliance and best practices most coal-fired and other industrial plants are required to use **scrubber** technology to curb emissions. Scrubbers are diverse technologies that are used to eliminate harmful pollutants from exhaust streams.

CRITERIA POLLUTANTS

The **EPA** identifies six classes of air pollution as **criteria pollutants**. These are pollutants found throughout the environment that can cause significant harm to human health and even damage property:

- **Carbon monoxide (CO)**—A colorless, odorless gas the comes from fossil fuel–burning vehicles, gas stoves, and volcanoes. Carbon monoxide is toxic, and when concentrated indoors may lead to death. In outdoor spaces carbon monoxide can pose a risk to people with preexisting health conditions.
- **Ground-level ozone (O_3)**—A secondary pollutant formed from the reaction of oxides and organic compounds. While Earth's stratospheric ozone layer is beneficial, ozone on the ground can cause harm. Ground-level, or tropospheric, ozone is often concentrated in urban areas due to the large amount of motor vehicles on highways and roads. For this reason, many news sources broadcast ozone alerts on a regular basis. Ozone causes breathing problems from moderate to severe and can harm the environment.
- **Lead (Pb)**—A heavy metal element that gets into the environment from metal smelting, transportation, utilities, and battery manufacturing. Lead is a neurotoxin and can also damage the kidneys as well as the cardiovascular system. Lead is persistent in the environment; it is generally immobile and nonreactive. Lead in soil can degrade the soil, causing damage to plants and animals in the soil.
- **Nitrogen dioxide (NO_2)**—Nitrogen dioxide in the environment comes from the burning of fossils fuels, particularly petroleum for transportation. Nitrogen dioxide is part of group of highly reactive chemicals called nitrogen oxides (NO_X). Nitrogen dioxide is often concentrated in urban areas due to the large amount of motor vehicles on highways and roads. Nitrogen oxide can cause breathing problems ranging from moderate to severe. It reacts with water vapor in the air, causing acid rain, something that causes damage to human infrastructure and the environment.
- **Particulate matter**—Airborne particles of varying size, from barely visible to microscopic, that are emitted into the environment from a variety of sources such as construction sites, power plants, motor vehicles, or outdoor fires. Particulate matter can cause breathing problems from moderate to severe. It can also be an environmental hazard.
- **Sulfur dioxide (SO_2)**—Sulfur dioxide in the environment comes from power plants, industrial facilities, large motor vehicles, and volcanoes. Sulfur dioxide is part of group of chemicals called sulfur oxides (SO_X),

but sulfur dioxide is found in much higher levels of concentration. Sulfur dioxide is a respiratory hazard, especially for the very young, the elderly, and those with preexisting conditions. It reacts with water vapor in the air, causing acid rain, something that causes damage to plants and entire ecosystems.

TOXIC AIR POLLUTANTS

Toxic air pollutants are known to cause cancer, reproductive harm, neurological disorders, and developmental disorders; disrupt the immune system; or cause respiratory problems. Some are produced naturally, such as hydrogen sulfide and other volcanic gases, but many are anthropogenic, produced by industry, smelting, and sewage treatment. These pollutants are not monitored as closely as the criteria pollutants, although there are standards for detection in place in most cases.

FOSSIL FUELS AND SMOG

Smog, a word that is a combination of smoke and fog, is a mixture of air pollutants that are usually visible due to the presence of particulate matter such as smoke and soot. This mixture of pollutants occurs in urban areas because of industrial production and the large amount motor vehicles in such areas. It became a problem at the beginning of the industrial revolution in England, and it continues to be a problem today. Smog tends to become trapped in low-lying areas due to weather conditions, topography, and thermal inversions.

There are two main types of smog. Smog from coal and oil-burning industries is **industrial**, or **gray air**, **smog**. It continues to be a problem especially in colder climates and places of stagnant air. In recent years China, one of the world's largest economies, has had major problems with this form of smog due to the large number of coal-fired power plants in that country. In general, governments have stringent conditions put upon power plants to keep emissions low in order to prevent industrial smog.

Photochemical, or **brown air**, **smog** is produced through light reaction with various chemical air pollutants such that are generally emitted by motor vehicle exhaust. This creates a variety of secondary pollutants and often forms a brown haze in the air that can cause irritation to the eyes and throat. This type of smog often occurs in urban areas within hot, sunny climates within basins. It is a common problem in Los Angeles and Mexico City.

GROUND TRUTH

The Great Smog of 1952

By 1952, smog had become a way of life for the inhabitants of London, England. Since the early industrial revolution in the late 18th century, mechanized progress fueled by steam and coal was a way of life in England, bringing clouds of steam, soot, and ash that often shrouded the city. The smog that crept across the streets of London in the evening of December 5, 1952, was different though.

It was especially cold that day, and most Londoners were burning large amounts of coal in their homes to keep warm. The smoke from this combustion did not escape very far into the air though. A high-pressure cell had formed over London, bringing calm conditions and very little wind. To make matters worse, a thermal inversion had formed over London, trapping both cold air and moisture near the ground. The inversion not only trapped coal smoke from homes near the surface of the Earth, but factory smoke as well.

During the morning of December 6, the ash, gases, and soot from combustion combined with the fog that had condensed from moisture and low temperatures to form smog. This smog was so thick that it was impossible to see in some places. Serious traffic problems occurred later in the morning as people attempted to go about their business.

By December 9, 1952, the damage had been done. Around 4,000 people died due to the smog, but the number may be higher. People continued to suffer respiratory problems after the smog had lifted. There were also reports of cattle dying from asphyxiation at Smithfield Market, London's largest livestock and meat market.

The Great Smog of 1952, as it became known, led the British government to pass legislation so that it would never happen again. These included the Clean Air Acts of 1956 and 1968. Though smog was still a problem in London for a few decades after the great smog, thankfully nothing like it has happened since.

Figure 12.7 The Great Smog of 1952, London, England

RURAL AIR POLLUTION

Rural air pollution comes from a variety of sources. Outdoor air pollution is unconfined, so pollutants from cities, factories, and industry can drift into nearby rural areas. The large volume of pesticides used on farms can become airborne. Cattle, chicken, and hogs in feedlots often emit large volumes of ammonia, hydrogen sulfide, and methane. These gases cause high rates of respiratory problems for people living or working in rural areas near these operations.

CFCS

Earth's stratospheric ozone layer is important because it blocks a great amount of harmful ultraviolet (UV) from reaching the Earth. **Chlorofluorocarbons (CFCs)** are a group of chemicals that damage the ozone layer. Prior to the late

1980s, the widespread use of CFCs had harmed the ozone layer to a point that a hole had formed in the ozone layer over Antarctica. CFCs also cause skin cancer, harms crops, and decrease productivity in the ocean.

In 1989, 20 nations began enforcing the Montreal Protocol, including the United States. This was an agreement to ban the use of CFCs. It worked, and the use of CFCs was all but eliminated. This reversed the damage to the ozone layer as well, although it is still recovering. The Montreal Protocol is an environmental success story and should stand as an example of what governments can do when they cooperate for positive change.

ACID RAIN

Acid rain is precipitation, usually rain, that contains acidic chemicals. Rainwater is often naturally acidic due to the presence of carbonic acid. Air pollutants, particularly nitrogen and sulfur oxides, decrease the pH of rainwater even further. These chemicals react with free hydrogen in water to form acidic chemicals, particularly nitric acid and sulfuric acid. When acidic precipitation falls to the ground, it forms **acid deposition** on surfaces. These acids cause harm in a variety of ways: Carbonate rocks used in building materials weather away due to dissolution, metal ions solubilize and pollute water, crops become damaged, nutrients become leached from soils due to chemical reactions with acidic species, fish die in large concentrations, and trees become damaged. Acid rain and acid deposition are difficult and persistent problems. Some legislation has been passed to tackle the issue, and industry has put some measures into place, but it remains work to be done.

INDOOR AIR POLLUTION

Indoor air pollution is often more severe than outdoor air pollution because air is concentrated in small spaces without much dispersal. People in many parts of the developed world spend most of their time indoors. This increases exposure time to a host of air pollutants. Many synthetic chemicals **volatilize**, or change phase from solid or liquid to gas. Many of these chemicals are not regulated. For decades buildings have been sealed off with insulation to improve energy efficiency. Until recently asbestos and other fibers have been used for insulation. Now these materials are being removed from most public buildings because asbestos is made from silicate minerals that are known to cause severe health problems such as **mesothelioma** (a type of cancer that damages organs, particularly the heart) and lung cancer.

In developing countries, people continue to burn dung, charcoal, crop waste, and wood indoors, often with little to no ventilation. These sources of combustion release dangerous pollutants like carbon monoxide and particulate matter, resulting in a variety of, at times severe, health problems. Millions of people in the developing world become sick and even die each year due to this problem.

Another source of indoor air pollution is secondhand cigarette smoke. This is a problem that has seen a sharp decline in recent years in most developed countries. In the United States, the use of cigarettes and other tobacco products has sharply declined. From 2005 to 2021 the use of tobacco decreased from 20.9% of the population to 11.5%. Cigarettes, especially most commercially available brands, contain thousands of different chemicals. Cigarette smoke irritates the eyes, nose, and throat and are known carcinogens.

Radon (Rn), a radioactive chemical element, is a colorless, odorless gas that is naturally emitted from rock, soils, and sediments. It can build up and become concentrated in enclosed spaces such as basements. Radon can damage lung tissues if inhaled and may lead to lung cancer. Most homes are now radon resistant but older buildings may not be. Radon testing kits are inexpensive or even free through state and local government programs.

Volatile organic compounds (VOCs) are a diverse class of pollutants that have only been recently recognized. VOCs are emitted from many common substances such as dry-cleaning chemicals, hydraulic fluids, paint thinners, and petroleum fuels. They are the components of many plastic products, which release VOCs through volatilization. VOCs are usually released in very small amounts but can become concentrated in small, enclosed spaces. They can be found in pesticides as well as in, formaldehyde, a chemical used in insulation and pressed wood board. It is also used as an embalming fluid. Some VOCs are known to have adverse health effects, but many are still poorly understood.

Organisms are also pollutants. This includes bacteria, fungi, and viruses. These disease-causing organisms can become airborne vectors for disease. From 2020 to 2022, the COVID-19 pandemic raged across the world. COVID-19 is caused by the SARS CoV 2 virus. COVID-19 caused at least 3 million deaths worldwide. COVID-19 is especially problematic because it is directly transmitted through respiratory droplets, which can become concentrated in enclosed spaces. Animal dander and dust mites are often airborne and can adversely affect people who have respiratory problems. Fungi in the form of mildew and mold can cause respiratory problems as well.

Sick building syndrome is a series of nonspecific illnesses that affect people who spend much of their time within specific buildings. It is often caused by one or more of the aforementioned indoor air pollutants, but sometimes the problem is not so easy to understand. Many problems associated with indoor air pollution can be solved to some extent by using natural building materials and making sure that buildings are well ventilated. Getting outside more can also help.

Earth's Climate

The climate of the Earth encompasses atmospheric conditions over large areas of the Earth's surface and over long time scales. Climate also implies long-term change. These conditions include moisture content, precipitation, temperature, wind patterns, and other components of Earth's atmosphere. Earth's climate is directly influenced, and influences, the biosphere, hydrosphere, and lithosphere. Tectonic change over large spans as well as ocean circulation are especially tied to changes in Earth's climate, as are astronomical changes in Earth's position within the solar system with respect to the Sun.

Global climate change is a term that simply refers to Earth's climate changing over time. These changes affect all weather and climate patterns. Climate change on a global scale occurs within geological spans of time far beyond human lifetimes. While forcing elements within the atmosphere, or the Earth system as a whole, may hasten climate change, the change is not sudden. It would take an event on the scale of an asteroid impact or the eruption of a supervolcano to enact sudden climate change.

Global warming is a term used to describe increases in Earth's average temperature. For that matter, **global cooling** has occurred in the past as well. Global warming and cooling are just one component of the global climate changes that have been occurring in cycles for at least the past 600,000 years, probably longer. These changes are cyclical and happen beyond human control. At the same time, human beings do have an effect on our planet. Earth's average temperature has increased by an average of 0.06° Celsius, 0.11° Fahrenheit, per decade since 1850, which is about 2° F in total. One can possibly dismiss some of these findings due to the veracity of recordkeeping over time, but the rate of warming since 1982 is a set of data points that are known to be accurate and robust. The rate of warming since 1982 is has accelerated to 0.36° F (0.20° C) per decade. The year 2023 was the warmest year on record since recordkeeping began in 1850.

Greenhouse Gases

When solar radiation strikes Earth's surface, some of that radiation is absorbed, and some of it is directed upward. This redirected wavelength of radiation is in the infrared spectrum. Greenhouse gases in the atmosphere absorb that radiation and redirect back to Earth's surface, heating it. The higher the concentration of greenhouse gases, the more radiation is redirected.

There are a number of greenhouse gases in the atmosphere. Carbon dioxide has the highest concentration and is the fourth-most abundant gas in Earth's atmosphere at 0.4% of total atmospheric composition. This is a small number, but the amount of carbon dioxide is increasing steadily, and has been since data began to be collected on the matter. Carbon dioxide levels have climbed from around 320 ppm in 1960 to around 420 ppm in 2020, a 24% increase.

Other greenhouse gases include methane, nitrous oxide, and others. Overall atmospheric methane and nitrous oxide concentrations in the atmosphere are currently fairly low, around 1911.8 parts per billion (ppb) for methane. Nitrous oxide concentrations are a bit lower than methane. These greenhouse gases, in comparison to carbon dioxide, re-emit solar radiation at higher levels. They have higher **global warming potential** than carbon dioxide. The global warming potential is the ability to re-radiate heat as measured against carbon dioxide's potential, which is 1. Methane has a global warming potential of between 27 and 30, and nitrous oxide has a global warming potential of around 273. The atmospheric concentrations of both methane and nitrous oxide are currently increasing.

Milankovitch cycles

The **Milankovitch cycles** are named after Serbian mathematician and scientist Milutin Milanković, who formulated them in the early 20th century. The Milankovitch cycles are series of cyclical changes in Earth's position with respect to the Sun based on data-driven mathematical calculations. Like all models, these calculations can be used to understand past conditions as well as potential future conditions.

The Milankovitch cycles consist of three basic components:

1. **Orbital eccentricity**—Changes in the shape of Earth's orbit around the Sun. The orbit of the Earth around the Sun is not perfectly circular but close to it. Currently, orbital eccentricity is slowly approaching its least elliptical, most circular position. This cycle for eccentricity is about 100,000 years.
2. **Axial tilt**—The tilt of Earth's axis with respect to its orbital plane. The current axial tilt of the Earth is 23.4°, and it is slowly decreasing. The axial tilt of the Earth is always between 22.1 and 24.5 degrees with respect to the orbital plane, which is perpendicular to the path that the Earth makes around the Sun. Axial tilt is a cycle that spans 41,000 years.
3. **Axial precession**—Also known as "axial wobble," the inclination of orbit with respect to movement around the Sun. The cycle of axial precession back to the Earth's original position is around 23,000 years.

Each of these large-scale, long-term factors are calculated together to give a complete picture of Earth's long-term, not short-term, climate change. Another factor for long-term climate change on a geological time scale is the position of Earth's continents as a function of tectonic plate movement. Tectonic activity influences ocean circulation. When large land masses occupy the polar regions, ice sheets build up, leading to cooling. The opposite can occur if there are not land masses at the poles and no polar ice sheets.

Long-term climate change is inexorable and occurs regardless of anthropogenic, or even biological, activity, to a certain extent. Currently, it appears that there is a long-term push toward the slow warming of the Earth, given the large-scale factors. These changes happen, again, over geological, even astronomical time scales. At the same time, there is a "spike" in increased average warming that should not be ignored. For that matter, biological activity is responsible for atmospheric oxygenation. This cannot be discounted, nor can the activity of human beings.

Other Influences

As we can see, influences on Earth's overall climate are manifold. In truth, these help illustrate the fact that science is inherently complicated and systematic changes are often subtle and hard to completely characterize in simple terms. For instance, scaling problems exist in which simple explanations that may apply to smaller systems often do not work for larger complicated systems.

The Sun is the primary driver of weather and climate on Earth. As such, it would seem that variability in solar output would be highly influential with regard to Earth's climate. As it is, this variation is not enough to cause great changes in Earth's climate.

The ocean is the largest reservoir for carbon, hence carbon dioxide, on Earth. The oceans hold more than 50 times more carbon than the atmosphere. Carbon absorption slows the warming of atmospheric carbon; it is not a preventative phenomenon. A positive feedback effect is currently occurring due to warming oceans in the tropics and in the polar regions. Warmer oceans absorb less carbon dioxide than cooler oceans because carbon dioxide is less soluble in relatively warmer waters. As less carbon dioxide is absorbed in the ocean, more carbon dioxide remains in the atmosphere.

Earth's albedo is also a positive feedback element. Around 2.5 million square kilometers of Arctic sea ice has melted since 1980. Mountain glaciers throughout the world have been melting as well. Ice sheets and glaciers increase Earth's total albedo, the amount of solar radiation reflected into space. When enough ice melts a positive feedback effect occurs in which more solar radiation is absorbed due to the absence of ice. As more ice melts, conditions become warmer, in turn melting more ice. This is an effect that can be observed. Geomorphologist and photographer James Balog started a group called the Extreme Ice Survey in the early 21st century in which he, and his team, ultimately set up 43 time-lapse cameras to document the increase or decrease in glacial and sheet ice. Every single camera took time-lapse images of ice slowly melting over a period of several years. In some cases, glaciers completely disappeared. His work can be seen in the documentary *Chasing Ice*.

Radiative Forcing

Radiative forcing is when energy entering the Earth's atmosphere is greater than the amount that leaves it. The Sun emits radiation, which enters Earth's atmosphere, making it to Earth's surface. If the net amount of radiation leaving the planet is less than that absorbed by the surface, the atmosphere will become warmer. Apparently this is what is currently happening. This effect is called radiative forcing because the energy difference can force changes in atmospheric conditions on Earth, the climate.

Scientists measure and calculate factors regarding incoming solar radiation versus outgoing energy. These factors are then used to create **box models**, which serve as a graphical simplifications of the calculated models, which are often quite complex. Box models offer explanations of associated components within systems and their relationships within systems. For instance, a box model can be used to explain the calculations behind water storage and release within a reservoir. Box models can be used in a similar way to model the input of solar radiation versus the output of energy from Earth's atmosphere, radiative forcing models (Figure 12.8). According to NOAA, overall radiative forcing for 2022 was a positive 3.4 watts per square meter (3.4 W/m2) on average for planet Earth. This is the net positive amount of power added, in heat, to one square meter on average for the entire planet. This increase is fairly rapid. In 1979 the amount of radiative forcing was 1.8 W/m2, a 43-year increase of 47%.

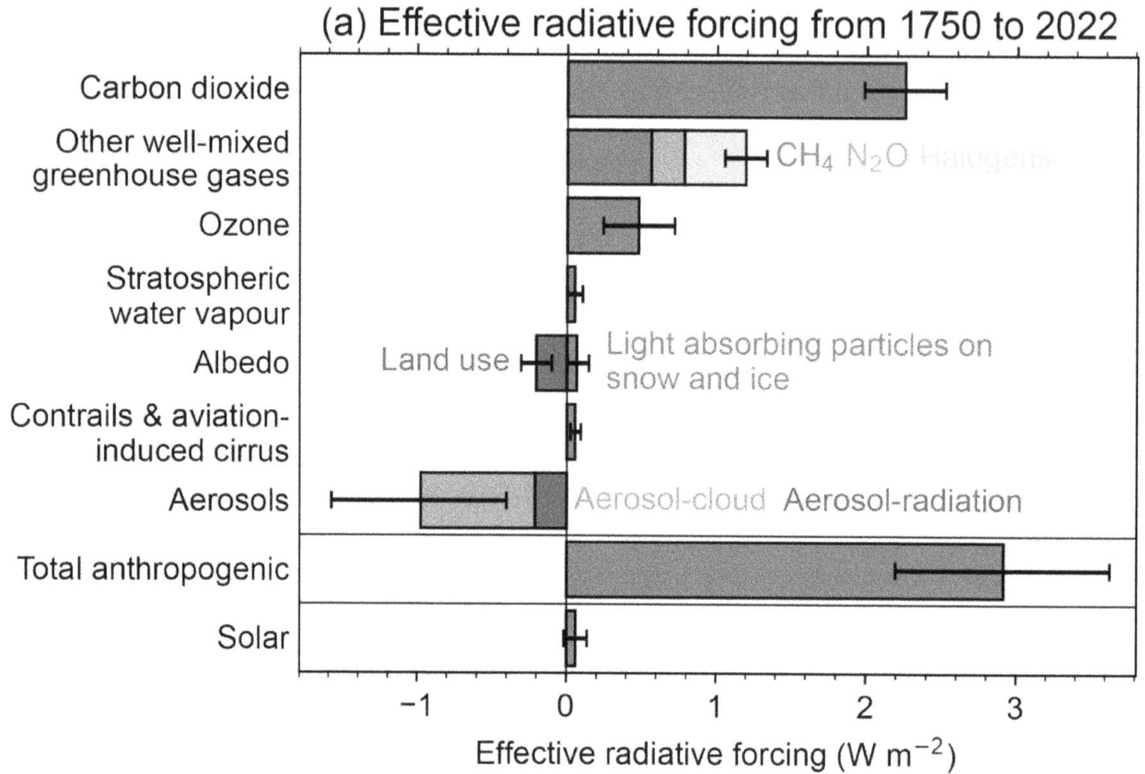

Figure 12.8 Radiative forcing model

Scientific Evidence

There is a body of evidence from the past that supports current analysis. One of the tenets of geology is that "the past is the key to the present." **Proxy indicators** are indirect lines of evidence that substitute for direct measurements when there is no direct evidence. These proxy indicators let us know what conditions were like in the past. In paleontology, fossils are a form of a proxy indicator. There are proxy indicators for climate science too, which is how we can understand Earth's overall climate and weather long before data on current conditions started to be recorded.

Evidence of past climates come from different sources. Ice cores offer very robust data on past climates. Glaciers and ice sheets in the polar regions are often thousands of years old. Some ice sheets in Antarctica may be 1,000,000 years old. These ice sheets can be sampled by drilling and extracting cores. Air bubbles trapped in these ice sheets can be analyzed to shed light on atmospheric conditions in the past such as atmospheric composition, gaseous components, frequency of wildfires, precipitation, snowfall, solar activity, and trends in temperature. This data can be placed in a timeline using radiometric dating and sedimentary techniques.

Cores taken from rocks and sediments contain fossils, microfossils, plant remains, and pollen grains. These can be analyzed using paleontological and sedimentological/stratigraphic techniques to obtain data from past climates as well. Fossilized tree rings indicate growth as well as data from the climate in which they grew. Fossils themselves are indicative of past climates. For instance, a large frog fossil will not be found with fossil beds from an ancient, dry tundra. In this regard coral reefs indicate past conditions, both fossilized and otherwise. For the sake of accuracy and clarity, scientific methods of exploration and analysis require multiple lines of evidence.

Direct Evidence

Current direct atmospheric sampling, from measurements of gas composition to temperature readings to measurements of melting glaciers, provide extremely robust data as evidence for climate change. In addition, simple observation can provide a wealth of evidence, even though those observations can be anecdotal. For instance, people who are older than the age of 50 may be able to remember colder winters and multiple snowstorms where they live. Although this direct evidence is valid, another person of that age group may dispute this. At that point, the argument is not a scientific one but one based on belief, culture, or worldview. Those sorts of arguments are best left out of science.

Weather and Climate Models

Models are constantly used for weather forecasting, whether for a scientific agency or a local news channel. Although these models are not always correct, much to our chagrin as we walk out into an unexpected rainy day, they are generally reliable. Climate models are similar, although they use much more computation power.

A prevalent type of climate model is a **coupled general circulation model**, a computer program that performs complex calculations on data such as atmospheric circulation, ocean circulation, atmosphere–ocean interactions, as well as feedback mechanisms to make climate process simulations. These and other models are becoming more accurate in analyzing current conditions and making predictions.

Scales of Change

As we all know, climate change, not to mention global warming, which is but one aspect of global climate change, has been a controversial subject for quite some time. Enormous resources are poured into attempting to sway the viewpoint of the public in one direction or another. Climate change denial? Alarmist environmentalism? Unfortunately, a lot of these efforts are poorly scientifically informed. Viewpoints on climate change are often based on culture or politics, from blind faith to oddly irrational, ill-informed skepticism. Many of the individuals and organizations that attempt to sway the public toward climate denial have either economic or personal interests in mind and use thinly veiled efforts at belief or pseudoscience to cover those facts up. To top it all off, we live in a world in which people seriously believe that the Earth is flat!

Healthy, informed skepticism is not generally a bad thing. For instance, climate science, much like other sciences that explore natural phenomenon on a grand scale like astronomy and geology, run into **scaling problems**. Scaling problems are like this: Investigating small systems does not always lead to an understanding of larger systems because of differential scaling. Large, complex systems are just that, complex, and it takes an enormous amount of data, analysis, and sheer work to come to a robust understanding of those systems. This work often takes decades.

At the same time, we have already amassed an enormous amount of evidence for rapidly accelerating climate change and global warming increase. Many organizations, from the Intergovernmental Panel on Climate Change (IPCC) to the WHO to NASA, all actively research climate change and advocate for measures to be taken.

Conclusion

So, what are we to do? Should we pursue **mitigation**, action to change the system, in the form of the reduction of greenhouse gas emissions as well as replacing greenhouse-emitting energy sources, such as coal, with non-emitting sustainable ones such solar power? Should we pursue **adaption**, an acceptance of the ongoing problem leading to what amounts as "dealing with it"? Perhaps the answer is a little of both in addition to a fair amount constructive cooperation, as well as critical thinking, scientific knowledge, and a little bit of common sense.

Discussion Questions

Directions: Review the chapter in order to completely and correctly respond to the questions and prompts.

Keep this in the back (or maybe even front) of your mind: Science is system of methods that allows us to understand not only our planet, and ourselves, but of the entirety of the universe. Answering the following questions correctly will lead to a more robust understanding of how to apply science to ENVS problems.

1. Has Earth's atmosphere always been the same? If not, how was it different in the past?
2. Which variable gas has the greatest concentration in the atmosphere? What are its benefits and impacts?
3. Describe the troposphere.
4. Using the formula for relative humidity on page 227, calculate the relative humidity under the following conditions: actual vapor density of 15.5 g/cm3 and saturation vapor density of 17 g/cm3. At what point will saturation be reached?
5. Would you be able to sail east in the ocean a few miles north of the equator? Why or why not?
6. Give some examples of both point-source and nonpoint-source pollution that you might have seen.
7. Have you ever seen the effects of acid rain? Where was it?
8. What can you do in your neighborhood to fight pollution? Describe it.
9. Do you have examples of climate change that you have experienced? What was it?
10. What is some other evidence for global climate change that you can think of?
11. Do you believe or think that Earth's climate is rapidly changing? Why or why not? Please use critical thinking, reason, and science to support your argument.

Bibliography

O'Hare, G., Sweeney, J., and Wilby, R., 2013, *Weather, Climate, and Climate Change: Human Perspectives*: Abingdon, Routledge, 403 p.
Stover, D., 2019, Marshall Shepherd: *Connecting atmospheric science and society: Bulletin of the Atomic Scientists*, vol. 75, no. 4, 205–209.
Withgott, J., and Laposata, M., 2019, *Environment: The Science Behind the Stories*: New York, Pearson, 7th ed., 768 p.

Credits

Fig. 12.1: NASA, "The Thin Line of Earth's Atmosphere and the Setting Sun," https://commons.wikimedia.org/wiki/File:Thin_Line_of_Earth%27s_Atmosphere_and_the_Setting_Sun.jpg, 2009.
Fig. 12.2: Brockert, "General Gas Composition of Earth's Atmosphere," https://commons.wikimedia.org/wiki/File:Atmosphere_gas_proportions.png, 2004.
Fig. 12.3: NOAA & Mysid, "Earth's Atmospheric Layers," https://commons.wikimedia.org/wiki/File:Atmosphere_layers.svg, 2014.
Fig. 12.4: Copyright © by pfctdayelise (CC BY-SA 2.5) at https://commons.wikimedia.org/wiki/File:Mount_Lushan_-_fog.JPG.

Fig. 12.5: Copyright © by Tyler Chow (CC BY-SA 4.0) at https://commons.wikimedia.org/wiki/File:Thermal_Inversion_in_Urban_Environment.png.

Fig. 12.6: U.S. Fish and Wildlife Service, "Point Source Air Pollution Smoke Rising from Plant Tower," https://commons.wikimedia.org/wiki/File:Air_pollution_smoke_rising_from_plant_tower.jpg, 2013.

Fig. 12.7: Copyright © by N T Stobbs (CC BY-SA 2.0) at https://commons.wikimedia.org/wiki/File:Nelson%27s_Column_during_the_Great_Smog_of_1952.jpg.

Fig. 12.8: Copyright © by Piers M. Forster, et. al. (CC BY-SA 4.0) at https://commons.wikimedia.org/wiki/File:ESSD_Radiative_Forcing_1750_to_2022.png.

Chapter 13

Energy Resources

Power for Civilization

Objectives

- Understand that most of Earth's energy comes from the Sun.
- Know about the three fossil fuels.
- Come to an understanding that fossil fuel reserves will not last forever.
- Know that all fossil fuels have an environmental impact and release greenhouse gases.
- Understand the benefits and drawbacks of nuclear power, including the possibility for catastrophe.
- Know what the benefits and drawbacks are of the alternatives to fossil fuels.

KEY WORDS AND TERMS

Combustion (n.): Rapid redox reaction involving oxygen that releases energy, chemical species, and light.

Energy (n.): Mass multiplied by velocity squared, the ability to do work.

Consider This . . .

Energy is required by all living things to function. What gives us, human beings, energy is food—substances that we, like all other heterotrophs consume in order to process for energy. This energy initially comes from photosynthetic autotrophs that are dependent on sunlight to manufacture their own food. Most forms of energy that we harness for technology are powered from fossil fuels. In a sense all processes on Earth are powered by the Sun, directly or indirectly. Even nuclear power uses materials that were already present in the proto-stellar disc 4.5 billion years ago.

Introduction

Energy powers human civilization. Nothing on the scale of entire nations, and a global civilization, could be accomplished without energy resources of enormous magnitude. We use energy for our living spaces, commercially, at work, for industry, for science and technology, for transportation, for conflict, and for exploration. Energy resources, on a colossal scale, have provided us with comfort and convenience beyond those of people of the past, at least from our perspective.

We obtain energy from a variety of sources. We obtain energy directly from the Sun, from within the bonds that hold together the nuclei of atoms, from the remains of plants, from wind, from running water in streams, and from heat that originates in the Earth's core. Currently, humanity is exploring even more sources of energy with varying results. Beyond all of the other sources of energy, we continue to obtain energy from the residues of ancient plants and microorganisms, otherwise known as **fossil fuel**, for most of our energy needs. As mentioned previously, we obtain energy directly from the Sun, solar power, but fossil fuels are energy sources that are the direct result of photosynthesis (Figure 13.1). That is energy from the Sun as well but in an indirect manner, chemical energy from biological residues and stratigraphic products.

Figure 13.1 The Sun provides us with power and energy

Fossil Fuels

For the past 300 years or so, fossil fuels have been the dominant source of energy used to power the machinery and infrastructure of civilization. Prior to the 18th century, biomass, the direct **combustion** of biological material, mostly plants, was the main source of energy for civilization. One of the reasons for the dominance of fossil fuels for civilization and economic development is that fossil fuels are relatively efficient at delivering high-energy content compared to the amount of energy that it takes to use them. Fossil fuels are relatively easy to store, transport, and burn on a large scale, unlike other forms of energy.

Fossil fuels are used to generate secondary energy, electricity, which is used for commercial, industrial, and residential purposes for both transportation and within the electrical grid. The electrical grid is a series of interconnected electricity sources used for infrastructure that are used in all developed nations and in many developing nations. Most of the energy used for electrical grids comes from fossil fuels.

Origin of Fossil Fuels

Fossil fuels are combustible energy resources that are derived from the remains of certain ancient organisms. The remains of these organisms provide residues, formed through stratigraphic and fossilization processes that are the base materials for fossil fuels. This is why they are called "fossil fuels": The organic residues are essentially fossilized remains of organisms that are anywhere from 350 million to several million years old. Some fossil fuels are younger, but these are anomalously produced deposits that do not represent the bulk of fossil fuel deposits.

The organic material that is the basis of fossil fuels is broken down in an **anaerobic environment**, one that is depleted of oxygen. The organisms that provide these materials made their homes in various past environments such as the bottoms of lakes, epicontinental seas, and wetlands. The source material for the three types of primary fossil fuel determines their occurrence (Figure 13.2):

Figure 13.2 The fossil fuels: coal, oil, and natural gas

- **Coal**—The remains of ancient plants that were buried during stratigraphic geological events, mostly within ancient wetlands. These beds of ancient plants were buried in anaerobic environments, ultimately undergoing diagenesis and lithification to become the organic rocks known as coal. Coal is the primary source for secondary electrical generation.
- **Petroleum**—Also known as crude oil, these are the remains of ancient microorganisms and algae that were buried beneath the sediments at the bottom of lakes and shallow seas under anaerobic conditions and constant temperature. This mostly fluid solution of hydrocarbon, a product of sedimentary diagenesis, is currently the most widely used energy resource on the planet.
- **Natural gas**—Also known as simply "gas." In a similar manner to petroleum, most natural gas in production formed from the remains of ancient microorganisms that were buried beneath the sediments at the bottom of lakes and shallow seas under anaerobic conditions. The difference with natural gas is that the hydrocarbons **volatilized**, turned to gas, due to variable heating. These gas deposits became trapped within rock formations and even rocks. Natural gas is mostly methane, a simple hydrocarbon plus minor amounts of other gases. Within the last few decades, natural gas has become a very widely used energy resource due to the discovery and exploitation of new deposits.

Fossil Fuel Reserves

The availability of fossil fuel deposits depends on the geologic history of any given region. Some regions have very few fossil fuel reserves, while others have substantial ones. For instance, Saudi Arabia reserves are within interbedded sedimentary strata, including evaporites, carbonates, and clastic sedimentary rocks. These formed in sequence over millions of years with multiple deposits of oil. This includes the most productive, extensive oil field in the world, Ghawar. Over 50% of the world's proven reserves are in the Middle East, including Saudi Arabia. **Proven reserves** are deposits of oil and natural gas that are available for current and future production.

Venezuela has the world's largest proven reserves within Miocene strata: high-porosity delta and coastal plain sedimentary rocks. Within the last few decades the United States, Canada, and other countries began using **hydraulic fracturing**, or **hydrofracking**, technology to extract oil and gas from previously unrecoverable reserves. Hydrofracking is the use of technology to fracture sedimentary rocks at depth, such as shale, using high-pressure fluids. This has made the countries less dependent on oil and gas from other nations.

The United States is the country with the most proven recoverable coal reserves, at around 249 million metric tons in 2020. Russia has the second largest known proven recoverable reserves. China has significant reserves of coal as well. Coal deposits are a source of natural gas. **Coalbed methane (CBM)** is methane gas that is released during coal mining. Because CBM is economically recovered and not released into the environment, it reduces the possibility of greenhouse gas emissions. Methane, or natural gas is often associated with oil reserves as well and can also be recovered from landfills. The United States is the world's largest producer of natural gas.

The length of time for a nation's fossil fuels to last depend on the size of proven recoverable reserves, the rate of production, the rate of use, and technological advances that enable recovery within previously unrecoverable reserves.

Fossil Fuels Across the World

The annual use of fossil fuels is far greater in developed nations than in developing countries. Developed, industrial nations use fossil fuels for commercial enterprise, homes, industry, transportation, and more. Fossil fuels are harder to obtain in the developing world. In developing countries and regions fossil fuels are generally used for subsistence activities such agriculture, cooking, and home heating, often on a limited basis. The governments of developing nations, as well as their wealthy inhabitants, generally make greater use of fossil fuels as opposed to the citizens of those countries, who often live far below the poverty level. Most people in developing nations and regions use animal energy or simply manual labor as day-to-day energy sources. They also extensively use combustible biological sources, such as wood or dung.

Fossil Fuel Production and Use

Fossil fuels are usually used for either electrical generation or transportation. Coal is usually used for electrical generation, although oil or natural gas can be used to generate electricity as well. This is often due to the regional availability of a particular energy resource, as well as the cost of transportation. For instance, a large number of electricity plants in Florida use petroleum or natural gas as opposed to coal because using petroleum and natural gas from the Gulf of Mexico is generally more cost-effective than shipping coal in from elsewhere.

Petroleum is widely used for transportation in the form of either gasoline or diesel for automobiles and trucks. **Kerosene** or a type of petroleum called **jet fuel** is used for airplanes. The shipping industry uses a suite of different types of petroleum to power boats, including **heavy fuel oil (HFO)**, **marine diesel oil (MDO)**, and **marine gas oil**

(MGO), depending on the type of vessel. All of these forms of petroleum are refined using **petroleum fractionating** or **petroleum distillation**.

Petroleum fractionating is done in **oil**, or **petroleum**, **refineries**, industrial facilities in which petroleum is chemically transformed into different products with differing properties. **Crude oil**, the raw, original form of petroleum from the ground, is heated within **fractionating columns** or **fractionating towers** (Figure 13.3). These facilities heat petroleum differentially within towers, lower heat products near the top, and increase heat products at the bottom. Different weights and compositions of the refined petroleum are collected at different heights. Each of these petroleum fractions have their own properties and uses. For instance, gasoline for automobiles is not subject to intense heat and is collected near the top of the tower. Kerosene for aviation fuel is collected near the middle of the tower, and **bitumen**, or **tar**, highly viscous petroleum products used for roads and building, are collected near the bottom at the highest heating point.

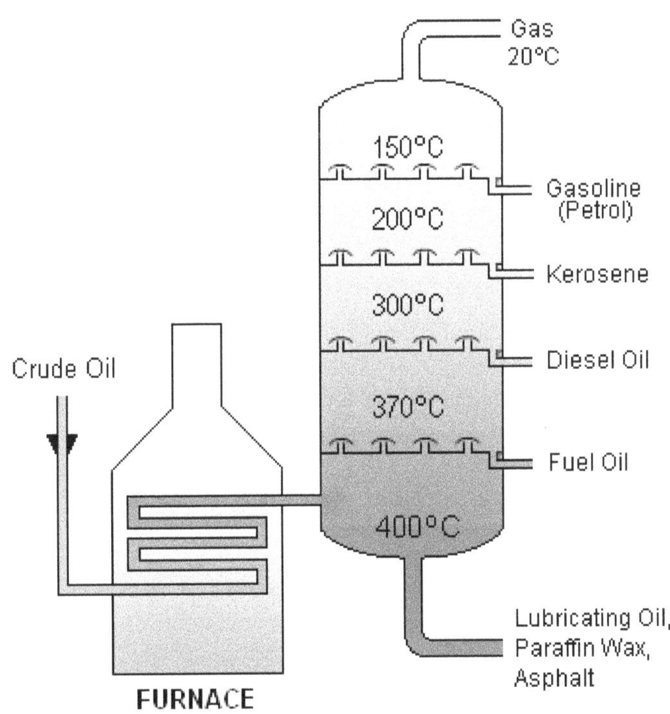

Figure 13.3 Diagram of a fractionating column

Coal does not undergo refinement like the petroleum does. Coal is generally processed with additives used to remove unwanted chemical content, such as mercury, in a process called **scrubbing**. The United States provides incentives for the removal of potential pollutants in coal such as mercury or sulfur. Natural gas is refined using a number of processes in order to remove unwanted chemical species, resulting in nearly pure methane.

Once refined, fossil fuels have to be transported out of the refineries and then be distributed to commercial, industrial, and residential markets. This requires time and energy, as well as efficient routes for distribution. Petroleum is distributed by rail and truck across land, by ships on the ocean, and by pipelines that crisscross entire continents. These pipelines are controversial due to the environmental impacts that they cause, including the potential for leaks along the often thousands of miles of length. Recently, the Keystone XL pipeline was canceled in the United States for this reason.

Natural gas, not unlike petroleum, is transported in a similar manner. To transport by truck rail, or ship, natural gas is cooled to the point of a phase change into a liquid state. This **liquefied natural gas (LNG)** is then suitable for transportation by vehicle. This, of course, poses a hazard should an accident occur. Coal is simply transported by rail, truck, or ships worldwide, with little need for much preparation since it is a solid product.

Energy Return on Investment

Generating energy within a fossil fuel extraction system requires initial inputs of energy. Initial capital investments are required for land and exploration. Once proven recoverable reserves have been discovered, extraction and processing are required. Once obtained, the resource has to be transported to market. All of this requires geophysical exploration, engineering, development, scientific equipment, storage tanks, wells, and other factors. The amount of logistical effort that is put into energy resources versus the capital return is known as **energy return on investment (EROI)**, which is essentially the energy or gains that are returned versus the energy that goes into the investment on a given fossil fuel. Fossil fuels generally have very high EROIs, around 20:1, which is why their use is so prevalent. This fluctuates from time to time though due to economic, logistic, transportation, and other factors.

Coal

Coal has been used as a fuel or heat source longer than any other fossil fuel, and it is also the most abundant fossil fuel. The abundance of coal is due to the enormous reserves beneath the ground within geological formations throughout the world. In addition, many of these coal reserves, or coal seams, are within stratigraphic layers of rock that are fairly close to the surface of the Earth (Figure 13.4). Since coal is a solid that generally requires little processing to mine, it is fairly easy to obtain. Worldwide, the supply of coal is estimated to be available for at least 200 years at current levels of production.

Coal is composed of plant matter that contains a lot of cellulose, woody plant material that was buried within stratigraphic layers in the distant past. Throughout many of the geologic periods in the Phanerozoic Eon, favorable worldwide climates supported the growth of enormous forests of trees and other large plants. As tectonic plates shifted, over millions of years and altering paleoenvironments, environments of the distant past, these plants became buried in sediments. Over time sediments were compressed deep within the upper crust under **anaerobic** conditions. Anaerobic refers to extremely low- to no-oxygen conditions. In this manner, organic plant matter is progressively transformed into a material that is mostly carbon.

Plant matter near the Earth's surface that is depleted of oxygen yet subject to very little pressure becomes **peat**, the lowest grade of coal. Peat is still used for burning these days. The lack of oxygen as well as burial greatly reduces the chance for decomposition. As peat becomes further buried and subject to further pressure as well as heating, the material turns to a higher carbon content and becomes **lignite**. Lignite that further matures as burial increases becomes **sub-bituminous** coal. At this point the coal is mostly carbon, along with minor amounts of hydrogen and other constituents. Maturation at further depths leads to the formation of the next highest grade of coal, **bituminous coal**. If this coal deposit is further buried, it moves past the window for **diagenesis**, sedimentary rock formation, and metamorphoses to the metamorphic rock **anthracite**.

Figure 13.4 Coal seam within strata

The presence of sulfur, mercury, arsenic, and other trace metals is indicative of coal formation at the edge of the ocean or a sea because these chemicals precipitate out of seawater. Coal that formed in ancient freshwater wetlands does not have as many of these impurities in it. Much of the coal in the eastern United States formed at continental margins next to the sea, so they contain the impurities. Coal in what is now the western United States generally formed in freshwater wetlands and does not feature impurities. Since impurities like mercury and trace metals are pollutants, these are taken out of the coal before shipping using technology like scrubbing.

Coal has been used for thousands of years, starting with China around 3,500 BCE. Coal was used for heating during the Roman Empire. With the advent of the steam engine and then coal-fired engines, coal began to become widely used. Coal was first mined extensively, commercially, in the 18th century. Coal drove the industrial revolution and steel production in the 19th and 20th centuries. Thomas Edison built the first power station in the late 19th century. Since then, coal has become the primary source for secondary electrocortical power throughout the world. It's secondary because electricity is produced by turbines from burning coal. The turbines generate electricity by spinning, like a motor, which induces an electrical current.

Coal is mined in either shafts through subsurface mining, which uses a network of tunnels underground, or through strip mining. Strip mining is the practice of removing topsoil to expose a deposit for extraction. The most extreme variation of strip mining is mountaintop removal in which large, heavy machinery is used to remove entire mountain peaks. Mountaintop removal is also done through detonation, blowing up entire mountain peaks. As one can tell, this is devastating to not only entire ecosystems but to people as well.

There is no form of coal mining or use that does not impact the environment, despite government regulations. Miners themselves are subject to extremely dangerous working conditions and often develop chronic ailments like black lung disease. Mining often takes place in the presence of water, resulting in acid mine drainage. Using coal to generate electricity requires combustion, which releases pollutants into the air. Even scrubbing technology does not remove all of the impurities that result in air pollution and acid rain. Combustion itself releases carbon dioxide and even methane into the atmosphere, both of which are greenhouse gases.

Petroleum

Petroleum is the world's most widely used fuel, and it's been that way since the 1960s, mostly due to its predominance as energy for transportation. Petroleum from crude oil is also used for other products such as synthetic chemicals, tar, heating fuel, and others. Despite ongoing controversy with its use, oil is still widely used even though its use has dropped off slightly in recent years, possibly due to the COVID-19 pandemic.

Crude oil, the source material for petroleum and petroleum products, comes from the tissues of ancient microorganisms that became buried beneath the substrate of ancient shallow seas. During very warm periods of the Earth's prehistory, such as the Cretaceous period, the seas, many of them epicontinental, teemed with microscopic life. As their remains collected at the bottom of theses seas, and as tectonic activity drove the development of new depositional basins, these microorganisms became further buried under layers of sediment and thus were subject to very little decay.

Deep within the Earth, under further anaerobic conditions and heat, the remains of the microorganisms turned into liquid as **kerogen** developed. Kerogen is a mixture of many organic compounds and is the source material for crude oil development. Crude oil, once it forms, is a solution of hundreds of hydrocarbons, compounds made of only carbon and hydrogen. These hydrocarbons form polymers, complex long-chain molecules. The formation of crude oil takes place within the oil window. The oil window is the range of depth in which oil forms, and it is between 2–5 kilometers below the surface of the Earth. Beyond that window, further down, oil volatilizes into gaseous form, natural gas. Crude oil also takes time to form, anywhere from thousands to millions of years.

Oil and natural gas reside underground because they are held there within geological formations known as **traps** (Figure 13.5). Traps are impermeable cap rocks that prevent highly mobile hydrocarbons, both petroleum and natural gas, from migrating upward. There are several different types of petroleum traps:

- **Anticlinal traps**—Anticlines are arched slabs of rock at or near the surface of the Earth. If the top of an anticline is impervious, it could potentially trap crude beneath it.
- **Fault traps**—Faults in the Earth's crust can form geometric features of impermeable cap rock that traps petroleum and natural gas.
- **Salt dome traps**—Deposits of impermeable salt rise up through fractures in clastic sedimentary rock, rising above oil and gas deposits, trapping them from above.
- **Stratigraphic traps**—Impermeable sedimentary rocks such as shale trap oil and gas from above.

Keep in mind that fossil fuels are only found in sedimentary rock strata.

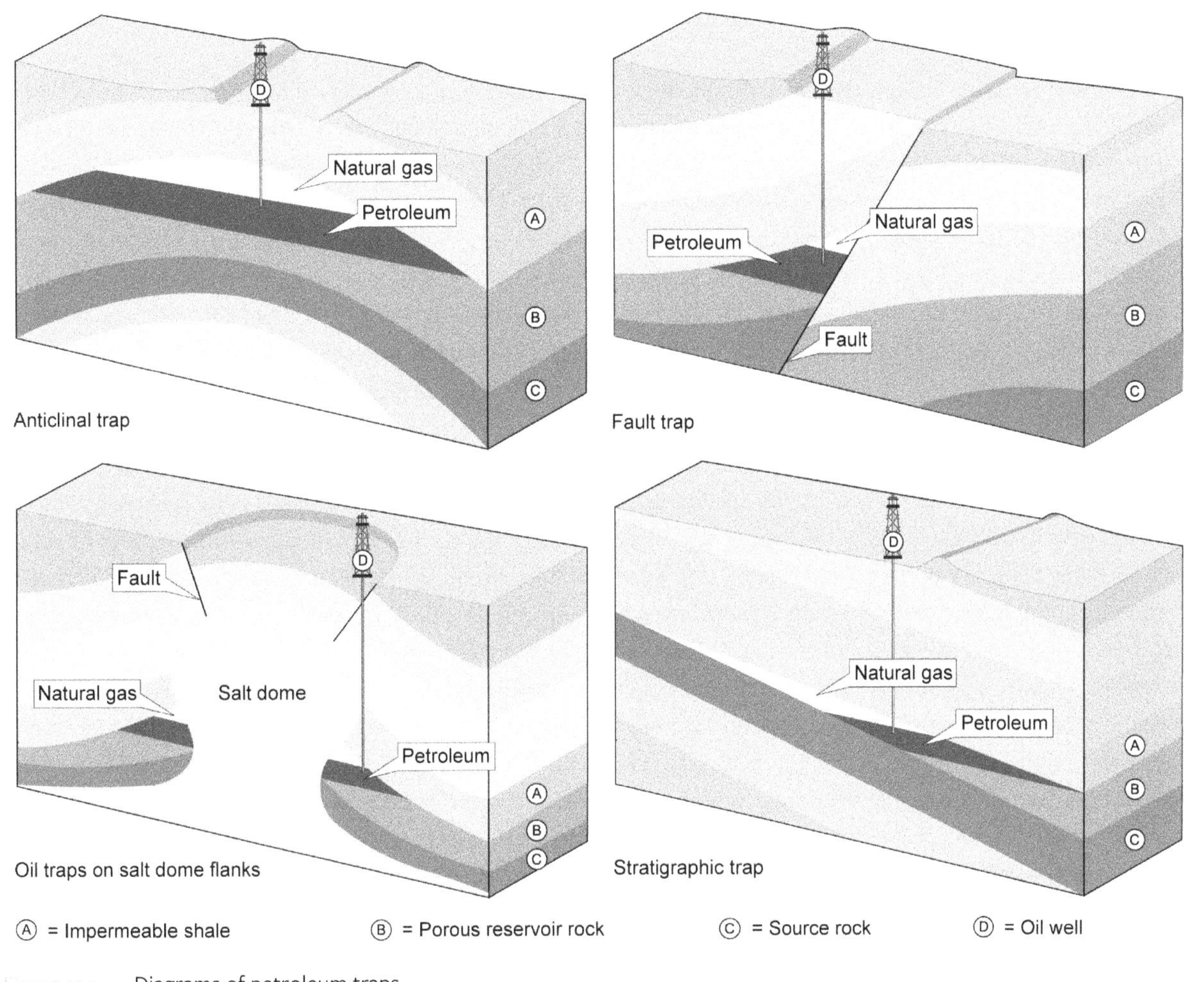

Figure 13.5 Diagrams of petroleum traps

Oil and gas in traps can rise above the water table. Extraction of this oil and gas can be problematic because it leads to groundwater contamination.

Where oil is plentiful, such as in the Middle East, it emerges from the ground in seeps. Seeps are also present in the relatively thin sediments at the bottom of the ocean. In the past oil has been readily available to people from seeps on land. People have been using bitumen or **asphalt**, particularly thick oil deposits, as adhesives since prehistoric times. There is evidence of its use by prehistoric humans. They may have used bitumen to attach the heads to spears.

The use of petroleum is first seen in the archeological record from the banks of the Euphrates River 6,000 years ago. Asphalt was quarried there for use as mortar and decoration. Bitumen was used around 4,000 BCE in Mesopotamia as caulk for building. It was also used in Ancient Egypt for mortaring the pyramids. Despite early use, oil had not been widely used throughout most of history. Until recently oil has historically been used as a sealant, a building material, "paint," a medicine. Native Americans used fossil fuels from seeps as medicine for hundreds of years before it was sold in the 19th century as a "tonic." Please do not drink petroleum!

In 1849 a Canadian geologist named Abraham Gesner distilled crude oil to create a new lamp oil called kerosene. Ultimately, kerosene replaced whale oil and Gesner would become known as the "father of the petroleum industry," although he neither achieved wealth or fame in his lifetime. The fact that petroleum products are readily combustible would become relevant a few decades later.

A few years later, in 1859, the Pennsylvania Rock Oil Company contracted Edwin L. Drake to drill a well near Titusville, Pennsylvania. The purpose of the well was to extract crude oil to be used as lamp oil. This was the world's first oil well, the beginning of the oil industry.

The oil industry truly began with the inception of John D. Rockefeller's Standard Oil of Ohio in 1870. In 1901 the first Texas "oil gusher" Spindletop came into production; then two years later Henry Ford began mass producing an automobile named the Model T. The Model T ran on a little-used byproduct of petroleum production, gasoline. Prior to the Model T, most crude oil that was in production went into the development of kerosene. Kerosene was primarily used as a lamp oil. Years later kerosene would be used for jet fuel. After rapidly increasing numbers of automobiles, like the Model T, started to be produced, it drove up the demand for gasoline. Gasoline became the predominant fraction of oil used around the world, and is today.

Petroleum, as well as natural gas, is found at depths between around 2 and 5 miles underground in stratigraphic layers of sedimentary rock. These occur in isolated deposits, collecting in porous layers beneath impermeable layers trapped within impermeable layers. Geologists who work for oil companies initially conduct seismic surveys to determine whether fluids are present within sedimentary rock strata. Seismic signals, essentially acoustic signals related to sound, convey information about Earth's depths. When a strong possibility for oil and gas has been determined exploratory drill cores are made to provide a visual record of the subsurface. Seismic exploration is initially done because it is not as expensive as "wildcat" randomly drilling to explore for oil. These locations are mapped regardless of whether oil and gas are discovered. If oil and gas are discovered, oil companies make a "play" for the deposit and begin production.

Oil is a fluid, so it is sometimes under pressure and rises to the surface due to the pressure head being higher than the surface of the Earth. These are called "gushers," and these types of wells are not very common anymore. Another type of **primary extraction** is the use of mechanical pumps to obtain oil. Some oil deposits are more difficult to extract, so **secondary extraction** methods must be used. Secondary extraction involves some means to extract difficult to get to oil by using solvents, steam, or water to induce extraction. These methods can often become expensive.

It is estimated that we have enough reserves, at present production rates, to last us for 25–40 years. This number might change as technology meets demands, for instance when fracking technology opened up previously thought unrecoverable reserves. The **reserve to production ratio** is the amount of total remaining reserves divided by the annual productions rate: R/P. This is, of course, a ratio that fluctuates annually and with some uncertainty since both reserves and rates are often not publicly disclosed by oil companies, especially oil companies from other nations. Politics, in addition to economics, has been part of the global oil market for some time.

Regardless, oil reserves are not infinite, and at some point the amount of global proven recoverable reserves will dwindle. As production declines the crises will likely begin. Prices will increase and supplies will drastically decrease. Geophysicist M. King Hubbard formulated a metric for oil reserve production called **Hubbard's peak**. It was predicted that we would reach peak oil production at around 1970. This was somewhat accurate, but he did not predict hydrofracking technology and how it has opened up new reserves in recent years.

Another series of reserves are in **oil sand**, also known as **tar sands**. These are sand deposits that contain 1%–20% bitumen. These deposits of oil are costly to process, both for the sand content and the lack of hydrogen in the deposits. These deposits are removed by strip mining in a manner that is ecologically devastating. Most of the world's deposits of tar sand are found in Alberta, Canada, and Venezuela.

Natural Gas

Natural gas has become the fastest growing sector of the fossil fuel industry. It accounts for about 30% of energy use in the United States. World supplies are projected to last about 52 more years.

Natural gas is essentially methane, CH_4, a simple hydrocarbon plus other minor chemical species that are often added during production. Most natural gas in production currently comes from stratigraphic deposits as the volatilized organic compounds from the remains of ancient oceanic microorganisms. There are other, minor sources of methane, such as biogenic gas, which is created from the decomposition of organic matter in deep, low-oxygen reservoirs like the bottom of wetlands, otherwise known as swamp gas. Many landfills capture biogenic natural gas from decay to sell. Thermogenic gas is inorganic in origin and is generated from compression and heat. Coalbed methane is found in coal seams. The source material for natural gas is kerogen, much like crude oil. Natural gas is often flared off, burned away in remote drilling areas. In other parts of the world, like Alaska, gas is captured during oil drilling and reinjected into the ground for future use.

Extraction of natural gas began in 1821 in Freedonia, New York, but it was only an isolated incident used locally. Natural gas was initially used for streetlamps, then for cooking and heating. Once production intensified in the 1920s, pipelines were laid for thousands of miles (Figure 13.6). Now the transport of natural gas is fairly economic and safe, including not only pipelines but transport by rail and truck using liquefied natural gas. The top reserves in the world are in Russia, Iran, Qatar, and the United States. Markets have opened in many parts of the world due to hydrofracking technology.

Fossil Fuel Impacts

Despite the obvious benefits that they have had for human civilization, the occurrence, extraction, and use of fossil fuels has a massive impact on ecosystems. The use of fossil fuels releases pollutants into the environment, the nature of which depends on the fossil fuel in question. These pollutants cause serious harm to human health. Fossil fuel fluids, petroleum, and natural gas contaminate ground- and surface water supplies. Mining involved with fossil fuel extraction through mining has an overwhelming negative impact on local ecosystems, including both the coal and tar sand mining.

Figure 13.6 Natural gas pipeline

Coal mining causes environmental impacts specific to its practice. Coal mining near the water table or in conjunction with water sources leads to acid mine drainage, which contaminates natural waters as well as municipal water sources through chemical runoff. Acid drainage comes from sulfur-bearing minerals and sulfur content in coal. Sulfur reacts to water to produce sulfuric acid (H_2SO_4). Mountaintop removal is mining on a catastrophic scale, causing harm not only to ecosystems but local communities of human beings. Coal mining is one of the most dangerous jobs on Earth due to the inherent risk of mine-shaft collapse. Coal miners risk fatal black lung disease from the inhalation of dust if they don't have the proper breathing equipment. Finally, coal seams can catch fire and burn for decades, polluting an entire region. A coal seam caught fire in Centralia, Pennsylvania, in 1962. It is still burning.

Natural gas is essentially methane, a greenhouse gas many times more potent than carbon dioxide. Atmospheric methane levels are currently very low, but they are rising worldwide on average. Much of this is attributable to human extraction. Natural gas is also a source of pollutants. The extraction of natural gas in southeast New Mexico has caused a regional problem with pollution. In addition to methane, which is odorless, areas near the gas fields, including Carlsbad, New Mexico, almost always smell like rotten eggs due to the release of sulfates from gas drilling. The entire region sits of bedrock that contains gypsum ($CaSO_4$), a sulfate mineral. Methane has several sources from agriculture. Methane is released from rice production and is produced by the digestive processes of cattle.

Public policy has been enacted to combat pollution and greenhouse gas emissions from fossil fuels, but we can make a difference as well, especially if we work together. One need not drive everywhere for one thing. Walking promotes physical activity, which is good for the human body. It is also a carbon-neutral activity. This may seem small, but it is an effort. On a larger scale, there are alternatives to the use of fossil fuels that are highly viable sources of energy.

Alternatives to Fossil Fuels

Fossil fuels provide around two-thirds of global electricity generation: oil, coal, and natural gas. There are alternatives to these sources of secondary energy generation. Given the widescale, almost unfettered use of fossil fuels and their obvious impacts on the environment, it is a good idea to explore these alternative forms of energy production. We need to start using alternative sources for much more energy production right now. We should have been using them decades ago. With that in mind, several alternative energy resources have been used for quite some time now.

Nuclear Energy

Figure 13.7 A nuclear reactor

Nuclear energy is energy obtained from splitting the nucleus of an atom, generally an isotope of **uranium (U)**. Uranium, element 92, is a radioactive element that is used for energy. It is also used in atomic weaponry. Isotopes are atoms that differ in atomic mass due to differences in the number of neutrons within the nucleus, usually the addition of extra neutrons. Splitting an atom is called **nuclear fission**. With nuclear fission the nuclear force that holds atomic nuclei together is overcome. This causes an enormous release of energy that can then be harnessed for nuclear power. There is estimated to be around 230 years of uranium in reserves that can be used for power at current rates of production.

Nuclear power has many benefits but several clear drawbacks. These drawbacks have constrained the use of nuclear power over the years. Regardless of the drawbacks, nuclear energy provides 18.2% of the United States' total power output. Nuclear power provides France with 62.6% of its power.

Nuclear Reactions

Nuclear reactors are facilities within **nuclear power plants**, power stations that use heat to turn turbines to produce electricity as secondary energy, much like coal-fired power plants, except in this case the source of heat is a nuclear reactor (Figure 13.7). Nuclear reactors are devices that initiate and control nuclear fission to deliver energy in the form of heat.

Uranium-235 (U-235), an isotope of uranium, is used to provide the energy for nuclear power through the nuclear fuel cycle, the life cycle of uranium from mining through spent uranium, nuclear waste. Uranium is a radioisotope, an isotope that emits subatomic particles and radiation as it decays into lighter radioisotopes, ultimately becoming stable **daughter products**, the final stage of radioactive decay in which the initial isotope decays into a non-radioactive element. Radioisotopes decay through a series of **half-lives**, in which half of the isotope decays into the daughter product. The daughter product of U-235 is **lead-207 (Pb-207)**.

The half-life of U-235 is 700 million years, so once U-235 as a fuel is used up as **spent fuel**, a lot of the **radioactive waste** is still around. Uranium's half-life means that it will take 700 million years for even half of a given amount to decay to Pb-207. This is the first drawback to nuclear energy because the radioactive waste must be disposed of safely.

To initiate fission, U-235 is bombarded with free neutrons. This initiates a **chain reaction** that causes all isotopes within the **nuclear core**, the mass of radioisotopes, to systematically split. This chain reaction must be controlled. In order to do this, substances called **moderators** are used. Water is often used, as are **control rods** made of graphite (C), a form of carbon. Water is generally used no matter what to help cool down the extremely hot nuclear reactions. All of the devices that pertain to nuclear reaction are housed in containment buildings that are constructed to prevent radioactive leaks due to accidents, natural hazards such as earthquakes or tsunamis, or sabotage.

Nuclear fusion is the nuclear process that is responsible for the Sun's vast energy output. Nuclear fusion, as opposed to nuclear fission, occurs when the nuclei of two atoms are fused, releasing energy. In the Sun this happens when two hydrogen (H) nuclei fuse, forming a helium (He) nucleus, a free neutron, and energy. Enormous amounts of energy are released through this process. Humanity achieved nuclear fusion in 1952, when the first hydrogen bomb was detonated by the United States in the Marshall Islands. Otherwise, to this date no one has achieved a controlled fusion reaction. The attempt to do this continues to this day, since in doing so humanity will have a limitless energy source fueled by the most abundant element in the universe, hydrogen, and hydrogen from water at that.

Nuclear Power

One of the key benefits of nuclear power is that it helps humanity avoid the emission of over 500 million metric tons of carbon each year. Nuclear power releases no greenhouse gases, and barely any pollutants other than some local thermal pollution. As for the dangers of working in nuclear power plants, they are generally safer than working in coal-fired plants.

There are only two drawbacks to nuclear power, but they are considerable. Again, radioactive nuclear energy must be disposed of somehow, and any accidents within nuclear facilities have the potential to be catastrophic. Today there are over 400 operating nuclear plants in the world. If these facilities are well maintained to the highest standards, they can continue to provide safe, clean energy for years. Unfortunately, this is not always the case.

Meltdown

A nuclear **meltdown** occurs when a severe accident or catastrophe causes the core to heat up uncontrollably. This can be due to damage to the reactor or a loss of moderators or coolant water. As the atomic core overheats, radiation is released into the environment. A meltdown does not necessarily result in an explosion, but the extreme heat can cause fires, which may initiate explosions.

Two of the most severe nuclear meltdown events occurred in the 20th century: Three Mile Island in the United States, and Chernobyl in Ukraine, at the time part of the Soviet Union. Three Mile Island occurred at the Three Mile Island Nuclear Generating Station near Harrisburg, Pennsylvania, on March 28, 1979. The accident, a minor meltdown, could have been a lot worse, as within less than a week it was contained. It caused a wave of reform in United States nuclear policy.

The Chernobyl disaster occurred on April 26, 1986, at the Chernobyl nuclear power plant, near the city of Pripyat, Ukrainian, SSR, in the Soviet Union. There was an initial explosion of the plant, then a catastrophic meltdown. Fires raged for 10 days until they were finally put out. Over 100,000 nearby residents were evacuated from the area. The accident killed 31 people, but many more perished in the years that followed the accident from illnesses and cancer. The landscape in the entire area became contaminated for years. Only recently has the semblance of healthy ecosystems returned to the still-abandoned area.

A few minor accidents have occurred over the years at nuclear power facilities, but nothing on the scale of Chernobyl, with the exception of the Japanese Fukushima nuclear accident of 2011 caused by a tsunami, resulting in multiple fires. When nuclear power plants age, they require more maintenance and more stringent safety measures. At times this can become too costly and nuclear plants are shut down.

Another concern is with sabotage or some other form of armed attack on a nuclear power plant. This recently happened. On March 4, 2022, the Zaporizhzhia nuclear plant in southeastern Ukraine was attacked by the Russian armed forces during the second Russian invasion of the Russo-Ukrainian war. As conflicts around the world continue, attacks on nuclear power plants remain a very real threat. On a related subject, stolen radioactive materials could be used by terrorist groups to stage attacks with small, atomic "suitcase bombs."

Nuclear Waste Disposal

The long half-lives of uranium and other radioisotopes ensure that they will continue emitting radiation for thousands to even millions of years. Radioactive waste has to be stored in extremely stable and secure containers in extremely stable and secure locations where the radioactive waste can do no harm. The containment vessels are usually composed of thick concrete, lead, and steel.

Nuclear waste in the United States is currently being held in decommissioned nuclear power plants at around 80 locations. As more nuclear waste is generated, new facilities will have to be found, which will be a challenge.

Biomass

Almost 3 billion people, worldwide, use biomass for most of their energy needs. This is mostly in the form of wood from trees, and the people in question usually live within the developing world. People within the developed world often take easy access to electricity and heat for granted, but people in the developing world must use more basic energy sources for survival. This **biomass**, energy taken directly from biological materials, usually plants, is used for cooking, heating, lighting, and work. Wood that is used for energy is called **fuelwood**. Other sources of energy and heat include charcoal and dried animal dung. Biomass, especially fuelwood, is renewable, but only as long as it is harvested sustainably and not overharvested. This is a challenge because deforestation, as well as soil erosion, continues to run rampant in certain parts of the world.

Although many in the developing world rely on biomass, many developing countries are implementing fossil fuels. In certain part of the world biomass itself is burned in power plants, resulting in secondary electrical generation.

Biofuels

Biofuels are biomass sources that are converted to liquid fuels for automobiles and trucks. Ethanol is a biofuel that is derived from biomass using carbohydrate-rich crops such as corn. Biofuels are widely used in the United States as additives to gasoline. Many vehicles can run on E-15, an 85% gasoline, 15% biofuel mixture. Some vehicles run completely on biofuels. Diesel engines can be converted to all-biofuel engines.

Hydroelectric Power

Hydroelectric power, or **hydropower**, requires the use of the kinetic energy of running water to turn turbines within a hydroelectric power plant, producing secondary electricity. The electricity that isn't immediately used is stored in impoundments, facilities that store the energy for later use.

Water for hydroelectric power is usually stored in a **reservoir** that is created when a river is **dammed**. Other, smaller hydroelectrical facilities use a **run-of-the-river** system in which a turbine is placed in a river without the obstruction of a dam.

Hydropower accounts for about 16% of the world's electricity production. This has caused **hydroelectric dams** to be placed across most of the world's largest rivers. This is problematic not only for the ecological damage that dams, and their resultant reservoirs, cause, but also because the world has essentially run out of rivers to dam for hydropower. The only remaining rivers that can be dammed are in remote areas, such as parts of the Amazon Basin and Siberia, where few people live. Paying for hydropower facilities in those areas would be infeasible due to the extremely high cost of development and maintenance.

Hydropower has two clear benefits to it. These are major advantages over the use of fossil fuels. Hydropower is clean, no greenhouse gases are emitted, and no pollution is generated, and second, it is completely renewable. As long as water flows in rivers downstream, energy is available to turn turbines. It also has an EROI of around 10:1, which is remarkable. At the same time, the lack of available river water may be a blessing in disguise because large hydropower facilities, such as China's Three Gorges Dam, cause immense damage to the environment despite the clean energy that is generated.

Solar Energy

The Sun is the source for all of Earth's energy at its surface. It powers all biological activity, the hydrologic cycle, and weather. In direct sunlight the Sun provides a little over 1.3 kilowatts of power per square meter. Solar energy is the great potential energy of our times.

Solar energy has always been used by humanity. The most common way to use solar energy is **passive solar energy**. Passive solar energy entails using direct sunlight most efficiently. The classic method for passive energy is building structures that collect heat as the sun shines through south-facing windows, opposite in the southern hemisphere, retaining that heat in **thermal mass**, materials that compositionally store heat such as clay. Other passive solar methods include planting vegetation, including trees, in strategic locations to cool or insulate buildings.

Active solar energy makes use of technologies that absorb solar energy, translating it to electricity, then either using it directly or storing it for later use, usually a combination of both. Active solar energy is not new. The first **solar panels**, flat solar-absorptive materials that collect solar energy, were built in 1881 by Charles Fritts in the United States. Fritts's idea was a good one, but his panels were prototypical and not extremely efficient. That would change with time.

A very prevalent type of active solar energy use is that of solar panels used to collect heat from the Sun. These panels are installed on rooftops. The panels themselves are made of dark, heat-absorbing materials. Air, antifreeze, or water is passed through the collectors and transferred throughout the building. This heat can be stored and used later.

The most common type of active solar collection is the use of **photovoltaic cells** that collect solar energy to be used as electricity. This electricity is stored in specialized batteries. Photovoltaic cells make use of the photovoltaic effect. This occurs when sunlight strikes a pair of metal plates in a PV cell, which releases electrons, inducing an electrical current. A PV cell consists of two layers, an n-type layer that is rich in electrons and a p-type layer that is electron poor. Sunlight striking the upper n-type layer causes electrons to flow to the p-type layer, inducing an electrical current. This current is then harnessed using technology designed to transfer the current as well as store energy for later use.

Around 4.2 million homes in the United States use solar energy for power (Figure 13.8). Solar power is a clean, greenhouse gas-free, pollutant-free method of energy use. Solar energy is also used for businesses and in some industries. Solar power provides benefits to people who live in remote areas since solar panels can be equipped to provide electricity and heat without relying on an outside facility. This is especially useful for medical facilities in remote areas.

Figure 13.8 Solar panels being installed

Many developed counties use solar power, and that continues to grow. One drawback to solar power that is often discussed is the use of it in areas that are not always brightly lit by the Sun, for instance in Germany and part of the East Coast of the United States. Another case is made against higher latitudes and the dispersal of sunlight in those regions. These arguments are currently invalid. The efficiency of current solar technology as well as the ability to store energy are usually more than enough to provide energy using this most clean and limitless energy source.

Wind Power

Wind power makes use of **wind turbines**, devices that harness the power of wind. Wind power is not new. Windmills have been used for hundreds of years to pump water for agriculture. Windmills have been used to generate electricity since the late 19th century. The global 1973 oil embargo initiated government-funded research into the development of wind power in the United States, as well as research in solar power. This funding ended in 1980 for little more than political reasons. Fast-track to the 21st century and wind is one of the fast-growing sectors of energy production.

The technology for wind power is harnessed using blades that rotate with the wind, turning a turbine. This rotates machinery inside a compartment called a **nacelle,** which generates secondary energy, electricity. This machinery is placed atop towers that are anywhere from around 40 to 100 meters high. Higher towers ensure less turbulence and higher wind speeds. Wind turbines are often very efficient. Small changes in wind speed can yield big jumps in electrical generation.

Wind power is currently widely used in agriculture, much as it has been historically. The wind turbines used for agriculture are constructed in groups of turbines called **wind farms**. These are sometimes collected in groups of close to 100. Wind power for municipal use makes use of wind farms as well. Wind farms are currently being built in offshore facilities that make use of the constant powerful winds of the ocean.

Wind power is now widely used and is growing. China currently leads the world in wind energy production, followed closely by the United States. The installed capacity of wind power installations in the United States is over 74,000 **megawatts** of power. A megawatt is a million **watts**. A watt is the metric unit of power equivalent of one joule of energy per second, making power a time-dependent unit of measure.

There are multiple benefits to wind power. Upon installation, wind produces no emissions and no pollution. Wind emits no greenhouse gases. It much more efficient than conventional energy resources. From an economic standpoint, landowners can lease land for wind farms for profit.

There are a few drawbacks to wind power. We have no control on when and how wind blows. Some regions are naturally windier than others. This drawback can be overcome by erecting turbines on high ground above the tree line or on top of tall buildings. Higher up in the atmosphere, even in regions that have relatively calm ground winds, wind is always blowing. Another drawback is the amount of space that wind farms take up. Residents may not want to have them near their homes. One of the major drawbacks to wind power is that flying animals can become sucked into wind turbines, harming them or killing them. Even with this drawback it seems like there could be a solution, for instance, some form of wire netting that could protect animals from being sucked into turbines.

Geothermal Energy

Geothermal energy is energy that is generated from within the Earth. It is closed system energy originating from heat deep within the Earth, the heat of the Earth's mantle. The heat of the mantle is the residual heat of Earth's formation and, more so, the heat emitted by radioactive elements deep within the Earth. These sources cause the formation of mantle plumes as well as magma within the Earth's crust. This deep heat from within the Earth powers the planet's plate tectonic system. It does not originate from the Sun, unlike all of the other forms of energy that are currently used. It is a renewable energy source since Earth's interior heat will not likely dissipate for another few billion years or so.

The Earth's interior heat is harnessed in geothermal power plants that use heated water to turn turbines to generated secondary electricity. Geothermal energy is also used to heat facilities and homes. In regions near geothermal heat sources, often areas of active volcanism, pipes are used to circulate heat from within the Earth into buildings to heat them.

Geothermal is completely renewable, but a limited amount of greenhouse gases and pollutants are released. Also, geothermal is very site specific. It can only be used in areas that sit atop geothermal heat sources, which limit their use. Globally, only a small percent of power is provided by geothermal energy. Only countries like Iceland, which is atop large-scale active volcanic and magma-producing crust, use geothermal extensively. Geothermal energy in the United States is mostly used by residents in the western states.

Energy Sources From the Ocean

Energy that uses the kinetic energy of ocean water, as well as the ambient heat off the ocean, is currently in development. Ocean energy is promising because it is completely renewable, non-polluting, and free of greenhouse gases.

The oceans tides are a constant throughout the world, displacing large amounts of water as the tides rise and fall. This **tidal energy** is capable of being harvested. The displaced water in tidal bays can be the perfect setting to harness energy. There are currently only a few tidal energy facilities, such as in France and South Korea. One of the problems with tidal power is the overall cost of development for use in many coastal zones. There are few coastal areas of the United States in which tidal power is feasible.

Ocean wave energy may be a less costly venture than tidal energy. Wind-driven wave motion on the surface of the ocean is harnessed, then converted from kinetic energy into electricity. This and other forms of energy harnessed from the kinetic motion of the ocean, including the use of ocean currents, are still in development. Costs remain high for any sort of widespread use.

Ocean thermal energy is a promising potential resource. Daily, the ocean in tropical regions absorbs enormous amounts of solar energy, equivalent to that of over 200 billion barrels of oil. This alone makes it a very promising source of renewable energy, but the costs are currently prohibitive. It, like other sources of ocean energy, remain in research and development.

Hydrogen Power

Hydrogen power is promising since it uses the most abundant element in the universe, hydrogen (H), to generate energy. Hydrogen used for municipal electricity does not currently exist. The use of hydrogen is still in development. Some methods for using hydrogen are free of greenhouse gases and pollutants; others emit them. Hydrogen needs to be harnessed safely because it can be explosive. Currently hydrogen is only used in some industrial applications, although a few hydrogen-powered buses are currently operating in Europe and Japan.

Conclusion

Energy is a given within the Universe. It is what drives change; it is "that which does work". Beyond the need of energy for biological processes, humanity requires energy, in vast amounts, for the continued operation of civilization. At the same time, harnessing some forms of energy are problematic. We are at a crossroads in development right now. Will we choose to use more advanced forms of limitless, sustainable energy, or will we continue to be almost completely dependent on energy from combustion?

Discussion Questions

Directions: Review the chapter in order to completely and correctly respond to the questions and prompts.

Keep this in the back (or maybe even front) of your mind: Science is system of methods that allows us to understand not only our planet, and ourselves, but of the entirety of the universe. Answering the following questions correctly will lead to a more robust understanding of how to apply science to ENVS problems.

1. Why can we say that most energy resources on Earth ultimately come from the Sun?
2. What are the three fossil fuels? Describe them.
3. What are the benefits and drawbacks in using fossil fuels for energy?
4. Is there a reason that anthracite is relatively rare? Why?
5. What are some of the public health hazards posed by fossil fuels?
6. Is nuclear power a safe alternative to fossil fuels? Why or why not?
7. Which of the alternative energy resources, other than nuclear power, do you think is the overall best alternative for energy production, safety, no pollution, no greenhouse gas emissions, and so on? Why?
8. Do you think fusion power will be available soon? Why or why not?
9. Describe the photovoltaic effect.
10. What are some drawbacks to building windfarms off the coast?
11. Can you think of some other possible energy alternatives not mentioned? What are they, and how would they work?

Bibliography

Smil, V., 2017, *Oil: A Beginner's Guide*: London, Oneworld Publications, 224 p.
Withgott, J., and Laposata, M., 2019, *Environment: The Science Behind the Stories*: New York, Pearson, 7th ed., 768 p.

Credits

Fig. 13.1: Copyright © by Aleksandr Serebrennikov (CC by 4.0) at https://commons.wikimedia.org/wiki/File:Sun_in_the_sands.jpg.
Fig. 13.2: Copyright © by Pixeltoo (CC BY-SA 4.0) at https://commons.wikimedia.org/wiki/File:Combustibles_fossiles.png.
Fig. 13.3: Copyright © by Mbeychok (CC BY-SA 3.0) at https://commons.wikimedia.org/wiki/File:Crude_Oil_Distillation.png.
Fig. 13.4: Copyright © by TrimmerinWiki (CC BY-SA 4.0) at https://commons.wikimedia.org/wiki/File:Austinmer,_NSW_-_Coal_seam_in_headland_at_Bell%27s_Point.jpg.
Fig. 13.5: Copyright © by MagentaGreen (CC BY-SA 3.0) at https://commons.wikimedia.org/wiki/File:Oil_traps.svg.
Fig. 13.6: Copyright © by ReubenGBrewer (CC BY-SA 3.0) at https://commons.wikimedia.org/wiki/File:14_06_29_Natural_Gas_Pipeline_Sign_Clearwater_FL_02.jpg.
Fig. 13.7: Copyright © by Myesd (CC BY-SA 4.0) at https://commons.wikimedia.org/wiki/File:TRR1-M1-Reactor-TINT.jpg.
Fig. 13.8: Copyright © by ArnoldReinhold (CC BY-SA 4.0) at https://commons.wikimedia.org/wiki/File:NNHS_solar_panels_lift_prep.agr.jpg.

Chapter 14

Waste Management

Handling Impacts

Objectives

- Understand that human activity produces waste, and that this is inevitable.
- Come to an understanding of how waste products pose a risk to human health and well-being.
- Know about the different modes of waste: municipal, industrial, and wastewater.
- Understand that waste needs to be managed and that regulations are required.
- Know that industrial waste is the most severely damaging waste and why.
- Understand that water is critical to human beings, as well as all life on Earth, and needs to be free of contaminants.

KEY WORDS AND TERMS

Geospatial (adj.): Refers to data or conditions that are specific to geographic locations.

Non-potable water (n.): Water that is not suitable for drinking or other use by human beings.

Potable water (n.): Water that is suitable for drinking or other use by human beings.

Consider This . . .

What have you thrown away today? Was the trash that you threw away unavoidable? Here's one more question: Could that trash have been something of value to somebody else? "One person's trash is another person's treasure" may seem like a cliché, but it is a very real one. For instance, oxygen is largely a waste product produced as a byproduct of photosynthesis. We certainly wouldn't consider oxygen to be waste, would we?

Introduction

Human activity, particularly industry, produces waste. This is inevitable as the process of use, in the physical and chemical sense, involves processing goods in which waste products are inevitably produced. These waste products are often toxic, or deleterious to human existence, and they must be processed and reclaimed in order to not cause damage to human health or harm the environment.

What Is Waste?

Waste is any unwanted material or substance that is the byproduct of anthropogenic activities. Waste products are also the discarded remnants of matter that had been used or spent. Waste products can be biological, chemical, or physical.

Waste can be classified demographically. **Municipal solid waste**, otherwise known as trash (or rubbish in the United Kingdom) is any non-liquid waste that comes from businesses, homes, institutions, and public facilities (Figure 14.1). **Industrial solid waste** is generated by agriculture, industrial production, mining, petroleum production, powerplants, and technology. **Hazardous waste** products are liquids or solids that are corrosive, flammable, reactive, or toxic. **Wastewater** is water used commercially, industrially, or residentially that has become **non-potable** from use. Wastewater is also **stormwater**, polluted runoff from streets and storm drains.

Figure 14.1 Rubbish in the UK

Waste Management Goals

There are three main goals when it comes to waste management:

1. **Reduction**: Reduce the volume of waste that is generated. This can be done through minimized packaging and general conservation. Reduction is the optimum strategy for waste management.
2. **Recovery**: Repurpose or reuse materials rather than discard them outright. Recover materials that can be used again or recycle them. Recovery is the next best solution for waste management after reduction.
3. **Disposal**: If materials need to be completely disposed of, do so in a way that is efficient and safe.

Methods of Management

Waste can be track through the **waste stream**, which is the flow of waste as it moves from the point at which it's been discarded to either recovery or disposal. The waste stream is time dependent and **geospatial** in nature.

Impacts caused by excessive movement within the waste stream, as well as the overall volume of waste itself, can be reduced by using materials more efficiently, as well as not overconsuming goods. Other solutions include goods that use less packing and recovery. Recovery includes **composting** and **recycling**. It is the next best method of waste management after reduction of initial use. Composting is the use of discarded organic material, usually plant food scraps, to be used as nutritive material for further plant growth. When materials are composted, natural decomposition is induced, resulting in high-carbon material that contains an abundance of plant nutrients. Recycling is the repurposing or refabrication of goods in order to obtain new goods.

At its best the recovery of goods and materials mimics the recycling of matter within ecosystems. Nature is extremely efficient; all materials are recovered within the biosphere.

Municipal Solid Waste

Municipal solid waste (MSW) is discarded consumer and commercial materials from common use. MSW is known as "trash" in the United States, and "rubbish" in the United Kingdom. "Garbage" refers to food, or household, waste.

Solid Waste

Solid waste in most nations, including the United States, consists of food scraps, packaging, paper and cardboard products, plastic products, and yard debris. These are the principal components of municipal solid waste.

Paper and cardboard products remain the largest component of MSW in developed nations, despite recycling efforts. Despite composting efforts, food scraps are the second largest component of MSW (Figure 14.2). Packing used for consumer products accounts for most disposal. Food scraps and organic materials are the primary components of MSW in developing countries. Municipal and commercial waste disposal is widely used in developed, wealthy countries. Such efforts are often absent in many developing regions.

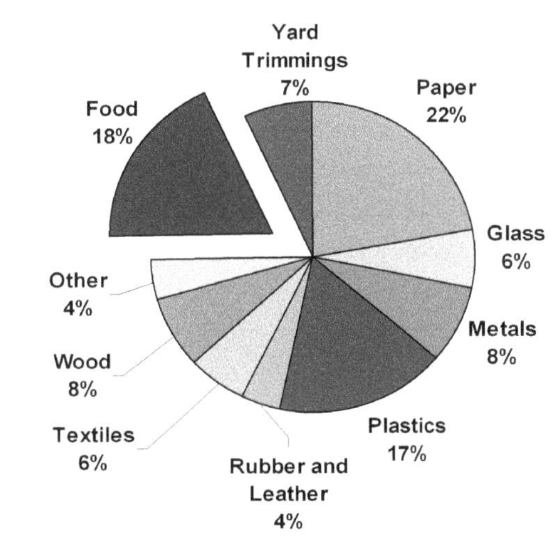

Figure 14.2 Food as the second largest component of MSW

Global Waste Generation

People with higher incomes in developed countries not only consume more goods, but they also discard more waste. At the same time, as developing nations vie for affluence, consumption increases.

Globally, there is a push for a higher standard of living and more economic affluence, leading to an overall rise in the standard of living. As wealth increases, individuals on a global scale tend to discard more goods. The wealthy often throw away goods that can still be used. Poor people in both the developing and developed world support themselves by scavenging discarded items, repurposing them, or selling them.

Disposal Methods

In the past, waste disposal occurred wherever it was convenient. Throughout history **middens** were public sites in which people discarded nearly everything: animal remains; human excrement and waste; mollusk shells and aquatic organism waste; plants, pots, and containers; discarded stonework; and many other items. These middens are a source of information for past living conditions and cultures. Burning has also been used throughout history as a method of disposal.

Throughout developing regions, to this day, open dumping at sites of convenience is still used, as is burning. This illustrates that waste disposal is not only a matter of convenience but a matter of economic affluence, or the lack thereof. Industrialized, developed nations use facilities in which waste is sorted, then buried in covered and lined facilities, or **incinerated**. Incineration is the burning of waste materials using extremely high heat sources. Recycling in many developed nations decreases the volume of waste in landfills.

Sanitary Landfills

Sanitary landfills are facilities in which large amounts of MSW, and other waste products, are sorted and buried underground or assembled into large piles. Sanitary landfills are engineered to dispose of, and reclaim, waste as efficiently as possible. In the United States, all sanitary landfills must meet standards set by the EPA under the Resource Conservation and Recovery Act (RCRA) of 1976.

In these facilities, waste is compressed under its own weight and subject to partial decomposition by bacteria. Soil is layered on top of waste to speed decomposition and reduce odor, although there is generally no escaping the smell of a landfill. Burying waste in soil reduces the infestation of pests. Liners composed of plastic and other impermeable materials such as clay are used to, in theory, stop leachate from entering the local environment, especially water sources. **Leachate** is contaminated water that infiltrates through layers of soil into the environment.

All landfills eventually close. When that happens, they must be capped with soil or some other material to further hasten decomposition. Even after closure, landfills must be maintained for some time to ensure that they do not contaminant the environment.

Landfills are generally located on the outskirts of cities and towns, away from residential areas. They must also be located away from wetlands or other environmentally sensitive areas. In addition, they must be located away from faults with a history of earthquake activity, and they must be located no closer than two miles from gradients that provide municipal waters. They must 20 feet or higher than the water table.

In the past there were many more landfills in the United States than the current 1,700 or so. In the mid-1980s there were around 7,600 landfills in the United States, which means that thousands of abandoned landfills dot the American landscape. Tougher government regulations were the reason for this. Today, in the 21st century, nearly half of all U.S. landfills are privately owned. Most landfills are gigantic, located far from municipal areas.

INCINERATION

Around 10% of all waste in the United States is incinerated. Incineration is done in specially constructed facilities that can handle the extreme heat. Although incineration can be an efficient method of waste disposal, there are some impacts. Ash from incineration must be disposed of in hazardous waste facilities. Incineration also creates and releases hazardous chemicals as air pollution. Scrubbers must be used to reduce the amount of these pollutants that enter the atmosphere.

Incineration can be used to generate electricity. The heat from incineration can be used to generate power in **waste-to-energy facilities (WTE)** (Figure 14.3). Only about 1% of all municipal electrical generation in the United States comes from WTE, but it is a reliable source of power at least. Another drawback to WTE is that it is expensive to operate and maintain, making it hard to turn a profit.

Figure 14.3 Waste to energy facility in Australia

LANDFILL GAS

Because of the high levels of decomposition present, landfills emit large amounts of gas, mostly carbon dioxide and methane. Methane can be harnessed from landfills to be used by the public. Gas that is not used is often burned off in flares to reduce odor.

Source Reduction

Waste can best be handled initially through **source reduction**. Source reduction keeps waste from being generated in the first place. All of the costs associated with the waste stream are avoided in this manner. Source reduction conserves resources, nearly eliminates pollution, and saves money.

An inordinate amount of municipal waste consists of packaging. This can be eliminated on the manufacturing level by reducing the amount of packaging for goods. Manufacturers can also use more recyclable materials in packaging. Consumers can respond by purchasing minimal, or no-packaging, goods. For instance, many grocery chains encourage the use of bags brought from customers' homes as opposed to plastic bags. Bulk packaging can also aid in the reduction of excess packaging.

Recovery and Reuse

Planned obsolescence is the practice of designing goods to lack durability or breakdown in the short-term. Although this might enable manufacturers and businesses to turn higher profits, it is disastrous with regard to the waste stream. It is also costly for consumers. Consumers can reduce the amount of waste materials that enter the waste stream due to planned obsolescence by purchasing durable goods that can be reused over and over again. In the long run this also saves money for the consumer.

Another action that can be done for the reduction and recovery of goods is that of donating to charitable organizations that maintain stores in which goods are sold. These stores are cost effective and are of great benefit to those with low incomes. One can also buy groceries in bulk, reuse food and drink containers, compost vegetable food scraps, purchase rechargeable batteries, and rent or borrow materials when appropriate.

Compost

Composting is beneficial to not only the consumer but to the environment as it supports sustainability and healthy plant growth. In order to compost, one need only set aside a container or an area somewhere nearby, for instance at the edge of a yard. A lid of some kind should be used to contain heat because heat supports the life cycle of decomposer organisms. When waste is added it hastens heat production and the decomposition process. Organisms like bacteria, earthworms, and mites are effective decomposers.

In many communities municipal composting programs are implemented. When utilized, these programs ensure that food and yard waste is diverted away from the waste stream to local composting facilities. This reduces pressure on landfills. Using compost for gardens and small farms promotes soil biodiversity, reduces the need to industrial fertilizers, and promotes healthy plants and gardens.

Recycling

On average, around 90 million tons of MSW are kept from landfills thanks to recycling efforts in the United States. The process of recycling is known as the **recycling loop**. There are three steps to the recycling loop:

1. The act of collecting and processing recyclable materials from homes using curbside recycling containers and designated local collection locations is the first step in the recycling loop (Figure 14.4). These materials are transported to **materials recovery facilities (MRFs)**, where workers use machines to sort items, clean them, and prepare them for reprocessing.
2. Recyclables that are used to produce new goods constitutes the second step of the recycling loop. Many manufacturers use recyclable goods. One can see this on a lot of labeling.
3. The purchase of recycled goods by consumers in the third step in the recycling loop. This is an essential step. After this occurs the recycled goods can be returned to step 1, thus closing the loop.

According to the EPA, recycling is a great environmental success. This may be true, but the rates of recycling across the United States, and the world for that matter, vary widely. Also, the types of materials that are recycled vary from place to place. Businesses are often not interested in recycling because of the costs. In this case, recycling should be encouraged through the cost-effective implementation of the practice and the actions of the consumer.

Figure 14.4 Recycling container next to a trash can

Industrial Waste

Industrial waste consists of materials used in construction, industry, large-scale agriculture, mining, and petroleum production (Figure 14.5). In the United States around 7.6 billion tons of industrial waste is produced per year. Most waste is industrial waste.

Figure 14.5 Barrels containing industrial waste in Washington state

Waste disposal and recycling efforts for industry are often similar to methods for municipal solid waste. In the United States, regulations for handling industrial waste vary from state to state, but as always, the federal government takes precedence with regard to regulating industrial waste. Many state and local government regulations are less stringent than those used by the federal government.

Industries do not always need to have permits regarding the handling of waste depending on the locality. This includes liners for leachate and ground- and surface water monitoring systems. Note, industry generates a lot of wastewater.

It might seem like handling waste on the industrial level in an efficient and sustainable manner would be cost-effective. This is true from a physical standpoint, but it is not always true from an economic one. While operationally reducing the volume of waste by being careful is efficient, it is often more profitable in the short-term to simply produce what is needed as quickly as possible. On the other hand, the cost of waste disposal continues to rise, so less waste generated may ultimately make the most sense.

Industrial Ecology

Industrial ecology is the practice of designing industrial systems to mimic those of ecosystems, inducing sustainability, reduced inputs, and maximized efficiency, while at the same time maximizing profits. Arrangements within industry are designed to operate within closed, somewhat self-sustaining systems. Through **life cycle analysis**, the life cycle of a product is examined, and ways are found to make the production process more ecologically efficient.

Using these methods, rather than discarding them, waste products can be reused for raw materials. Since the emphasis is on mimicking ecosystems, environmentally harmful materials are phased out. Products are made to be more sturdy, recyclable, or reusable. Businesses and industry that use the industrial ecology approach do things in environmentally sound ways with less impact on human health. They also reduce waste. This ultimately saves money in many ways, one of them being the need to avoid any fees and fines that are often part of industrial processes. If the problem isn't there in the first place, there is no need to spend money to fix it. Artificial wetlands are an example of industrial ecology since they are like natural features.

Hazardous Waste

Hazardous waste is waste that is harmful to human health and/or the environment. These waste products are a large feature of industry and seemingly unavoidable in many cases. The following are the main types of industrial waste. All are directly toxic to living things to some extent. Some types of hazardous waste, such as radioactive waste, occupy many of these categories:

- **Corrosives**—Substances, particularly chemicals species, that can corrode materials especially metals
- **Ignitables**—Substances that are prone to combustion and catching fire
- **Reactives**—Unstable compounds and chemical species that react readily with other chemical species, often explosively or by producing dangerous fumes
- **Toxins**—Substances that are directly harmful to human health through direct contact, ingestion, and/or inhalation

By its very nature, industry is the largest source of hazardous waste. Due to the nature of hazardous waste, it is highly regulated. Industrial sources of hazardous waste include agriculture, construction, mining, technology, and utilities. Chemical manufacturing, which includes petroleum, coal, and general hydrocarbon production, is the largest source of hazardous waste, accounting for over 80% of all hazardous waste generated in the United States alone. This includes products that end up being used in households and small businesses. The military also counts as a producer of hazardous waste. There are federal programs in place that address this issue.

Organic Compounds

Organic compounds, or **organic chemicals**, are loosely defined as any chemical that contains carbon. A more specific definition would be that organic compounds are chemicals that contain the **organic** carbon-carbon or carbon-hydrogen bond. Volatile organic compounds, used for many products such as plastics, are organic compounds (Figure 14.6).

Figure 14.6 Chamber for measurement of volatile organic compounds emitted from furnishings

Organic compounds are particularly hazardous due to their persistence over time, they tend to persist in the environment a long time because their toxicity persists over time. Synthetic organic compounds resist decomposition in the environment. These compounds are often found in the plastics that are commonly available. These compounds are made to persist since they are often used to keep even buildings from decaying. They are also used to store products and many of them are used in pesticides. Unfortunately, their very function as persistence substances causes them to become persistent pollutants as well as persistent waste products. These products are readily absorbed into human tissues. Many are known carcinogens, endocrine disruptors, mutagens, and teratogens.

Heavy Metals

Heavy metals are a class of metals that are particularly dense as well as toxic, or potentially harmful to human health, even at low concentrations. These metals, while normally thought of as large and heavy, are often dissolved in other substances or part of chemical compounds. These include arsenic (As), chromium (Cr), copper (Cu), lead (Pb), nickel (Ni), and tin (Sn). These metals are used extensively in industry for use in industry for dyes, electronics, metal fabrication, pigments, and wiring.

They easily enter the environment in many ways but often do so simply when they are not properly disposed of. Some heavy metals, such as mercury, are fat soluble, breaking down slowly in human tissue: They bioaccumulate and biomagnify.

Hazardous Waste Disposal

Prior to the last 50 years, hazardous waste disposal was not strictly regulated. Throughout the first half of the 20th century the public was generally not aware of the extreme hazards that hazardous waste poses to human health. It was assumed that hazardous waste, and waste in general, would just "disappear" back into the environment.

Local municipalities have been designating sites or specific methods for the disposal of hazardous household waste. States are required to safely manage the disposal of hazardous waste through RCRA. Entities that generate large amounts of hazardous waste have to obtain permits and track waste throughout its entire waste cycle. Much of this is intended to prevent illegal dumping. Illegal dumping is a problem since hazardous waste disposal is expensive, resulting in illegal dumping. Of course, fines are expensive as well, as is going to court. Beyond the costs and legalities, the dumping of hazardous waste is unethical. It creates health risks and poses extreme risks to the environment.

With ethics in mind, the **Basel Convention** is an international treaty that specifically targets the disposal of hazardous waste. This has not always stopped industrial nations from dumping waste into developing nations.

Hazardous waste can be disposed of properly using several methods that are not difficult to implement, or so it would seem. Impervious liners are required in landfills and industrial leachate removal systems. The design and construction standards for these methods are stricter than that for general municipal sanitary landfills. These must always be located far from aquifers and the water table.

Liquid hazardous waste can be stored in **impoundments**. Shallow depressions are lined with clay, plastic, or some other impervious substance. Wastewater evaporates, leaving behind the solid residue of hazardous waste, which is then transported elsewhere. This method is often used for mine spoil, which often contains organic compounds of other toxins. A problem with this method is that the liner can crack, causing leachate to leak into the environment. Storms can also cause overflow, which causes the waste to become entrained in runoff.

Deep well injection is a method that involves drilling deep beneath the water table into rock, sediments, or regolith, then injecting waste into the well that was formed. This is a long-term disposal method that is not without its problems. It is intended to isolate waste far from human habitation, yet the well casings can corrode. Hazardous waste can become mobile through groundwater transport, ultimately making its way into soil and surface water.

Electronic Waste

Since the mass proliferation of computers and computerized technology since the early 1980s **electronic waste**, or **e-waste**, has been a problem. E-waste includes all sorts of commercial, industrial, and residential computational and electronic devices: cell phones, computers, DVD players, fax machines, printers, VCRs, and so on. Many of these devices have long since become obsolete. These are often simply disposed of, making their way to landfills, but due to their components they should be treated as hazardous waste.

Some businesses and public organizations, such as colleges and universities, sponsor programs to recycle and reuse e-waste. Many secondhand stores throughout the country sell supposedly obsolete electronic items that are perfectly functional and useful.

The Superfund

On a global level, thousands of former military and industrial sites are contaminated with hazardous waste. Dealing with this problem is expensive and time-consuming, so often nothing gets done. In the United States the **Comprehensive Environmental Response Compensation and Liability Act** of 1980 **(CERCLA)**, also known as the **Superfund**, addresses this problem, as well as other problems associated with hazardous waste.

In 1980 the Superfund was established as a federal program, under the guidance of the EPA, to clean up hazardous waste and sites polluted with hazardous waste. The EPA employs scientists and other experts to identify contaminated sites, then take action to protect groundwater near these sites and clean up the pollution. Amendments to the Superfund later charged the EPA to clean up brownfields, lands that are contaminated to the extent that they are difficult or impossible to use post-contamination.

EPA scientists, once a site is identified, assess how close people live to the site, whether wastes are confined or likely to spread, and whether the sites currently threaten drinking water supplies. Harmful sites are put on the EPA's National Priority List and ranked according to the level of risk they pose to human health. These sites are cleaned up on the basis of evaluation as well as the availability of funds. It is required that the EPA hold public hearings to inform residents in the area of its findings and to allow the public to have a voice through feedback.

One of the tenets of the Superfund is that the "polluter pays the principle." Violators are charged with paying for clean-up. Sites are often shut down and inoperable for a long time due to this because many violating entities cannot pay for clean-up. Cases often end up in costly litigation. Sometimes the responsible parties cannot be found due to the particular company no longer existing, or some other reason. Sometimes responsible parties flee the country.

Chemical and petroleum industries are subject to a federal tax that was established through a trust fund. The EPA also has Superfund special accounts set up to deal with specific polluting entities, such as chemical and steel producers. On average, Superfund clean-ups cost over $40 million and can take many years to complete. Entities and business interests often offer this as an excuse to resist action. This reflects self-interest and the push for their own profits and is not in the best interest of the public at large, which is composed of people who vastly outnumber these business interests.

Wastewater

Wastewater consists of any water that has been used by the public, becoming non-potable to some extent. This water is in the form of agricultural wastewater, commercial wastewater, residential wastewater, industrial wastewater, and sewage. While sewage remains a problem, industry remains the largest source of wastewater. Wastewater is critical since water is of prime importance to human beings and life itself.

Residences and small businesses, especially in rural areas, make use of **septic systems** to deal with wastewater, especially sewage. These systems consist of underground tanks that separate solids and oils from wastewater. The water drains into a drainage field beyond the tank, in which microbes decompose the waste in the water, These tanks need to be periodically pumped to remove the volume of waste, which is then transported to a landfill.

In populated areas, municipal wastewater and sewage systems are employed. Wastewater from businesses, homes, and industries are sent through the systems for reclamation in **water treatment** plants so that the water can become potable again. Industry often requires more steps to do this (Figure 14.7).

Figure 14.7 Wastewater treatment plant, Czech Republic

There are several steps to treating municipal wastewater:

1. **Primary treatment**: Water is physically separated from solid wastes in settling tanks, known as clarifiers. These feature-filtering mechanisms are used in large numbers in larger urban areas. The typical retention time in settling tanks is two to six hours. Once separation has occurred, water is sent to secondary treatment.
2. **Secondary treatment**: Water is stirred to aerate it, allowing for decomposition of remaining wastes to be consumed by aerobic bacteria.
3. **Tertiary treatment**: Water is further treated by passing through a series of filters, treated with chlorine or ozone, and/or irradiated with ultraviolet light. Tertiary treatment is used by most municipalities, although it is not always required. Tertiary treatment removes especially difficult-to-deal-with organic compounds or pollutants.

Treated water from municipal wastewater plants is periodically evaluated by the government agencies for up to several dozen primary contaminants. If found in violation, water treatment plants are required to fix the problem. In the past, many municipal sewage and stormwater pipes were combined. This caused periodic severe, costly, and damaging overflows. This has largely been phased out in recent years.

The use of **artificial**, or **constructed**, **wetlands** have been implemented in municipal systems, businesses, and industry in recent years (Figure 14.8). Natural wetlands are proven to be effective at cleaning up contaminated water through natural processes. After some form of treatment at a facility, or after secondary or tertiary treatment at a water treatment plant, water is pumped into a wetland. Microbes in the wetland consume any contaminants that remain further aided by plants and soil or sediment. The **potable water** is then released into waters or allowed to drain into the subsurface. These wetlands can also serve as habitats for wildlife and viable sites for recreation. Thousands of artificial wetlands are in use, worldwide.

Figure 14.8 Artificial wetland, Australia

Conclusion

The use of energy and materials has always resulted in the production of unwanted matter and energy, waste. Amplified by the scale of municipalities, regions, and even nations one can see how waste generation is problematic. In order to take-on these problems innovation is required; in fact, not only innovation, but innovation in sustainability. Finding sustainable solutions for waste management mirrors the need for it in our ongoing relationship with Earth's environment.

Discussion Questions

Directions: Review the chapter in order to completely and correctly respond to the questions and prompts.

Keep this in the back (or maybe even front) of your mind: Science is system of methods that allows us to understand not only our planet, and ourselves, but of the entirety of the universe. Answering the following questions correctly will lead to a more robust understanding of how to apply science to ENVS problems.

1. What exactly is waste, and why is it critical that we deal with it?
2. What are the types of waste that can be classified demographically?
3. Described the three waste management goals.
4. How was municipal solid waste dealt with in the past versus the present?
5. What is waste incineration? What are its benefits and drawbacks?
6. Why is recycling important?
7. Describe the three steps to recycling.
8. What are the impacts of industrial waste? How severe is the problem?
9. Describe sources of industrial waste and why they are dangerous. Describe as many as possible.
10. Describe the steps that are used in municipal water treatment plants.
11. What are the benefits of artificial wetlands?

Bibliography

Newton, D., 2020, *Waste Management: A Reference Handbook*: Santa Barbara, California, ABC-CLIO, 324 p.
Withgott, J., and Laposata, M., 2019, *Environment: The Science Behind the Stories*: New York, Pearson, 7th ed., 768 p.

Credits

Fig. 14.1: Copyright © by Acabashi (CC BY-SA 4.0) at https://commons.wikimedia.org/wiki/File:Wheelie_bins_overflowing_rubbish_Dongola_Road_Tottenham_London_England_1.jpg.
Fig. 14.2: USEPA Environmental-Protection-Agency, "Food as the Second Largest Component of MSW," https://commons.wikimedia.org/wiki/File:How_Much_Of_Your_Waste_Is_Food%3F_(3679929496).gif, 2009.
Fig. 14.3: Copyright © by Calistemon (CC BY-SA 4.0) at https://commons.wikimedia.org/wiki/File:East_Rockingham_Waste_to_Energy_Plant,_January_2023_01.jpg.
Fig. 14.4: Copyright © by Bart Everson (CC by 2.0) at https://commons.wikimedia.org/wiki/File:Recycling_Bin_and_Trash_Can,_Mid-City_New_Orleans.jpg.
Fig. 14.5: Copyright © by Robert Ashworth (CC by 2.0) at https://commons.wikimedia.org/wiki/File:Waste_barrels_in_the_Duwamish_industrial_area_(6111564488).jpg.
Fig. 14.6: Copyright © by Tracey Nicholls, CSIRO (CC by 3.0) at https://commons.wikimedia.org/wiki/File:CSIRO_ScienceImage_2906_Measurement_of_volatile_organic_compounds.jpg.

www.ingramcontent.com/pod-product-compliance
Ingram Content Group UK Ltd.
Pitfield, Milton Keynes, MK11 3LW, UK
UKHW061702190726
13853UKWH00008B/2360